T0143290

Cognitive Systems Engineering

The Future for a Changing World

EXPERTISE: RESEARCH AND APPLICATIONS
Robert R. Hoffman, K. Anders Ericsson, Gary Klein,
Michael McNeese, Eduardo Salas, Sabine Sonnentag, Frank Durso,
Emilie Roth, Nancy J. Cooke, Dean K. Simonton
Series Editors

Hoc/Caccibue/Hollnagel (1995)—Expertise and Technology: Issues in Cognition and Human–Computer Interaction

Hoffman (2007)—Expertise Out of Context

Hoffman/Militello (2008)—Perspectives on Cognitive Task Analysis

Mieg (2001)—The Social Psychology of Expertise Case Studies in Research, Professional Domains, and Expert Roles

Montgomery/Lipshitz/Brehmer (2004)—How Professionals Make Decisions

Mosier/Fischer (2010)—Informed by Knowledge: Expert Performance in Complex Situations

Noice/Noice (1997)—The Nature of Expertise in Professional Acting: A Cognitive View

Salas/Klein (2001)—Linking Expertise and Naturalistic Decision Making

Smith/Hoffman (2017)—Cognitive Systems Engineering: The Future for a Changing World

Schraagen/Chipman/Shalin (2000)—Cognitive Task Analysis

Zsambok/Klein (1997)—Naturalistic Decision Making

Cognitive Systems Engineering
The Future for a Changing World

Edited by
Philip J. Smith
Robert R. Hoffman

CRC Press
Taylor & Francis Group
Boca Raton London New York

CRC Press is an imprint of the
Taylor & Francis Group, an **informa** business

CRC Press
Taylor & Francis Group
6000 Broken Sound Parkway NW, Suite 300
Boca Raton, FL 33487-2742

First issued in paperback 2019

© 2018 by Taylor & Francis Group, LLC
CRC Press is an imprint of Taylor & Francis Group, an Informa business

No claim to original U.S. Government works

ISBN-13: 978-1-4724-3049-6 (hbk)
ISBN-13: 978-0-367-87940-2 (pbk)

This book contains information obtained from authentic and highly regarded sources. Reasonable efforts have been made to publish reliable data and information, but the author and publisher cannot assume responsibility for the validity of all materials or the consequences of their use. The authors and publishers have attempted to trace the copyright holders of all material reproduced in this publication and apologize to copyright holders if permission to publish in this form has not been obtained. If any copyright material has not been acknowledged, please write and let us know so we may rectify in any future reprint.

Except as permitted under U.S. Copyright Law, no part of this book may be reprinted, reproduced, transmitted, or utilized in any form by any electronic, mechanical, or other means, now known or hereafter invented, including photocopying, microfilming, and recording, or in any information storage or retrieval system, without written permission from the publishers.

For permission to photocopy or use material electronically from this work, please access www.copyright.com (http://www.copyright.com/) or contact the Copyright Clearance Center, Inc. (CCC), 222 Rosewood Drive, Danvers, MA 01923, 978-750-8400. CCC is a not-for-profit organization that provides licenses and registration for a variety of users. For organizations that have been granted a photocopy license by the CCC, a separate system of payment has been arranged.

Trademark Notice: Product or corporate names may be trademarks or registered trademarks, and are used only for identification and explanation without intent to infringe.

Library of Congress Cataloging-in-Publication Data

Names: Smith, Philip J. (Philip John), 1953- editor. | Hoffman, Robert R., editor.
Title: Cognitive systems engineering : the future for a changing world /
edited by Philip Smith, Robert R. Hoffman.
Description: Boca Raton : Taylor & Francis, CRC Press, 2017. | Series:
Expertise : research and applications | Includes bibliographical references and indexes.
Identifiers: LCCN 2017015928| ISBN 9781472430496 (hardback : alk. paper) |
ISBN 9781315572529 (ebook)
Subjects: LCSH: Systems engineering. | Cognitive science. | Human-machine systems. |
Expert systems (Computer science)
Classification: LCC TA168 .C597 2017 | DDC 620.001/17–dc23
LC record available at https://lccn.loc.gov/2017015928

Visit the Taylor & Francis Web site at
http://www.taylorandfrancis.com

and the CRC Press Web site at
http://www.crcpress.com

Dedication

What is really new about CSE that sets it apart from the specializations and focus areas that came before? Task analysis was always "cognitive," in that it recognized the importance of cognitive processes in the conduct of tasks, even ones that were ostensibly mostly physical (see Hoffman and Militello 2008). Human factors, as it emerged in industrial psychology toward the end of World War I and matured during World War II, was always about human–machine integration. Then came the computer and the concepts of cybernetics. These were game changers and, as they changed the work, they changed the scientific study of work. And as work became progressively more and more cognition-dependent, systems engineering had to become progressively more and more cognition-focused.

CSE considers the creation of complex macrocognitive work systems. The contributions of Woods are ideas and methods for dealing with the complexity, just as important as the early notions of cybernetics, control theory, and information theory were for the World War II-era work in human factors.

David D. Woods received his BA (1974) at Canisius College and his MS (1977) and PhD (1979) at Purdue University. He was a senior engineer at the Westinghouse Research and Development Center from 1979 to 1988. He has been on the faculty at The Ohio State University since 1988.

David is past president of the Resilience Engineering Association, past president and a fellow of the Human Factors and Ergonomic Society, and a fellow of the Association for Psychological Science and the American Psychological Association. He has been the recipient of the Westinghouse Engineering Achievement Award

(1984), a co-recipient of the Ely Award for best paper in the journal *Human Factors* (1994), and the recipient of the following: the Laurels Award from *Aviation Week and Space Technology* (1995) for research on the human factors of highly automated cockpits, the Jack Kraft Innovators Award from the Human Factors and Ergonomics Society (2002) for advancing Cognitive Engineering and its application to safer systems (2002), an IBM Faculty Award (2005), the best paper award at the 5th International Conference on Information Systems for Crisis Response and Management (ISCRAM 2008), a Google Faculty Award (2008), and the Jimmy Doolittle Fellow Award for Research on Human Performance, Air Force Association, Central Florida Chapter (2012).

David has investigated accidents in nuclear power, aviation, space, and anesthesiology, and was an advisor to the Columbia Accident Investigation Board. He has served on National Academy of Science and other advisory committees including Aerospace Research Needs (2003), Engineering the Delivery of Health Care (2005), and Dependable Software (2006). He has testified to the U.S. Congress on Safety at NASA and on Election Reform. He has worked extensively at the intersection of engineering and health care as a board member of the National Patient Safety Foundation (1996–2002) and as associate director of the Midwest Center for Inquiry on Patient Safety of the Veterans Health Administration. He is co-author of *Behind Human Error* (2010), *A Tale of Two Stories: Contrasting Views of Patient Safety* (1998), *Joint Cognitive Systems: Foundations of Cognitive Systems Engineering* (2005), and *Joint Cognitive Systems: Patterns in Cognitive Systems Engineering* (2006).

His chapter in this volume retraces his career path, and thus, in this dedication, we focus on the *honoris causa* for this book. Woods' impact on human factors, cognitive systems engineering, and engineering broadly has been profound. Everyone who has met him and who has heard his presentations, or has read his works, has been affected, often to the core of their beliefs and premises about the science of cognitive systems. For ourselves as editors of this volume, we continue to consider ourselves students of David whose work has benefited immeasurably from David's touch. Beginning with impactful achievements such as those involved in display and workstation design for the nuclear industry, and continuing with recent innovative concepts of the resilience and adaptivity of cognitive work systems, David has provided a legacy of ideas, methods, and students.

David's manner of thought and expression is notorious. Indeed, his friends compiled a "Daveish Dictionary" of mind-bending expressions. Daveish expressions include the following: Negotiate trade-offs in domains of complexity, collaborative envisioning exercise, shortfall in coordinative mechanisms, breaking the tyranny of the 2d screen, dynamic trajectories between intertwined processes, anticipatory displays, performing at the edges of the competence envelope, placing human actors in the field of practice, change triggers processes of expansive adaptation and transformation, multiple parties and interests as moments of private cognition punctuate flows of interaction and coordination, context gap complementary vector, escalating demands as situations cascade, projecting trajectories and side effects from courses of action, degrees of commitment ahead of a rolling horizon, and extending human perception over wider ranges and scopes. These are of course taken out of context. But even in context, Daveisms bend one's mind. And they are not just butter-thick wordplay. They spread across serious concepts and important issues. Anyone who encounters a Daveism and does not grok it is well advised to think until they do.

David's reach never exceeds his grasp. Spanning industrial and utilities safety, patient safety and health care, teamwork analysis, human–robot coordination, aviation human factors, space mission operations, and other venues as well, David's work is never, ever that of the dilettante who has spread too thin. David always spreads it on thick. And we love him. We are better scientists because of him, and better people as well.

REFERENCES

Cook, R.I., Woods, D.D. and Miller, C. (1998). *A Tale of Two Stories: Contrasting Views on Patient Safety*. Chicago IL: National Patient Safety Foundation.

Hoffman, R.R. and Militello, L.G. (2008). *Perspectives on Cognitive Task Analysis: Historical Origins and Modern Communities of Practice*. Boca Raton, FL: CRC Press/Taylor & Francis.

Hollnagel, E. and Woods, D.D. (2005). *Joint Cognitive Systems: Foundations of Cognitive Systems Engineering*. Boca Raton, FL: Taylor & Francis.

Woods, D.D., Dekker, S.W.A., Cook, R.I., Johannesen, L.L. and Sarter, N.B. (2010). *Behind Human Error* (2nd Edition). Aldershot, UK: Ashgate.

Woods, D.D. and Hollnagel, E. (2006). *Joint Cognitive Systems: Patterns in Cognitive Systems Engineering*. Boca Raton, FL: Taylor & Francis.

Woods, D.D., Patterson, E.S. and Cook, R.I. (2007). Behind human error: Taming complexity to improve patient safety. In P. Carayon (Ed.), *Handbook of Human Factors and Ergonomics in Health Care and Human Safety*. Boca Raton, FL: CRC Press/Taylor & Francis.

Contents

Part I   The Evolution and Maturation of CSE

Part II   Understanding Complex Human–Machine Systems

Part III   *Designing Effective Joint Cognitive Systems*

Part IV   Integration

List of Figures

List of Tables

Editors

Philip J. Smith, PhD, is a professor in the Department of Integrated Systems Engineering at The Ohio State University and a fellow of the Human Factors and Ergonomics Society. His research and teaching focus on cognitive systems engineering, human–automation interaction, and the design of distributed work systems. This research has been supported by the FAA; NASA; the National Heart, Lung and Blood Institute; the U.S. Army; and the U.S. Department of Education. Of particular significance has been his work on the following:

- The influence of brittle technologies on human–machine interactions
- Constraint propagation as a conceptual approach to support asynchronous coordination and collaboration in the National Airspace System, including work on the design of airspace flow programs, coded departure routes, collaborative routing, and the use of virtual queues to manage airport surface traffic
- Interactive critiquing as a model to support effective human–machine cooperative problem solving through context-sensitive feedback and the incorporation of metaknowledge into machine intelligence
- Continuous adaptive planning

Dr. Smith, his students, and his colleagues have won numerous awards, including the Air Traffic Control Association David J. Hurley Memorial Award for Research in Collaborative Decision Making, the Airline Dispatchers Federation National Aviation Safety Award, and best paper awards in *Human Factors* and *Clinical Laboratory Science*. (E-mail: smith.131@osu.edu)

Robert R. Hoffman, PhD, is a senior research scientist at the Institute for Human & Machine Cognition (IHMC) in Pensacola, Florida. Dr. Hoffman is a recognized world leader in cognitive systems engineering and human-centered computing. He is a fellow of the Association for Psychological Science, a fellow of the Human Factors and Ergonomics Society, a senior member of the Association for the Advancement of Artificial Intelligence, a senior member of the Institute of Electrical and Electronics and Engineers, and a Fulbright Scholar. His PhD is in experimental psychology from the University of Cincinnati. His postdoctoral associateship was at the Center for Research on Human Learning at the University of Minnesota. He was on the faculty of the Institute for Advanced Psychological Studies at Adelphi University. Dr. Hoffman has been recognized internationally in psychology, remote sensing, human factors engineering, intelligence analysis, weather forecasting, and artificial intelligence—for his research on the psychology of expertise, the methodology of cognitive task analysis, HCC issues for intelligent systems technology, and the design of macro cognitive work systems. Dr. Hoffman can be reached at rhoffman@ihmc.us. A full vita and all of his publications are available for download at www.ihmc.us/users/rhoffman/main.

Recent Books

Hoffman, R.R., LaDue, D., Mogil, A.M., Roebber, P., and Trafton, J.G. (2017). *Minding the Weather: How Expert Forecasters Thinks*. Cambridge: MIT Press.

Hoffman, R.R., Hancock, P.A., Scerbo, M., Parasuraman, R., and Szalma, J. (2014). *Cambridge Handbook of Applied Perception Research*. Cambridge University Press.

Hoffman, R.R., Ward, P., DiBello, L., Feltovich, P.J., Fiore, S.M., and Andrews, D. (2014). *Accelerated Expertise: Training for High Proficiency in a Complex World*. Boca Raton, FL: Taylor & Francis/CRC Press.

Hoffman, R.R. (Au., Ed.) (2012). *Collected Essays on Human-Centered Computing, 2001–2011*. New York: IEEE Computer Society Press.

Moon, B., Hoffman, R.R., Cañas, A.J., and Novak, J.D. (Eds.) (2011). *Applied Concept Mapping: Capturing, Analyzing and Organizing Knowledge*. Boca Raton, FL: Taylor & Francis.

Hoffman, R.R., and Militello, L.G. (2008). *Perspectives on Cognitive Task Analysis: Historical Origins and Modern Communities of Practice*. Boca Raton, FL: Taylor & Francis.

Crandall, B., Klein, G., and Hoffman, R.R. (2007). *Working Minds: A Practitioner's Guide to Cognitive Task Analysis*. Cambridge MA: MIT Press.

Ericsson, K.A., Charness, N., Feltovich, P.J., and Hoffman, R.R. (2006). *Cambridge Handbook of Expertise and Expert Performance*. New York: Cambridge University Press.

Selected Recent Publications

Hoffman, R.R., Henderson, S., Moon, B., Moore, D.T., and Litman, J.A. (2011). Reasoning difficulty in analytical activity. *Theoretical Issues in Ergonomic Science*, 12, 225–240.

Hoffman, R.R., and Woods, D.D. (2011, November/December). Beyond Simon's slice: Five fundamental tradeoffs that bound the performance of macrocognitive work systems. *IEEE: Intelligent Systems*, 67–71.

Moore, D.T., and Hoffman, R.R. (2011). Sensemaking: A transformative paradigm. *American Intelligence Journal*, 29, 26–36.

Hoffman, R.R., Lee, J.D., Woods, D.D., Shadbolt, N., Miller, J., and Bradshaw, J.M. (2009, November/December). The dynamics of trust in cyberdomains. *IEEE Intelligent Systems*, 5–11.

Contributors

Kevin B. Bennett, PhD, earned a PhD (applied–experimental psychology) from the Catholic University of America in 1984. Dr. Bennett is a professor in the Department of Psychology of Wright State University (Dayton, Ohio), and a member of the Human Factors/Industrial Organizational PhD Program. His research focuses on using interface technologies to design effective decision-making and problem-solving support systems. He has more than 36 years of experience in applying the Cognitive Systems Engineering/Ecological Interface Design framework to meet the associated challenges. He is a fellow in the Human Factors and Ergonomics Society and has served as an editorial board member of the journal *Human Factors* for more than two decades. He has more than 50 grants and contracts, 50 primary publications, and 110 secondary publications, and he has one co-authored book. This work has received several international awards. (E-mail: kevin.bennett@wright.edu)

Zarrin K. Chua, PhD, earned a doctorate in aerospace engineering with a focus on human–automation interaction in 2013 from the Georgia Institute of Technology. Her MS and BS degrees are also in aerospace engineering, from Georgia Tech and Virginia Tech, respectively. She has worked at research institutions such as Ecole Nationale de l'Aviation Civile, Institut Supérieur de l'Aéronautique et de l'Espace-SUPAERO, NASA Johnson Space Center, the Charles Stark Draper Laboratory, and the Technische Universität München. In addition to cognitive engineering, her research interests include cognitive modeling and simulation, experiment design, and neuroergonomics. (E-mail: zarrin.chua@gmail.com)

Richard I. Cook, PhD, is a research scientist in the Department of Integrated Systems Engineering and a clinical professor of anesthesiology at The Ohio State University in Columbus, Ohio. Dr. Cook is an internationally recognized expert on patient safety, medical accidents, complex system failures, and human performance at the sharp end of these systems. He has investigated a variety of problems in such diverse areas as urban mass transportation, semiconductor manufacturing, military software, and Internet-based business systems. He is often a consultant for not-for-profit organizations, government agencies, and academic groups. His most often cited publications are "Gaps in the Continuity of Patient Care and Progress in Patient Safety," "Operating at the Sharp End: The Complexity of Human Error," "Adapting to New Technology in the Operating Room," and the report *A Tale of Two Stories: Contrasting Views of Patient Safety* and "Going Solid: A Model of System Dynamics and Consequences for Patient Safety." (E-mail: cook.16@osu.edu)

Sidney W. A. Dekker, PhD (Ohio State University, 1996) is a professor of humanities and social science at Griffith University in Brisbane, Australia, where he runs the Safety Science Innovation Lab. He is also a professor (Hon.) of psychology at the University of Queensland. Previously, he was a professor of human factors and system safety at Lund University in Sweden. After becoming full professor, he learned to fly the Boeing 737, working part-time as an airline pilot out of Copenhagen. He is author of, most recently, *The Field Guide to Understanding "Human Error"* (CRC Press, 2014), *Second Victim* (CRC Press, 2013), *Just Culture* (CRC Press, 2012), *Drift into Failure* (CRC Press, 2011), and *Patient Safety* (CRC Press, 2011). His latest book is *Safety Differently* (CRC Press, 2015). (More at sidneydekker.com; E-mail: s.dekker@griffith.edu.au)

Elizabeth (Beth) P. DePass is a senior scientist at Raytheon BBN Technologies. She has spent the last 25 years developing a variety of military planning and logistics decision support systems. Her current research interests are applying cognitive engineering principles in the design and development of complex decision support systems that effectively promote human–machine collaboration. Together with her co-authors, she has merged cognitive engineering and design, work-centered analysis, and software engineering to design joint cognitive systems in support of military planners from Air Mobility Command and USTRANSCOM. DePass earned a BA in physics from the College of Charleston (South Carolina). She is a member of the Human Factors and Ergonomics Society. (E-mail: edepass@bbn.com)

Cindy Dominguez, PhD, is a principal cognitive engineer at the MITRE Corporation, brings cognitive engineering methods to bear in the development of fielded systems and is actively conducting and applying research toward improving human–machine teaming. Her current research passions include creating elegant designs for operators of complex systems, enabling collaboration at the intersection of humans and autonomy, and making cognitive engineering systematic within systems engineering. During 20 years as an Air Force officer and 10 years in the industry, she has led studies of cognitive work in settings that include command and control from tactical to strategic, submarine operations, intelligence analysis, and health care. Recent work in collaboration with design professionals has emphasized combining cognitive engineering and design thinking. (E-mail: cdominguez@mitre.org)

Karen M. Feigh, PhD, earned an BS in aerospace engineering from the Georgia Institute of Technology, Atlanta; an MPhil in engineering from Cranfield University, United Kingdom; and a PhD in industrial and systems engineering from the Georgia Institute of Technology. Currently, she is an associate professor in the School of Aerospace Engineering at the Georgia Institute of Technology. Her research interests include cognitive engineering, design of decision support systems, human automation interaction, and behavioral modeling. Dr. Feigh serves on the National Research Council's Aeronautics and Space Engineering Board and is the associate editor for the *Journal of the American Helicopter Society* and the *AIAA Journal of Aerospace Information Systems*. (E-mail: karen.feigh@gatech.edu)

John M. Flach, PhD, is a professor of psychology at Wright State University. He earned a PhD (human experimental psychology) from The Ohio State University in 1984. Dr. Flach was an assistant professor at the University of Illinois from 1984 to 1990 where he held joint appointments in the Department of Mechanical and Industrial Engineering, the Psychology Department, and the Institute of Aviation. In 1990, he joined the Psychology Department at Wright State University. He served as department chair from 2004 to 2013. He currently holds the rank of professor. He teaches graduate and undergraduate courses in the areas of applied cognitive psychology and human factors. Dr. Flach is interested in general issues of coordination and control in sociotechnical systems. Specific research topics have included visual control of locomotion, interface design, decision-making, and sociotechnical systems. He is particularly interested in applications of this research in the domains of aviation, medicine, highway safety, and assistive technologies. In addition to more than 175 scientific publications, John is a co-author of three books: *Control Theory for Humans* (with Rich Jagacinski; Erlbaum, 2003), *Display and Interface Design* (with Kevin B. Bennett; Taylor & Francis, 2011), and *What Matters?* (with Fred Vorhoorst; Wright State University Library, 2016). (E-mail: john.flach@wright.edu)

P. A. Hancock, DSc, PhD, is Provost Distinguished Research Professor in the Department of Psychology and the Institute for Simulation and Training, as well as at the Department of Civil and Environmental Engineering and the Department of Industrial Engineering and Management Systems at the University of Central Florida, where he directs the MIT^2 Research Laboratories. (E-mail: peter.hancock@ucf.edu)

Gary Klein, PhD, is known for (a) the cognitive models he described, such as the Recognition-Primed Decision model, the Data/Frame model of sensemaking, the Management by Discovery model of planning in complex settings, and the Triple Path model of insight; (b) the methods he developed, including techniques for Cognitive Task Analysis, the PreMortem method of risk assessment, and the ShadowBox training approach; and (c) the movement he helped to found in 1989—Naturalistic Decision Making.

The company he started in 1978, Klein Associates, grew to 37 employees by the time he sold it in 2005. He formed his new company, ShadowBox LLC, in 2014. The five books he has written, including *Sources of Power: How People Make Decisions* (MIT Press, 1998) and *Seeing What Others Don't: The Remarkable Ways We Gain Insights* (PublicAffairs, 2013), have been translated into 12 languages and have collectively sold more than 100,000 copies. (E-mail: gary@macrocognition.com)

Gavan Lintern, PhD, earned a PhD in engineering psychology from the University of Illinois, 1978. In his recent research, he has employed cognitive work analysis to identify cognitive requirements for complex military platforms. Dr. Lintern retired in 2009. He now works occasionally as an industry consultant and runs workshops in cognitive systems engineering, otherwise filling in as minder of the home pets and general home roustabout. He has published two books: *The Foundations and Pragmatics of Cognitive Work Analysis* in 2009 (Cognitive Systems Design), and *Joker One: A Tutorial in Cognitive Work Analysis* in 2013 (Cognitive Systems Design). (E-mail: glintern@cognitivesystemsdesign.net)

Jean MacMillan, PhD, is a leading expert in understanding, maximizing, and assessing human performance in complex sociotechnical systems. Her 25-year career has spanned a broad range of accomplishments in simulation-based training, human–machine interaction, and user-centered system design.

Dr. MacMillan is chief scientist *Emerita* at Aptima, Inc. and a member of its board of directors. Before joining Aptima in 1997, Dr. MacMillan was a senior scientist at BBN Technologies and a senior cognitive systems engineer at Alphatech (now BAE Systems). She has been a frequent contributor and strategic advisor to workshops and expert panels on human engineering issues for organizations such as DARPA and the military services. As an industry leader, Dr. MacMillan co-chaired a three-year National Research Council study on military needs for social and organizational models, which resulted in the publication of *Behavioral Modeling and Simulation: From Individuals to Societies.*

Dr. MacMillan holds a PhD in cognitive psychology from Harvard University. She is currently a member of the editorial board of the *Journal of Cognitive Engineering and Decision Making* and is associate editor for cognitive systems engineering for the online journal *Cognitive Technology.* (E-mail: macmillj@aptima.com)

Laura Militello is the chief scientist of Applied Decision Science, LLC, the company she helped to found seven years ago. She applies cognitive systems engineering to the design of technology and training to support decision making in complex environments. Her specialty is supporting decision making in health care. She is also acknowledged as one of the masters of advanced Cognitive Task Analysis methods, and one of the leaders of the Naturalistic Decision Making movement. (E-mail: l.militello@applieddecisionscience.com)

Christopher A. Miller, PhD, is the chief scientist and a co-owner of a small research and consulting company called Smart Information Flow Technologies, based in Minneapolis, Minnesota. He was formerly a research scientist, rising to the rank of fellow, at the Honeywell Technology Center.

Dr. Miller earned a BA in cognitive psychology from Pomona College and an MA and PhD in psychology from the Committee on Cognition and Communication at the University of Chicago. He has managed more than 40 research projects including multiple BAAs for DARPA, the U.S. Navy, Army, and Air Force, and NASA.

Dr. Miller has authored or co-authored more than 100 articles in the fields of computational etiquette, adaptive and adaptable automation, human performance modeling, cognitive work analysis and associated modeling techniques, linguistic analysis, and knowledge management approaches. (E-mail: cmiller@sift.net)

Amy R. Pritchett's research examines safety in dynamic multi-agent environments. Her current studies focus on effective distribution of authority between agents in general and air–ground and human–automation interaction in particular, especially in civil aviation. She also specializes in translating these methods into safety analyses and in identifying implications for air traffic operations, training, and regulatory standards. Pritchett has led numerous research projects sponsored by industry, NASA, and the FAA. She has also served via IPA as director of NASA's Aviation Safety Program, responsible for planning and execution of the program, and has served on several executive committees, including the OSTP Aeronautic Science and Technology Subcommittee and the executive committees of CAST and ASIAS. She has published more than 170 scholarly publications in conference proceedings and scholarly journals. She has also won the RTCA William H. Jackson Award and, as part of CAST, the Collier Trophy, and the AIAA has named a scholarship for her and selected her for the Lawrence Sperry Award. She served until 2014 as member of the FAA REDAC and chaired the Human Factors REDAC Subcommittee. She has also served on numerous National Research Council (NRC) Aeronautics and Space Engineering boards and numerous NRC committees, including chairing a recent committee examining FAA Air Traffic Controller Staffing. She is editor-in-chief of the Human Factors and Ergonomics Society's *Journal of Cognitive Engineering and Decision Making*. She is also a licensed pilot in airplanes and sailplanes. (E-mail: amy.pritchett@isye.gatech.edu)

Emilie M. Roth, PhD, is the owner and principal scientist of Roth Cognitive Engineering. She has pioneered the development and application of cognitive analysis methods in support of design and evaluation of advanced technologies, including applications in military command and control, railroad operations, health care, and process control. Together with her co-authors, she has participated in the design and evaluation of a series of work-centered support systems and collaborative automated planners for air transport mission planning for the Air Mobility Command and USTRANSCOM.

Dr. Roth earned a PhD in cognitive psychology from the University of Illinois at Urbana–Champaign. She is a fellow of the Human Factors and Ergonomics Society, is an associate editor of the *Journal of Cognitive Engineering and Decision Making*, and serves on the editorial board of the journal *Human Factors*. She is also a member of the Board on Human-Systems Integration at the National Academies. (E-mail: emroth@mindspring.com)

Penelope Sanderson, PhD, is a professor of Cognitive Engineering and Human Factors with the Schools of Psychology, ITEE, and Clinical Medicine at the University of Queensland. She leads the Cognitive Engineering Research Group where her current focus is the development of principles for the design of visual and auditory displays, and the development of approaches for understanding and designing workplace coordination.

Dr. Sanderson earned a BA (Hons I) in psychology from the University of Western Australia and an MA and PhD in engineering psychology from the University of Toronto. She is a fellow of HFES, IEA, and the Academy of the Social Sciences in Australia.

Her work has been recognized by HFES's Paul M. Fitts Education Award (2012), Distinguished International Colleague Award (2004), and Jerome Ely Best Paper Award (1990 and 2005); the Human Factors and Ergonomic Society of Australia's Cumming Memorial Medal (2014); and the American Psychological Association's Franklin V. Taylor Award (2012). (E-mail: p.sanderson@uq.edu.au)

Nadine Sarter, PhD, is a professor in the Department of Industrial and Operations Engineering and the director of the Center for Ergonomics at the University of Michigan. She earned a PhD in industrial and systems engineering, with a specialization in cognitive systems engineering, from The Ohio State University in 1994. Dr. Sarter's primary research interests include (1) types and levels of automation and adaptive function allocation, (2) multimodal interface design, (3) attention and interruption management, (4) decision support systems, and (5) human error/error management. She has conducted her work in a variety of application domains, most notably aviation and space, medicine, the military, and the automotive industry. Dr. Sarter serves as associate editor for the journal *Human Factors*. (E-mail: sarter@umich.edu)

Ronald Scott, PhD, is a lead scientist in the Information and Knowledge Technologies group at Raytheon BBN Technologies. Dr. Scott earned a BS in mathematics from MIT, and an MS and PhD in mathematics from the University of Chicago. His research is focused on design and development of decision support systems with a particular emphasis on two techniques: (1) dynamic visualizations to directly support the cognitive needs of the decision maker, and (2) development of algorithms, or tailoring of existing algorithms, to better allow the human decision maker to work collaboratively with the algorithm. (E-mail: ron.scott@raytheon.com)

Daniel Serfaty, PhD, as Aptima's principal founder, has established and implemented an ambitious vision for Aptima as the premier human-centered engineering business in the world. In addition to his management and governance duties, his work continues to involve the leadership of interdisciplinary projects for U.S. government agencies and private industries. These efforts aim at optimizing factors that drive cognitive, behavioral, and organizational performance as well as integrating humans with technologies to increase resilience in complex sociotechnical systems. Recently, he led the launch of a sister company to Aptima, named Aptima Ventures, dedicated to the commercialization of Aptima's intellectual property in human performance. Initial ventures in the new company's portfolio are in the fields of learning technologies, corporate training, predictive medical analytics, and cognitive health. His many industry activities include the leadership of various technology leadership forums and serving on the board of directors of several technology businesses and on senior advisory boards of academic, educational, and nonprofit institutions.

Serfaty's academic background includes undergraduate degrees in mathematics, psychology, and engineering from the Université de Paris and the Technion, Israel Institute of Technology, an MS in aerospace engineering (Technion), and an MBA in International Management from the University of Connecticut. His doctoral work at the University of Connecticut pioneered a systematic approach to the analysis of distributed team decision making. He is the recipient of the University of Connecticut Distinguished Service Award and has been inducted in its Engineering Hall of Fame. (E-mail: serfaty@aptima.com)

Colonel Lawrence G. Shattuck, PhD (U.S. Army, Retired), is a senior lecturer in the Operations Research Department at the Naval Postgraduate School where he directs the Human Systems Integration Program and serves as the chair of the Institutional Review Board.

Col. Shattuck graduated from the United States Military Academy in 1976 and served in the U.S. Army in a variety of positions and locations for 30 years. During his last 10 years on active duty, he was the director of the Engineering Psychology Program and Laboratory at the United States Military Academy. He holds an MS degree in human factors psychology from Rensselaer Polytechnic Institute and a PhD in cognitive systems engineering from The Ohio State University. (E-mail: lgshattu@nps.edu)

Stephen F. Smith, PhD, is a research professor of robotics and director of the Intelligent Coordination and Logistics Laboratory at Carnegie Mellon University. His research focuses broadly on the theory and practice of next-generation technologies for automated planning, scheduling, and coordination. He pioneered the development and use of constraint-based search and optimization models and has developed systems for collaborative decision making in a range of transportation, manufacturing, and space mission planning domains.

Dr. Smith earned a PhD in computer science from the University of Pittsburgh. He is a fellow of the Association for the Advancement of Artificial Intelligence (AAAI), is an associate editor of the *Journal of Scheduling*, and serves on the editorial boards of *ACM Transactions on Intelligent Systems and Technology, Constraints*, and the *International Journal on Planning and Scheduling*. He is also a member of the Executive Council of AAAI. (E-mail: sfs@cs.cmu.edu)

Robert Truxler has contributed to a number of work-centered support systems for the Air Mobility Command and USTRANSCOM as an engineer, designer, and technical lead. He earned a BA in computer science from the College of the Holy Cross and an MS in computer science from Boston University. Truxler currently leads an engineering team at TripAdvisor LLC where he applies cognitive engineering to tools designed for restaurant owners. (E-mail: rtruxler@gmail.com)

Jeffrey L. Wampler is a senior human factors engineer for the Air Force Research Laboratory's 711 Human Performance Wing at Wright–Patterson Air Force Base and serves as a research and development program manager for work-centered support system applications to transportation logistics command and control.

Wampler earned a BS in systems engineering and an MS in human factors engineering from Wright State University. Wampler has managed several highly successful research efforts for Air Mobility Command and USTRANSCOM and earned numerous awards for technology transition in the Air Force. Wampler has authored and/or co-authored more than 30 papers on human–computer interface design and specification, work-centered support systems, and collaborative automation. (E-mail: jeff.wampler@wpafb.af.mil)

Shawn A. Weil, PhD, is an executive vice president at Aptima, Inc. His research interests include analysis of social media, developing new concepts for command and control, investigating advanced training methods, and assessing communications for better organizational understanding. In his business strategy capacity, he provides corporate guidance regarding customer engagement, strategic planning, and internal business processes for the Federal Science and Technology market. In this role, he works with a broad range of Aptima's staff and consultants to ensure alignment between Aptima's technical capabilities and the real-world needs of Aptima's customers.

Dr. Weil's earned a PhD and MA in cognitive/experimental psychology from The Ohio State University with specializations in cognitive engineering, quantitative psychology, and psycholinguistics. He previously earned a BA in psychology/music from Binghamton University (SUNY). He is a member of the Human Factors and Ergonomics Society, the American Psychological Association, the National Defense Industrial Association, and the Cognitive Science Society. (E-mail: sweil@aptima.com)

David D. Woods, PhD, is a professor in the Department of Integrated Systems Engineering at The Ohio State University. He was one of the pioneers that developed cognitive systems engineering, beginning in 1980 and continuing for the next 35 years, with studies of systems of people and computer in critical risk, high-performance settings including nuclear power emergencies, pilot automation teams, anomaly response in space shuttle mission operations, critical care medicine, replanning military missions, and professional information analysis. The empirical results and design implications can be found in his books *Behind Human Error* (Ashgate, 1994; 2010, Second Edition), *A Tale of Two Stories: Contrasting Views of Patient Safety* (National Patient Safety Foundation, 1998), and Joint Cognitive Systems (*Foundations of Cognitive Systems Engineering*, Taylor & Francis, 2005 and *Patterns in Cognitive Systems Engineering*, Taylor & Francis, 2006). His more recent work founded resilience engineering—how to make systems resilient to improve safety—as captured in *Resilience Engineering: Concepts and Precepts* (Ashgate, 2006). Dr. Woods is past president of the Resilience Engineering Association (Ashgate, 2011–2013) and of the Human Factors and Ergonomic Society (1999). (E-mail: woods.2@osu.edu)

Part I

The Evolution and Maturation of CSE

1 Introduction

Philip J. Smith and Robert R. Hoffman

CONTENTS

This book has a number of goals. Foremost, the contributors have been asked to distill what they see as the major contributions to the science and practice of cognitive systems engineering (CSE), using concrete examples that illustrate how to apply this guidance. The target audiences include students and practitioners specializing in cognitive engineering, as well as the wide range of system designers coming from disciplines such as computer science, industrial design, information science, instructional design, cognitive psychology, and a variety of engineering fields such as aerospace, electrical, industrial, mechanical, nuclear, and systems engineering.

A second goal has been to provide a glimpse into the history and evolution of the field as it was driven by a push to design increasingly complex human–machine systems that were enabled by advances in a variety of technologies. This practical need, along with the *Zeitgeist* of the early 1980s, brought together an interdisciplinary mix of researchers from the psychological sciences such as cognitive psychology (and kindred fields like anthropology and social psychology) with fields focused on the invention of new technological capabilities (such as artificial intelligence) and fields focused on systems engineering, cybernetics, and control theory.

The demands from fielding complex sociotechnical systems, often with significant safety ramifications, forced recognition that overall work system performance emerged from interactions of human operators with highly coupled technological components. This focus on human–technology integration was then further extended to recognize that a focus at the level of the individual operator was not sufficient. Teamwork matters, as do interactions within broader distributed work systems where, while the individuals have to be characterized as part of a "team," in reality they act as members of distinct, separate interacting organizations. Group dynamics and organizational influences (as discussed in Chapter 7 by Weil, MacMillan, and Serfaty) thus come into consideration as a dimension of CSE, as do the impacts of broader economic and political forces.

As a third goal, the authors have been asked to highlight contributions by David Woods in recognition of his seminal role in helping to develop the framework defining the field of CSE and in contributing important details, as well to provide insights into their own journeys. This underlying theme provides interesting insights into science and technology development as human endeavors. It also highlights how the external demands that arise from developing and fielding functional systems can shape efforts and even affect implicit moral judgments as highlighted in Chapter 5 by Hancock.

1.1 THE EMERGENCE OF CSE

The first part of this book focuses on the evolution of CSE. Interviews with Dick Pew, Neville Moray, and Tom Sheridan provide an exceptional opportunity to understand how early work in cognitive human factors provided the foundation that supported the rise of CSE as a field. This early work included researchers with a focus on cognitive psychology, even before it was called that, using concepts from control engineering to model important constructs such as attention, information processing, and human performance, as well as researchers and practitioners concerned with modeling human performance on vigilance and psychomotor tasks directly relevant to the effectiveness of sonar operators in submarines and pilots flying aircraft. Much of this work was subsumed under the label of "human error" (as discussed in Chapters 11 and 12 by Sarter and Dekker, respectively).

From its beginning, CSE was driven by an understanding that in large complex work systems, human operators need to be considered as important components within tightly coupled human–machine systems, and that work system performance emerges as a result of the interactions of these components. This perspective emphasized a broader systems engineering perspective that highlighted a number of important concepts.

In a narrow, sense, CSE as a basic science thus studies the psychology of how cognitive tools shape human performance (cognitive work). In a broader sense, it focuses on how a designed work system, introduced into a broader social, organizational, and physical environment, influences individual performance and group dynamics as the humans interact with the natural, technological, and social environment, resulting in the emergence of global system behavior.

CSE is also, however, a field focused on the practice of design. The literature and practice of CSE explicitly translates this basic science into a set of prescriptive goals, guidelines, and methods for effective system design. It then goes a step further and illustrates the application of the guidelines and methods in a variety of functional systems (see Bennett and Hoffman 2015; Bunch et al. 2015; Eskridge et al. 2014).

1.2 RESEARCH AND DESIGN METHODS AND GUIDANCE

Whether for research to extend the science of CSE or for the design of operational work systems, CSE places a premium on working with experienced practitioners performing complex "real-world" tasks. Core techniques include cognitive task analyses and cognitive work analyses, as discussed in Chapter 6 by Sanderson and in Chapter 16 by Klein, Militello, Dominguez, and Lintern. The products of such analyses can also be used to structure cognitive walkthroughs to evaluate proposed system designs, making these techniques useful for both descriptive and predictive analyses. These techniques represent important extensions of traditional task analyses not only because they incorporate representations for cognitive work instead of just for externally observable performance but also because they provide a range of conceptual frameworks and "analytic templates" that help the designer to structure the analysis process.

However, the CSE enterprise goes well beyond providing such structured approaches to analyze and predict the impacts of particular design decisions. As reviewed in Chapter 10 by Smith and further emphasized in Chapter 15 by Miller, Chapter 14 by Roth et al., and Chapter 3 by Woods, numerous studies have provided insights into the influences of alternative system designs on operator performance. Examples include the impacts of brittle technologies on cognitive processes, the benefits of designing to support direct perception of the states of technologies or of work systems, or the need to design to support the adaptive behaviors of people and organizations when new cognitive tools are introduced.

Other chapters translate these findings from the science of CSE into specific principles and guidelines support design. They further provide illustrative applications of these design concepts (as discussed in Chapter 9 by Flach and Bennett, Chapter 15 by Miller, Chapter 13 by Pritchett, Chapter 11 by Sarter, and Chapter 14 by Roth et al.) and look forward to future challenges and opportunities confronting CSE (see Chapter 4 by Cook and Chapter 18 by Feigh and Chua).

REFERENCES

Bennett, K.B. and Hoffman, R.R. (November/December 2015). Principles for interaction design, part 3: Spanning the creativity gap. *IEEE Intelligent Systems*, 82–91.

Bunch, L., Bradshaw, J.M., Hoffman, R.R., and Johnson, M. (May/June 2015). Principles for human-centered interaction design, part 2: Can humans and machines think together? *IEEE Intelligent Systems*, 68–75.

Eskridge, T.C., Still, D., and Hoffman, R.R. (July/August 2014). Principles for human-centered interaction design, part 1: Performative systems. *IEEE Intelligent Systems*, 88–94.

2 Many Paths, One Journey: Pioneers of Cognitive Systems Engineering

Robert R. Hoffman

CONTENTS

This volume on the progress and state of the art in cognitive systems engineering would not be complete without some reflection on the history of the field. Lewis Hanes, Neville Moray, Thomas Sheridan, and Richard Pew were asked about the major influences on their work and ideas: *What do you regard as the most important lessons you learned over your career, with regard to the major concepts, theories, methods, and accomplishments in cognitive systems engineering?* The narratives presented in this chapter describe distinct career paths that chart a single journey, a journey leading to CSE.

To begin this narrative by extending the metaphor, someone had to clear an opening to the path that others might begin their journey.

2.1 LEWIS F. HANES

Lewis F. Hanes (Figure 2.1) served for 10 years in the U.S. Air Force and Air National Guard as a pilot flying aircraft ranging from single-engine jets to four-engine refueling tankers. He became interested in human factors engineering (HFE) based on experiencing problems with the way aircraft instruments were designed and used in the aircraft he piloted. He performed research and development (R&D) on military aircraft instrumentation for the first seven years of his professional career including in-flight and simulator evaluations. He received his BS, in agricultural sciences, and

FIGURE 2.1   Lewis F. Hanes.

then MA and PhD degrees with emphasis in HFE from The Ohio State University. From 1973 until 1992, he was a manager at the Westinghouse Science and Technology Center, responsible for a Human Sciences section and later for a department involving four sections (Intelligent Systems, Mathematics and Statistics, Decision Sciences, and Human Sciences). Although he provided R&D support to many Westinghouse Divisions (e.g., defense, CBS, appliances, elevators, and escalators), he became heavily involved in supporting the Westinghouse Nuclear Division beginning in 1974 as it designed new products, including the AP 600 reactor, the Safety Parameter Display System (SPDS), and the Computerized Procedures System.

From 1994 to 2004, Lewis was Human Factors Engineering Project manager in the Nuclear Power Division of the Electric Power Research Institute (EPRI). He continued to work on EPRI projects after retirement from EPRI until 2015. He was responsible for performing and also contracting and supervising a number projects focused on human factors guidance for control room design, operating and maintenance procedures, operator interfaces, tacit knowledge elicitation and capture (which is related to CSE), and training.

Lewis has served as a consultant for Duke Energy, the Department of Energy, and the U.S. Nuclear Regulatory Commission. He is a Fellow of the Human Factors and Ergonomics Society, and served as its president in 1975–1976. He was a senior member of the Institute for Electrical and Electronics Engineers, and served as the chairman of the IEEE Human Factors Engineering Subcommittee of the Nuclear Power Engineering Committee. Although Lewis has been involved in many sorts of projects, his involvement in human factors for the nuclear power industry is crucial for the history of CSE.

I started a human factors activity at the Westinghouse Research and Development Center in 1973. I became very busy with support requested by several Westinghouse Divisions, and found it necessary to add HFE personnel to the activity, which became a section. The first person hired in 1974 was Dr. John O'Brien, and his initial assignment was to support a project with the Nuclear Power Division of Westinghouse. This was five years before the nuclear plant accident at Three Mile Island (TMI). John developed a very strong rapport with some key people in the nuclear division, and he was helping them with a variety of projects related to new products and reactor designs. In 1979, John decided to join EPRI and so he resigned. I needed somebody to replace John, and went to the 1979 (I believe) Annual Human Factors Society meeting to recruit. I interviewed Dr. Dave Woods, who had just finished his PhD degree at Purdue. I was impressed and offered him a job with the Westinghouse Research and Development Center. An initial assignment was to replace John O'Brien and begin supporting the Nuclear Power Division. One of his early projects involved helping them develop an SPDS instrument. After TMI, the NRC required every nuclear plant to have one integrated display that showed control room operators the critical safety parameters associated with the plant. Operators could view this display and quickly understand the state and "health" of the plant. EPRI developed a separate SPDS design. EPRI wanted to compare and evaluate its SPDS design to the Westinghouse design. Our group was selected by EPRI to perform the evaluation because we had access to a full-scale control room simulator, access to control room operators from several nuclear utilities, and HFE personnel qualified to perform evaluations. No other organization had these capabilities at that time. Dave Woods was heavily involved in this evaluation study. I believe that this evaluation project and the previous project to help design the Westinghouse SPDS had a strong influence on Dave's thinking about the cognitive-CSE area.

There were two other activities that started near the completion of the SPDS evaluation project that had major impacts on Dave's thinking and writing about CSE. John O'Brien, who then was at EPRI and had contacts with nuclear utilities around the world, and I established an information exchange activity involving European, Japanese, and other nuclear utilities and research organizations interested in HFE as it applied to nuclear plant control rooms. Participants included EdF (the French utility), the Risø National Laboratory from Denmark (with participation from Jens Rasmussen, Erik Hollnagel, and others), the Halden Reactor Project based in Norway, and research laboratory personnel from Italy, Germany, and Finland, among others. We would usually meet in Europe and conduct workshops at which attendees would discuss their projects and the entire group would identify major challenges. Dave became involved in these workshops and met and became heavily involved with Erik Hollnagel and Jens Rasmussen. The second activity involved a joint four-party agreement that included Westinghouse, the French equivalent of the U.S. Nuclear Regulatory Commission, EdF (the French utility), and Frametome (at the time a French company licensed by Westinghouse for nuclear plants). Human Factors was identified by the group as a topic of interest. I was responsible for the initial HFE participation, but became busy with other responsibilities and Dave took over. One of his activities involved helping EdF evaluate the design for a new plant control room that had a different design and operating philosophy than found in control rooms designed by vendors located in the United States.

The R&D work for the Nuclear Power and other Westinghouse Divisions grew rapidly in the 1980s. I added several personnel to the Human Sciences section to support the Nuclear Power and other Divisions, including Dr. Emilie Roth in 1982 (who had a PhD in cognitive psychology from University of Illinois and now heads Roth Cognitive Engineering), Dr. Randy Mumaw (who had a PhD in cognitive psychology from Carnegie-Mellon University and now is at the NASA Ames Research Center), and Dr. Kevin Bennett (who had a PhD in applied–experimental psychology from the Catholic University and is now a professor and co-director of the Cognitive Systems Engineering Laboratory at Wright State University). Dave interacted with these personnel and had a major impact on their thinking, and to a certain extent, their thinking had to influence his thinking. He and Emilie Roth worked very closely together and co-authored many reports and papers. We were fortunate in that several Instrumentation & Control (I&C) engineers in the Nuclear Power Division recognized the importance of HFE early in our work with them. The continued support of our work was attributable to the quality and value of support provided by John O'Brien first and then the results provided by Dave Woods, Emilie Roth, and others. The key I&C engineers were Dr. John Gallagher, Mr. Jim Easter, and Mr. Bill Elm. Dave definitely influenced their thinking about designing Human System Interfaces taking into account CSE concepts. Although Dave worked closely with all of the I&C engineers, he and Bill Elm worked very closely together. Bill subsequently left Westinghouse and established a company named Resilient Cognitive Solutions. His company has been very successful in applying CSE and related concepts to intelligence and also defense command and control applications. His company recently completed a three-year contract for EPRI to develop decision-centered guidelines for upgrading existing and developing new control rooms for nuclear plants.

My major roles in helping Dave, Emily, and others develop CSE and related concepts were to establish and obtain funding for several of the initial nuclear-related projects, as mentioned above, make sure funding was available for their work, add additional professional and support personnel to help them, and establish a leading-edge HFE laboratory to perform evaluations. My responsibility was to provide an environment and resources for them to succeed. I believe I provided a field that was fertile enough that they were able to grow in it. They had the talent but a fertile field is needed to take advantage of the talent.

2.2   NEVILLE MORAY

Neville Moray (Figure 2.2) was born in London in 1935. He studied medicine, physiology, philosophy, and psychology at Oxford University, and received a PhD in 1960 in experimental psychology. There, he researched the "cocktail party phenomenon" of attentional capture using the dichotic listening paradigm. At Oxford, he had taken the second course ever offered in computer programming.

I became very interested in computing, by reading the work of McCulloch and Pitts, von Neuman, Feldman, and Feigenbaum. In 1968 (at the University of Sheffield), I introduced the first computer for experimental control in a UK psychology department. We built the interfaces for it ourselves.

Figure 2.2   Neville Moray.

He continued to research the psychology of attention, teaching courses in experimental psychology and, as he says, what would come to be called cognitive psychology.

I was asking questions about how attention mechanisms worked in the brain, using behavioral methods and modeling. Then, I went for a sabbatical to MIT in 1968, and during that year, I met engineers for the first time who were interested in human–machine systems: Tom Sheridan and John Senders. I felt that the modeling they were doing was much more powerful than the kind of modeling I had got used to in behavioral attention research. I got interested in manual skills, control skills, and how people use their attention. So I stopped trying to model attention mechanisms in the brain and began working on how people pay attention to displays, systems they are using in real-world settings, applied settings.

This interest was confirmed and encouraged by my being made the UK representative to the NATO Science Committee Special Panel on Human Factors in the 1970s, which gave me the chance to see a very wide range of Human Factors problems. I think this experience was the accident that led me definitively away from "pure" psychology and into human factors. A later year at MIT with Tom Sheridan led to my getting involved more deeply with human factors work and in particular with the nuclear industry (Moray 2016).

In 1980, he was invited to join the Industrial Engineering Department of the University of Toronto to teach human factors. For the rest of his career, he worked in engineering departments rather than psychology departments. He served on the human factors panel of the U.S. National Research Council. As his research progressed into our age of ever more automation, he pioneered the study of mental workload, human–machine interaction, and trust between humans and automation.

I think the most important person was Jens Rassummseen, who invented the Skills–Rules–Knowledge trichotomy. I saw a lot of Jens though I never went and worked in Denmark. I served on many committees with him. National Research Council Committees, NATO committees, and various meetings in

Europe on human error, along with Jim Reason. What I saw happening was we had very powerful models of manual control and manual skills, generated by the engineers, and later on by the optimal control theory people at Bolt Beranek and Newman. I spent another year working at MIT with Tom. I knew Bill Verplank quite well also. (Verplank was co-author of the technical report that introduced the notion of levels of automation.)

What I saw happen was that the basic psychological research was best suited for skill-based behavior, manual control, automatic tasks, highly practiced. What began to appear, and this is where I met Dave Woods for the first time at a meeting after Three Mile Island, we began to move over to considering that in real human–machine systems, people work in teams, very complicated environments, very large numbers of displays, and I began to think that understanding eye movements was important, because that really is the paradigm of how attention is used by people in most real-life tasks. So I did some work on eye movements on radar operators in the UK and I slowly moved over to thinking about how people behave and distribute their attention dynamically in real-world situations.

People like Dave were also coming over to [the Rasmussen approach because it] opened the door to how we could think about and model real-time behavior in complex systems where people interact a lot with one another and also [use] these very large control and display systems, like in power plants.

I was still doing laboratory research, quite a lot of it was on mental workload, funded by NASA, and that was a nice bridge because you could to some extent quantitatively model workload but at the same time it was close to classical behavioral psychology.

One of the things I discussed a lot at the time with Dave Woods [was that] the behavioral psychologists were still rather unhappy about using the term "mental model." But the engineers were perfectly happy. I think that was because [of the development of] optimal control theory. And since the engineers were used to this in computers, it came quite natural to talk about mental models in humans as well. I did not suffer too much from Behaviorism because it was never as strong in Europe as it was in the United States.

Arising out of the workload and mental models work, and the work of Hollnagel, Woods, and Jim Reason, Tom Sheridan has always had this deep understanding of the importance of the relation of humans and machines. I always enjoyed the story that Tom went to his head of faculty at MIT and said, "Look, I should not really be called a mechanical engineer because I am interested in the human side." So they gave him the title of something like the Chair of Cognitive and Engineering Psychology.

Tom's work on automation and on models has always been a great inspiration in those areas, and completely at ease with the idea of applying engineering approaches to real human situations, and once you do that, you get directly involved with knowledge-based behavior, rich cognitive behavior, human error, and it feeds into all these problems.

John Senders at Toronto was a big influence. He started during the information theory stage of development of this kind of work. He did important work on eye movements and information as a model of dynamic visual attention. Was tremendously important. I learned a lot from him about how to be an eye movement researcher. The two great figures were Tom Sheridan and Jens Rasmussen. Including Dave Woods, we were all involved in nuclear human factors.

2.3 THOMAS B. SHERIDAN

Thomas Sheridan (Figure 2.3) was born Cincinnati, Ohio. He received his BS degree from Purdue University in 1951, his MS degree in engineering from the University of California–Los Angeles in 1954, his ScD degree from the Massachusetts Institute of Technology in 1959, and an honorary doctorate from Delft University of Technology in The Netherlands. He served as president of the Human Factors and Ergonomics Society, president of the IEEE Systems, Man and Cybernetics Society, and chair of the National Research Council on Human Factors. He received the Paul Fitts Award and the President's Distinguished Service award of the Human Factors and Ergonomics Society, the Centennial and Millennium Medals, the Joseph Wohl and Norbert Weiner Awards of the Institute for Electrical and Electronics Engineers, and the National Engineering Award of the American Association of Engineering Societies. He was elected to the National Academy of Engineering in 1995. He is currently professor emeritus in the Departments of Mechanical Engineering and Aeronautics–Astronautics at MIT. He is a pioneer of telerobotics and human-automation systems, and introduced the concept of "levels of automation."

> In high school I thought I might want to be an industrial designer. I went off to Chicago where there were a bunch of industrial design firms, and I found out that it was mostly marketing that they were interested in. I didn't get the sense that the most important thing was some kind of functional relationship with people that they were after.
>
> I was in ROTC during the Korean War and was at Wright–Patterson Air Force Base. I was there as an experimental subject and an assistant to a test pilot who was doing high-speed ejections from airplanes. I went through jump

Figure 2.3   Thomas B. Sheridan.

school and I rode centrifuges. It was a sort of discovery of the human body and cognition. I had fiddled with industrial design but I didn't know quite what I wanted. I wanted to do technical stuff but I wanted to help people. Combining engineering with human capabilities seemed interesting, and that opened my eyes to the notion that psychology and physiology and engineering could work together. There was a psychology branch there but I was kind of on the fringes, in what was at the time called the Integration Section. I did a two-year stint, got out of the service, went to UCLA and worked with John Lyman. He ran a biotechnology laboratory. I had interviewed with Paul Fitts but I did my Master's degree at UCLA.

I had worked one summer at MIT with John Arnold, who had a program called Creative Engineering, which sounded strange. The MIT Department Head was hard-nosed about the traditional disciplines, and didn't like what we were doing. But John gave me freedom to sort of start my own mini-course in human factors, because I'd been exposed to that at Wright–Patterson. So that was great opportunity, so I went back to MIT, as an instructor, allowed me to teach courses while doing my graduate work. So I started some courses in human factors.

MIT was mostly engineering and architecture, those kinds of things. But J.C.R. Liklider was there, and some people were trying to get a Department of Psychology started and they were turned down flat and they all went off in a huff to other places, like Bolt, Beranek and Newman, and DARPA. I got permission to go down to Harvard and get involved in their first year psychology program. At that time, B.F. Skinner, George Miller, E.G. Boring, and S.S. Stevens the psychophysicist, were all there. I had not done much reading, because I grew up as an engineer and I did problem sets. You did not have to read books. So it was a kind of a rude shock. I got this stack of books I had to read in that first semester. That was eye-opening.

I finally landed with an MIT professor in mechanical engineering who took me on even though what I was interested in was not exactly in his bailiwick, but he was a very smart guy and I learned a lot from him about systems. He was a systems guy, control theory and modeling systems, analogies. The big idea that I think I ended up with was how you take a few things and you can build analogies with … social systems and physical systems and economic systems, and so on. They are all pretty much the same thing, at some level. So that was an eye-opener.

The Sheridan and Verplank 1978 technical report is a classic, but not widely read or known today by human factors specialists. It stunned us both that that (the levels of automation) rating scale took off. I know Dave Woods has some negative reactions, like it was supposed to be an algorithm of some kind. I said no, it was just a vague idea of a suggestion. Something to think about. It was never intended to be a recipe or an algorithm. Other people have their own interpretations. The original report, those levels were just one part of one table, a much richer analysis.

There were other ideas in that. The notion that humans and computers, if they are going to work together, have to have models of each other. My example is athletic players on a team, think of a basketball team. The players have to have models of each other, not only physical models of where they are going to be several seconds hence, but also what they are going to do. Now we talk about human mental models … we think we know what we mean, I don't think we quite know, but that's a difficult issue. But for a computer to have a model of a person, or an animal or something else that's changing, I think is fairly doable.

What kind of questions were driving you that you wanted to work on?

Well, I guess it was the human–technology relationship I think I sort of latched onto very early on, it was style and fancy marketing stuff. That turned me off a little bit, and got me a little bit more oriented toward the engineering side. So, in terms of your question, I do not know. I think just sort of I worried a lot about social issues of technological progress. We're good at making technology and things seem to be getting better, but are they really? There is a Luddite component that has a lot of validity. I've always been a bit of a skeptic on the technology side of things.

With regard to theories, I consider myself a modeler, and from that perspective, I see CSE as currently weak on models that will influence engineers and technology decision makers. True, we have some nice qualitative models, but we must improve the quantification and ability to predict. We do respect data and experimentation, and that's good. But translating experiments into generalizable, denotative, and predictive models remains a challenge. "Situation awareness" is a case in point. The efforts are brave, but attention, alertness, cognitive models, and many other connotative mental constructs get mushed together.

2.4   RICHARD PEW

Richard Pew (Figure 2.4) received his BS degree in electrical engineering at Cornell University in 1956, his MA in psychology at Harvard University in 1960, and his

Figure 2.4   Richard Pew.

PhD in psychology at the University of Michigan in 1963. He served as president of the Human Factors and Ergonomics Society in 1977–1978 and president of Division 21 (Engineering Psychology) of the American Psychological Association in 1985–1986. He received the Paul M. Fitts Award of the Human Factors Society for outstanding contributions to Human Factors Education in1980, the Franklin V. Taylor Award of Division 21 of the American Psychological Association for outstanding contributions to Engineering Psychology in 1981, the U.S. Air Force Decoration for Exceptional Civilian Service in 1993, and the Arnold M. Small President's Distinguished Service Award of the Human Factors and Ergonomics Society in 1999.

I studied electrical engineering (EE) at Cornell and became particularly interested in the part of EE that studies design and operation of control systems. In my fifth year, I read a paper by Birmingham and Taylor (1954) on "Man-operated Continuous Control Systems." It immediately captured my attention because I didn't want to be an "engineering geek." I wanted to do something that had to do with people. The paper described the notion that control engineering concepts could be used to describe some aspects of human performance. Now that really appealed to me! More of that after a brief interruption...

During my Cornell career, I wanted to have some athletic activity. Just before high school, I had a serious illness (suspected of being rheumatic fever) and was told that I could not participate in athletics for several years. By the time I finished high school, I was cleared so that at Cornell I wanted to play a sport. Because I had no high school activity in any sport, I had no prior experience to draw on. As a sophomore, I picked fencing since few students who were interested in it, like me, had any prior fencing background (that is definitely not true today). And I could be on an equal footing, so to speak ... I took to it immediately and became a member of the Cornell Fencing Team for three years and was quite successful. In my fifth year, I was no longer eligible for the team but continued to fence with them and entered two national competitions. These were AAU sponsored rather than college meets. These competitions were the qualifications to make the U.S. Olympic Fencing Team in 1956. Much to my surprise (and the organizers'), I finished first in both competitions and, as a result, qualified for the Olympic Fencing team for the Games held in Melbourne, Australia, in November of 1956.

At Cornell, I was enrolled in the U.S. Air Force ROTC Program (Reserve Officers Training Corps) and, as a result, was obligated to spend two years on Active Duty after graduation. Since I had already qualified for the Olympic Team, my first duty assignment in the Air Force was to be based at Mitchell Air Force Base, in Garden City (very near where I lived with my Father), and to "train for and participate in the Olympic Games." I spent the summer living at home in Garden City, New York, and traveling into New York City to the New York Fencer's Club every day to train. I was actually a 2nd Lieutenant in the Air Force when I went to Melbourne in November (November was Melbourne summer, since it was in the Southern Hemisphere).

Participation in the Olympics was just amazing, from marching in the opening ceremony, to living in the Olympic Village, to having meals with many other U.S. participants who were luminaries with recognized names and reputations. And, of course, participation in the events themselves. I competed in the fencing Epee Team event and the Epee Individual event. In the Team event, I did not do well and we did not do well as a team. However, in the Epee Individual event among perhaps 40 participants, I qualified for a final round of eight and finished fourth behind three Italians, Pavasi, Delfino, and Mangiarotti.

When I returned from Australia, I sat around Mitchell Air Force Base doing almost nothing for 6 weeks before I was reassigned to Wright–Patterson Air Force Base in Dayton Ohio, my "dream assignment," since I had read about the Psychology Branch of the Aeromedical Laboratory as a place were I could learn more about the application of experimental psychology in engineering settings. I was assigned to the Controls Section, which fit my background in electrical engineering and servomechanisms (now called automatic control). It was a small group that did human factors research on various aspects of cockpit design. Since they were mostly psychologists with little knowledge of engineering, they were delighted to have me. When this project had been active for about six months, we decided that the team should visit all the labs around the country that did this kind of work. We went to the Cornell Aeronautical Lab, Princeton University, the Franklin Institute, and some west coast aerospace companies. This was a marvelous opportunity to get educated in how to build models of human performance, although it was all focused on control engineering.

I was at Wright–Patterson Air Force Base for about a year and a half, and then my Air Force tour was up. One of my colleagues in the Air Force said, "I know somebody in Boston, his name is J.C.R. ("Lick") Licklider, he just moved from MIT to Bolt, Beranek and Newman, Inc. (BBN), and he's looking for someone like you." Lick was a very senior scientist who was trained as a psychologist, and a self-taught computer scientist. Much to my surprise, he hired me solely on the basis of my very limited resume. As a result, I went to Boston and I started working for BBN. Jerry Elkind had been one of "Lick's" students at MIT. I worked with Elkind more than I did with Lick, but Lick certainly was an influence on me, to move in this direction, of looking at human performance from a scientific, engineering perspective.

Starting back when I was at Wright–Patterson, they told me that if you really want to do this, you've got to go back to school and get some psychology. Licklider wrote a letter of recommendation for me, to S.S. Stevens at the Harvard Psychology Department and I was accepted as a "Provisional Student" with the idea that if it worked out, I might stay and finish a PhD but I didn't like Harvard and they didn't particularly like me. What I was learning was theory, but I wanted to deal with practical things. At the end of that school year, the Harvard Experimental Psychology Department decided that they weren't going to invite me to continue. I decided to go back to BBN for a year and figure out what I wanted to do next. However, during my year at Harvard, I had met Elizabeth (Sue) Westin, a schoolteacher in Newton who was a Cornell graduate and I fell in love. We dated most of that year, were married in August 1959, and set up housekeeping in Cambridge, Massachusetts, while I worked at BBN and she in Newton. During that year at BBN, we contemplated what to do next, and concluded that it was going to be important for my career to obtain a PhD. I applied at the University of Michigan in psychology (to work with Paul M. Fitts) and UCLA in engineering (to work with John Lyman). Both had strong interests in the scientific study of human performance. I was accepted at Michigan and turned down at UCLA.

In contrast to Harvard, I just loved Michigan. I remember sitting in Prof. Fitts' first class, and as I walked out of the class, I thought to myself, "This is what I really want to know." The people that influenced me the most were Arthur W. Melton and Paul Fitts but also Robert M. Howe, a control engineer who taught me graduate-level control engineering and who later co-chaired a NASA research grant with me.

I had told BBN that I wanted to get my degree and come back, but it took me 14 years to get back because I wound up on the faculty at Michigan. I was in the Psychology Department but I teamed with people in the Control Engineering Department. My thesis was on motor skill timing. I decided to stay on one year at Michigan kind of as a postdoc working in Paul Fitts' lab. And then Professor Fitts passed away suddenly of a heart attack. The chairman of the department asked me to stay on, not just as a postdoc but as an assistant professor. And that's how I wound up staying for 11 years.

After 11 years, I finally decided that I really didn't like the academic world and I wanted to focus more on the things that were more applied and research oriented. I decided I'd go back to BBN. I've been at BBN ever since. I informally retired from BBN in 2000, but am still "on the books" and have an office.

The paper that had the most influence on me, partly because I was monitoring the work when they did it, was a monograph that was written by McRuer and Krendel (1957) called "The Dynamic Response of Human Operators." It's an absolute classic and it was a review of all the work in both engineering and psychology that had been done up to the 1950s that could be considered to be related to the issue of perceptual-motor control.

2.5   CROSSING PATHS

As these narratives suggest, all the paths crossed. Each pioneer mentioned their crossed paths with the others. For example, Lewis Hanes said:

> Dick Pew was a consultant on appliances at Westinghouse because they were just beginning to make the change from electro-mechanical to electronic appliances, like clothes dryers and refrigerators and stoves. They required more support than Dick could provide. I had known Dick from the Human Factors Society, he recommended me for the job, and I got the job there.

And it should not pass without notice that all the pioneers expressed respect for David Woods, even though they did not always concur.

Neville Moray: I've always enjoyed Dave's company and stimulation. He has had lots of good ideas, and I'm delighted to be part of a book in his honor.

Thomas Sheridan: I really respect the breadth of Dave Woods' thinking. For myself, I've always been interested in the social side of human factors. CSE folks, as with Human Factors folks in general, are not always welcomed by engineers because the CSE/HFE types are so often brought in to systems design too late and end up being critical of what the engineers have already designed, engendering ill will. Another feeling I have is that CSE types are reluctant to undertake any role in tackling the "cognitive factors" of larger social issues of our time—that impinge on economic and ethical values that are at least as important as human performance.

Richard Pew: I've always admired Dave Woods' work very greatly. I first met Dave when we worked together on a project for EPRI, in which we were looking at the nuclear industry's control panels. A [control] panel that had to be usable by anybody that walked up to it. The state of the art at the time was horrendous. He and I worked for EPRI for two or three years and produced a report on the design of control panels for the electric power industry. Dave was always great. He seemed to look up to me though I was never his mentor. We got along famously because we thought a lot alike.

2.6   OLD WINE IN NEW BOTTLES?

Paths taken by other pioneers who embarked "on the journey to CSE" have been charted, their lives and careers documented, and more yet could be charted. Although the pioneers quoted here noted the "leap" along the path—the leap of the 1980s due to the advance in computerization—all of the pioneers said in one way or another that "CSE is what we were doing all along." Thus, it is no surprise in hindsight that the pioneers balked when asked "What were the 'grand questions' you started with?" and "What resolutions did you finally achieve?"

Neville Moray: I don't think that I was trying to understand deep and fundamental questions. I sort of wriggled my way out from that by going into applied work. And it is not that I decided that the methods or approaches I was using were wrong, it was that progressively as I worked on new and different problems, I was led by the work and the phenomena to think in different ways and move over to the mix of engineering and psychology.

Richard Pew: My driving question was modeling the human operator in perceptual-motor control tasks. But it evolved. It wasn't anything that I started out thinking that's what I wanted to do. Although it wasn't any direct influence, I think the fact that I was fencing successfully influenced me to think about motor skills. And also Paul Fitts had worked in motor skills some … When I was an undergraduate, we never talked about models of anything. I was never influenced to think about the fact that a mathematical equation was a model, and most of engineering is based on mathematical models that represent the behavior of mechanical and electrical systems. But that's really what it is. So, the most fundamental thing that I learned was that you could think about mathematical models and that those models could represent the behavior of people as well as the behavior of mechanical and electrical things. Everything from Hick's law to control engineering models was all in that spirit. So that would be the most fundamental thing.

What were your thoughts when people some years ago started talking about cognitive engineering and CSE? My guess is that your response would have been "Well that's what we've been doing all along."

Correct. That's exactly what it was. We'd been doing it and they just gave it a name.

Thomas Sheridan: As far as cognitive engineering is concerned, I guess Dave Woods and others have championed that as something new and different, but I never quite have. I always saw the human–technology relationship as something that kept getting named something different. As we all know, it's got about 10 different terms … and it's true it did start out more or less way back in the late 1950s and 1960s with a more ergonomic reach and grasp, what you can see, and it has kind of evolved more cognitively. There's no question about that. But I don't see it as something crisp and clean and different from old-fashioned human engineering or human factors engineering.

What is really new about CSE that sets it apart from the specializations and focus areas that came before? Task analysis might be traced back to Denis Diderot in his encyclopedic description of the skills and methods of the craft guilds (1751; see Lough 1954). The analysis of tasks was conducted by early experimental psychologists in the 1800s. Task analysis methods matured by industrial psychologists in the decades of the two World Wars, leading to the notion of hierarchical task analysis (Annett and Duncan 1967) and various methods in use today. Across this entire evolution, task analysis was always "cognitive" in that it recognized the importance of cognitive processes in the conduct of tasks, even ones that were ostensibly or primarily physical–motoric (see Hoffman and Militello 2008). Human factors, as it emerged in industrial psychology toward the end of World War I and matured during World War II, was always about human–machine integration. Then came the computer and the concepts of cybernetics. These were game changers, and as they changed the work, they changed the scientific study of work. And as work became progressively more and more cognition dependent, systems engineering had to become progressively more and more cognition focused.

Lewis Hanes: In the 1950s and 1960s, when I was involved in performing R&D on military aircraft instrumentation, the major methods we used were task analysis, link analysis, work load analysis (sometimes using information theory), scenario analysis based on analytic methods and simulators, and so on. We tended to emphasize the physical actions pilots took while performing flight scenarios, which the then current task analyses methods supported. Although we did not have very good tools to analyze cognitive processes (e.g., decision-making, situation awareness) influencing pilot actions, we were certainly aware of the importance of these activities. I recall when we were designing new instruments to support improved pilot performance we used the analogy of not seeing the forest for the trees with the then current instruments (what we now would identify as a situation awareness problem). At that time, there was basically one sensor for each displayed parameter, for example, sensor measured altitude and the results were displayed as altitude. A separate display showed rate of altitude change even though these two parameters both show an aspect of altitude. The pilot had to develop and maintain a scan pattern to look at each of these individual instruments ("trees") to develop an understanding of the overall situation (mental integration to view the "forest"). We were developing integrated instruments that combined the parameters into a meaningful arrangement better to permit the pilot to see the "forest." The U.S. Air Force in the late 1950s had a research project funded and managed by the Flight Dynamics Laboratory at Wright–Patterson Air Force Base. (My introduction to HFE involved working for several years on this Air Force contract.) The contract objective was to develop a cockpit design (conceptual in nature since electronics, automation, and computers were not available to dynamically implement the concepts) that showed how individual sensor data could be integrated into presentations that enabled pilots to better and more accurately understand the situation and with less mental workload.

The increased interest in understanding cognitive processes and taking into account these processes during the early 1980s was stimulated by (1) the technology that became available, for example, distributed control systems, digital computers supporting automation, and electronic-based display systems; (2) recognition that human errors with serious consequences were attributable to cognitive (mental) errors; (3) the descriptions of CSE and related concepts by Dave Woods and others; and (4) the developing availability of methods and tools to support cognitive analysis (e.g., cognitive task analysis, applied cognitive task analysis, critical decision method, cognitive mapping, cognitive work analysis). The underlying interest in cognitive processes (mental models, decision making, etc.) had been around for many years, but in the 1980s, more formal recognition was given to these processes.

Changes in task analysis have been incremental, but CSE takes additional things into account, such as the environment. Fleishman invented task analysis as we know it (Fleishman and Quaintance 1984), but the basic task analysis is what it is, about the same. Cognitive analysis is incrementally better but has not made a leap forward. The tools are difficult to use, and I suspect this is a major reason that cognitive analyses are not being applied more widely today.

Neville Moray: Now, for the newer generation of people like Dave Woods and John Lee, there is a much greater acceptance of the idea of using cognitive science in applied engineering settings in the real world. And engineers are beginning to listen and see that we can actually do quite powerful modeling, at least of certain kinds of systems. My generation was lucky, it was a golden age. It was certainly fun. I never expected when I started doing work on auditory attention and the cocktail party problem that I would years later be consulting to the Army on building factories to destroy chemical weapons. I got a lot of satisfaction out of moving to applied problems. These are good problems.

Thomas Sheridan: In a funny way, we're coming back to task analysis in the robotics world. You can have the computer look at something and sort out a lot of different relationships in terms of their doing things. You can't quite get into their head, but you can infer … the business of inferring what somebody is intending is a tough one. So in a funny way, having humans and robots work together side by side is coming back to old-fashioned task analysis, I think.

Whether old wine or new, old bottles or new, the pioneers are proud of their contributions to the development of applied psychology and human factors in several countries. The pioneers are proud of having helped launch careers of outstanding students.

Richard Pew: I let students figure out what they wanted to do as much as I could. Though we had a lab and they more or less decided that they had to do something related to the lab. I was not a rigid faculty member who said "You have to work on my problems." There were people who worked that way.

Your approach is one of the defining features of a good mentor … letting students figure it out on their own, giving a little guidance and not imposing your own projects…

Right.

Thomas Sheridan: I learned a lot from the graduate students. I always felt I learned more from the graduate students than they learned from me. That was great. I just loved working with graduate students. Sheer fun. I guess I was lucky. The pioneers were finding paths while pursuing them. Like each and every one of us.

REFERENCES

Annett, J., and Duncan, K.D. (1967). Task analysis and training design. *Occupational Psychology*, 41, 211–222.

Birmingham, H.P., and Taylor, F.V. (1954). A design philosophy for man–machine control systems. *Proceedings of the IRE*, 41, 1748.

Fleishman, E.A., and Quaintance, M.K. (1984). *Taxonomies of Human Performance: The Description of Human Tasks*. New York: Academic Press.

Hoffman, R.R., and Militello, L.G. (2008). *Perspectives on Cognitive Task Analysis: Historical Origins and Modern Communities of Practice*. Boca Raton, FL: CRC Press/Taylor & Francis.

Lough, J. (1954). *The Enclycopedie of Diderot and d'Alermbert*. Cambridge: Cambridge University Press.

McRuer, D.T., and Krendel, E.S. (1957). *Dynamic Response of Human Operators. WADC Technical Report No. 56-524, Project 1365, Flight Control Laboratory, Aeromedical Laboratory*, Wright Air Development Center, Wright–Patterson Air Force Base, OH.

Moray, N. (2016). *Fellows Profile. Human Factors and Ergonomic Society* (downloaded 18 May 2016 from http://www.hfes.org/web/awards&fellows/fellowprofiles/profile_moray.pdf).

Sheridan, T.B., and Verplank, W.L. (1978). *Human and Computer Control of Undersea Teleoperations. Technical Report on Contract N00014-77-0256, Engineering Psychology Program*, Office of Naval Research, Arlington, VA.

3 On the Origins of Cognitive Systems Engineering
Personal Reflections

David D. Woods

CONTENTS

3.1   ORIGINS

The early 1980s were a time of great intellectual ferment, and many concepts and approaches that we take as standard today emerged in this period of exploration and interdisciplinary interaction. One major driver was the Three Mile Island (TMI) nuclear power plant accident in the spring of 1979. This event stimulated an international and multidisciplinary process of inquiry about how complex systems fail and about cognitive work of operators handling evolving emergencies (Woods et al. 2010).

Several other trends converged in the same time frame. One of these was the growing interest in artificial intelligence (AI), with advances that promised to automate or mechanize reasoning. The merger of AI with cognitive psychology and kindred disciplines (e.g., philosophy of mind) was a flash point for calls for "interdisciplines," one being cognitive science (Abelson 1976). As part of the discussions about cognitive science and given the context of safety of complex systems, Don Norman suggested that the new field of cognitive science also needed a companion field of cognitive engineering to apply findings about the fundamental nature of cognition whether carried out by machine, human, or a combination of the two (Norman 1980, 1986, 1987). This mix of pressing problems, new technological capabilities, and intermixing of different disciplinary approaches produced a variety of insights, concepts, methods, and techniques that challenged conventional thinking at that time.

This chapter recounts the emergence of some of these new concepts and approaches through the lens of my personal participation in the rise of cognitive systems engineering (CSE).

3.2 INITIAL COLLABORATIONS

I interacted with many people in this period. Some were engineers and designers; others were colleagues and mentors in research and development (R&D). Some were team players. I battled with others, ferociously at times, especially those who could not apprehend the importance of complexity and adaptation and who sought solace in the belief that human error was the illness (though only other people's human errors) and strict rule following (whether embedded in procedures, automation, or the new automation of machine expert systems) was the cure.

My interactions in the early 1980s centered on three collaborations. One was as part of the design team at Westinghouse, developing our approach to computerizing control rooms. Another collaboration was with my colleagues in the human sciences group at the Westinghouse R&D Center, notably Emilie Roth, John Wise, Lewis Hanes, and Kevin Bennett, as we carried out studies of how operators handled emergencies and how people interacted with expert systems and computerized support systems. Third, I worked closely with colleagues at Risø National Laboratory in Denmark where Jens Rasmussen had assembled a powerful group including Erik Hollnagel, Morten Lind, and Len Goodstein. There was cooperation between Westinghouse and the group at Risø, and we jointly explored issues about safety, complexity, cognition, computers, visualization, and collaboration.

There was also a great deal of interaction with veterans of human–machine systems who were also mentors to me such as Dick Pew, John Senders, Neville Moray, Don Norman, and others in the context of a series of international meetings on these topics. In particular, there were meetings in Italy in 1983, 1985, and 1986 on Human Error, on Intelligent Decision Support, and on Cognitive Engineering.

3.3 EXCITEMENT AND FRUSTRATION

This was a time of great excitement and energy as we felt that something new was afoot both practically and scientifically. What would emerge and stand the tests of time was not really clear at all to any of us. It was also a very lonely time as these problems fell outside of standard disciplines, challenging conventional boundaries as well as conventional wisdom. It felt like a fight for legitimacy with the outcome highly uncertain. One sign of this was the difficulty of getting work published in traditional outlets. Many of the developments, concepts, techniques, and findings of the early 1980s only appeared in print 5 to 10 years later and often in less visible outlets, at least, less visible to mainstream academics in experimental psychology, organizational factors, or engineering.

I entered this mix in the fall of 1979 after I finished my PhD in cognitive psychology where I had focused my studies on human perception and attention. I joined the Westinghouse R&D Center to carry out studies of operator cognitive work during nuclear power plant emergencies and to design computerized systems that would support operators during these emergencies. This field of designing systems to fit and support human capabilities had been labeled human factors. But at this time, concepts and techniques in human factors lagged far behind the findings and theories in cognitive psychology that I had been studying. As I observed operators handling simulated nuclear power plant emergencies, and as I participated in retrospective analyses of other actual nuclear emergencies, it was easy to see a variety of phenomena of human perception, attention, diagnosis, and decision making going on in these events (e.g., Woods 1982, 1984b; Woods et al. 1981, 1986). The human factors approaches at that time were unable to see, let alone gain traction on these topics.

Going on at the same time was the explosion of interest in AI technology and in particular the new attempts to build expert systems. Expert systems promised to mechanize reasoning so that organizations and designers need only collect relevant rules from the current human expert practitioners as a kind of "fuel" to power the new computerized inference engines. Knowledge engineering was a popular phrase to refer to the effort to collect a set of rules for the new machine expert systems. The advances in AI also encouraged the popular myth that these machine experts would take over cognitive tasks previously performed by people and carry them out faster, better, and cheaper by themselves (Hoffman 1992). But the workplaces exhibited complexities of cognitive work that seemed invisible to the computational analysis of expertise.

I was struck by the gap between the extremely spartan laboratory settings I had studied in my graduate work and the rich combination of factors at work in control room settings. Walking into a room full of engineers responsible for developing new computerized visualizations of the safety state of the power plant, they said to me, "Ah, here's our new cognitive psychologist, will you please tell us why operators made poor decisions and how we can use computers to help them do it better the next time." I was, and not for the first time in 1980, in shock—I was supposed to connect cognitive psychology to a complex setting where things could go wrong critically and use those connections to design new kinds of intelligent computerized interactive machines. I knew that what I had learned in school was based on tasks that had no direct connection to the nature of work in control rooms during emergencies and no connection at all to my new responsibility to assist the engineers in redesigning control rooms. I think I mumbled something about having left that book back in my office and that I would be right back when I knew full well there was no such book.

More than a little stunned by my new role, my first task was to figure out how to make progress. In the search for definitive results, the cognitive psychology methods that I was trained in had been focused on eliminating all extraneous variables and distilling stimuli down to the most elemental remains. This tactic was impossible for such a complex setting as control rooms and emergencies. I had to confront complexity and not run away from it.

3.4   CLEARING A NEW PATH

I well understood, based on the history of investigations into the mechanisms behind the tremendous power of human perception, that the way forward would begin with intense observations to identify the phenomena of interest. The good news was that several of my initial projects included lengthy periods where I could observe operators handling simulated emergency situations. In another project that followed the TMI accident, I was part of a team interviewing operators who had been involved in actual emergencies at nuclear power plants (needless to say, TMI was not the only adverse event in the industry, though these other events did not lead to severe consequences or garner the same levels of notoriety).

Prepared in this way, I began to see and characterize the differences between spartan experimenter-designed laboratory tasks and the rich set of interacting variables present in the control room when looked at as a kind of natural laboratory (Woods 2003). I was seeing a wide range of adaptations directed at coping with various forms of complexity. Activity was distributed across multiple agents, not isolated in only a single individual. Interconnected streams of activities evolved over time across varying phases and tempos. Individual behavior was embedded in, and could not be completely separated from, the context of larger groups, professions, organizations, and institutions. There were multiple interacting and often conflicting goals at stake so that managing trade-offs was both fundamental and hidden behind many actions and decisions. Change, especially in technological capabilities, was rampant. Tools and artifacts to aid cognition and coordination were everywhere. In fact, cognition without tools, while completely normal in the experimenter-designed laboratory, almost never existed in these workplace settings. Observing people use the existing control room technologies and examining potentially new designs revealed that clumsily designed technology was the norm, and that operators had to work around many kinds of gaps to accomplish their work (Cook et al. 2000).

3.5   THE IMPORTANT PHENOMENA ARE RELATIONAL

Many around me (including my sponsors) assumed uncritically that assessing human performance was a simple and straightforward matter of counting up successes and errors starting with the benchmark of standard procedures. But watching operators deal with complex emergency situations revealed many challenging forms of cognitive work where expertise and failure were intermingled. Digging behind the veil of procedure-following and everyone's narrow focus on major failures, I was able to begin to see how practitioners create safety normally despite conflicting goals, pressures, and dilemmas. Human adaptability compensated for many kinds of surprises that occurred due to the fundamental limits of procedures, plans, and automata (Woods et al. 1990b).

In other words, a large variety of phenomena were going on together as a "wrapped package." Untangling these required new methods and approaches. Traditional disciplines isolated out one or another of these as their topic of study and their methods were only able to assess each facet in isolation. In the natural laboratory of control rooms and emergencies, the phenomena of interest were interdependent and interconnected so that most of the action going on and the opportunities to learn depended on studying the relationships. Working on these new approaches was one of the issues in my collaboration with the Risø researchers (Hollnagel et al. 1981; Woods 2003).

One of these key relationships is the connection between problem demands and problem solving. This mutual relationship was central in the work of James J. Gibson about perception and action (1979), and the work of Jens Rasmussen about safety (Rasmussen and Lind 1981). The idea of problem demands would be central in the way I approached cognition in systems at work (Woods 1988, 1994; Woods and Patterson 2000).

Ten years later, people might comment that I simply had an opportunity to learn on the job how to carry out an ethnography of work. However, at this time, an ethnography of work was rare and novel. Work by the Tavistock Institute in the 1950s on "sociotechnical systems" was unknown to us at the time. Ed Hutchins had just begun his research on what would later be called "cognition in the wild" (1995). In addition, what we were doing was much more than ethnography because we had some additional levers that we could use to study a complex set of relationships. In studies of human perception and attention, the key is the design of the stimulus set; in studies of problem solving, the key is design and selection of the problem so that it poses the right challenges for the purposes of the investigation. I was not simply observing operators handle emergencies in the training simulator. I also had the opportunity to help design the simulated emergencies in order to challenge the operators in specific ways. I was able to deliberately design the problems that the operators would face to include specific cognitive demands, and I was able to present those problems as relatively full scope simulations that allowed repeated observations of how actual practitioners handled the specific probes and challenges built into each problem's design. Though it was never written down in this way, I still maintain today that the core skill of an effective cognitive systems engineer is the ability to design problems that challenge the boundaries of plans, procedures, and technologies. And the inverse holds as well. All plans, procedures, and technologies have bounds; these are hard to find and they move around over time; plus designers overestimate the range of situations the artifacts they create can handle (Perry and Wears 2012; Woods 2015).

3.6 GAINING CONFIDENCE

I was a part of the debates about the impact of technologies that automated reasoning, while at the same time I was observing how systems of people and various kinds of artifacts handled anomalies in concrete real situations with real consequences for any breakdowns. The contrast made crystal clear that previous notions such as Fitts' List—which contrasts what people can do versus what machines can do better—were completely irrelevant to the science or engineering challenges.

Instead, Erik Hollnagel and I in 1981 said that the unit of analysis is a *joint cognitive system* to focus attention on how cognitive work involves interaction and coordination over multiple roles of both human and machine (Hollnagel and Woods 1983). Ed Hutchins would later emphasize the same point using the label *distributed cognition* (Hutchins 1991, 1995).

To Hollnagel and me, the original core idea of CSE is that the base unit of analysis is the joint cognitive system—how multiple human and machine roles interact and coordinate to meet the changing demands of the situations they face as they attempt to achieve multiple and conflicting goals. We abandoned the language of machines compensating for human limits (after all, it is an obvious truth that all agents—human or machine, or combinations thereof—have finite resources and therefore limits). Instead, we focused on how people are adaptive agents, learning agents, collaborative agents, responsible agents, and tool-creating/wielding agents. As such, responsible people create success under resource and performance pressure at all levels of sociotechnical systems by learning and adapting to information about the situations faced and the multiple goals to be achieved.

3.7   SLOGANEERING

Four slogans captured the agenda for CSE. First, we began by studying and modeling *how people adapt to cope with complexity*, borrowing this phrase from the title of a paper by Rasmussen and Lind (1981). Second, we examined *how to make intelligent and machine agents team players* with others, noting that automata, while able to do some specific tasks very well, were fundamentally poor at coordinating with other roles—as we would put it later—they are strong, silent, and difficult to direct (Klein et al. 2004; Malin et al. 1991; Woods 1985). Third, we sought to utilize the power of good representations as captured by Don Norman's phrase—*Things that make us smart (or dumb)*—as one key method to support cognitive work (Hutchins et al. 1985; Norman 1987, 1993). And there was a fourth commitment as well—signaled by Ed Hutchin's phrase *Cognition in the wild* (Hutchins 1995): we sought ways to observe and study practitioners at work in natural laboratories of control centers, operating rooms, flight decks, mission control, and other settings where the phenomena of joint cognition in systems at work exist (Woods 2003).

Much more could be said about the key ideas in CSE, but these are available in many works, particularly the compilations Hollnagel and I prepared (Hollnagel and Woods 2005; Woods and Hollnagel 2006). For the remainder of this chapter, I will to highlight just a few of the early findings and concepts that arose from my work during the early 1980s and the initial rise of CSE.

3.8   FAILURES TO REVISE ASSESSMENTS AND FIXATION

One of my opportunities in the early period (1980s) was to observe operators in emergencies—in full-scope simulated emergencies where we could design the cases to challenge operators in particular ways and in retrospective analyses of actual cases to provide converging evidence. The result was a substantial set of data, but it was very easy to get lost in the variety of cases and in the details of the specific domain.

The challenge was how to step out of the details and richness of the setting and see general patterns. I started by looking at cases of situation assessment gone wrong, because situation assessment was a way to start to look at cognitive factors and because situation assessment seemed a relatively neutral conceptual framing.

For each case, I set up a general minimal pattern to lay out the practitioners' process based on Neisser's perception-action cycle—what data or event was noted and what knowledge did that data call to mind, and then how did the framing that resulted guide further exploration of the available data and modify the sensitivity to pick up on changes and new events (Hollnagel and Woods 2005; Neisser 1976; Woods 1993). I treated the control room team as if it were a single mind in the process tracing unless the interplay among the team members became important to the mis-assessment. For a cognitive psychologist having studied attention and problem solving, contrasting how mind-set evolved as a situation evolved—doing a process tracing—seemed an appropriate way to take previous work and use it in a new context (Woods 1993).

What I found was that, most of the time, the case began with a plausible or appropriate initial situation assessment, in the sense of being consistent with the partial information available at that early stage of the incident. In other words, people did not just get the wrong assessment; they started out assessing the situation correctly given the information and knowledge available at that point in time. What happened was that the incident evolved—new events occurred, faults produced time-dependent cascades of disturbances, new evidence came in, and uncertainty increased in some ways and decreased in other ways. In other words, the situation kept changing. Mis-assessments arose when the practitioners *failed to revise* their situation assessment and plans in a manner appropriate to the data now present. In 1988, I represented the general finding in the diagram shown here in Figure 3.1, to capture how the situation moved away from the initial assessment as the situation evolved. In general, mis-assessments occurred when practitioners could not keep pace with the changing events and evidence over time. In extreme cases, the failure to revise took on the quality of a fixation on the original assessment as the personnel involved discounted discrepant evidence and missed repeated opportunities to revise.

This finding was replicated in my latter work on Space Shuttle Mission Control and other settings. For example, Veronique De Keyser and I combined our data from different domains on cases that appeared to be fixations and were able to identify several different subpatterns for fixations (De Keyser and Woods 1990). The Columbia Space Shuttle accident provides a notable example at a broader managerial level. In that case, management failed to revise their assessment of safety risks as evidence about foam strikes during launch built up over multiple orbiter launches (CAIB 2003; Woods 2005).

The results on situation mis-assessment identified the kinds of problem demands that can make revision difficult. One of these is *garden path* problems where revision is inherently difficult because the data available early in the incident strongly suggest a plausible but ultimately incorrect situation assessment, while later or weaker cues point to the actual situation and conditions. One can then find or create a garden path problem in many different domains to observe and test how a joint cognitive system of various configurations performs at revising assessments as incidents evolve over time (e.g., Roth et al. 1992).

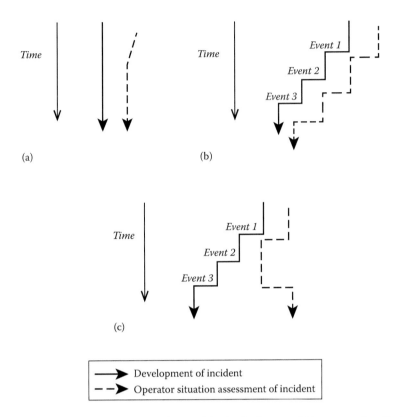

Figure 3.1 The original diagram used to explain the findings on failures to revise situation assessments. (a) A myth about the development of incidents—Operator has a single diagnosis followed by straight-through execution of response with some stereotypical feedback limited usually to action implementation. (b) Actual incidents develop and evolve due to propagation of disturbances, additional faults and failures of human and machine elements to respond appropriately. The operator must track the development of the incident and be sensitive to the next event. (c) When operational problems occur, the staff fails to update or adapt to new events. The event continues to evolve, while they are fixed on an earlier assessment of the situation. This erroneous assessment can then lead to misinterpretation and erroneous response to further developments in the incident. (From Woods, D.D. (1988). Coping with complexity: The psychology of human behavior in complex systems. In L.P. Goodstein, H.B. Andersen, and S.E. Olsen (Eds.), *Mental Models, Tasks and Errors* (pp. 128–148). London: Taylor & Francis, Figure 2, p. 133. Reprinted by permission of publisher.)

3.9   ANOMALY RESPONSE

The studies of control rooms during emergencies examined a major class of cognitive work—*anomaly response*. In anomaly response, there is some underlying process, an engineered or physiological process that will be referred to as the monitored process, whose state changes over time. Faults disturb the functions that go on in the monitored process and generate the demand for practitioners to act to compensate for these disturbances in order to maintain process integrity—what is sometimes referred to as "safing" activities. In parallel, practitioners carry out diagnostic activities to determine the source of the disturbances in order to correct the underlying problem.

Anomaly response situations frequently involve time pressure, multiple interacting goals, high consequences of failure, and multiple interleaved tasks (Woods 1988, 1994). Typical examples of fields of practice where this form of cognitive work occurs include flight deck operations in commercial aviation, control of space systems, anesthetic management under surgery, process control, and response to natural disasters.

In the early 1980s, the dominant view assumed a static situation where diagnosis was a classification task, or, for AI, heuristic classification (Clancey 1985). Internal medicine seemed archetypal where a set of symptoms was classified into one of several diagnostic categories in the differential diagnosis. However, this approach seemed limited in its capacity to capture the dynamism and risk that I was observing in anomaly response. Among many problems, classification missed the complications that arise when events can cascade. Plus, waiting until the classification was complete guaranteed that responses would be too slow and stale to handle an evolving situation.

In anomaly response, incidents rarely spring full blown and complete; incidents evolve. Practitioners make provisional assessments and form expectancies based on partial and uncertain data. These assessments are incrementally updated and revised as more evidence comes in. Furthermore, situation assessment and plan revision are not distinct sequential stages, but rather they are closely interwoven processes with partial and provisional plan development and feedback that lead to revised situation assessments. As a result, it may be necessary for practitioners to make therapeutic interventions, which have the goal of mitigating the disturbances before additional malfunctions occur, even when the diagnosis of what is producing the anomaly is unknown or uncertain (or even to just buy time to obtain or process more evidence about what is going on). It can be necessary for practitioners to entertain and evaluate assessments that later turn out to be erroneous. To reduce uncertainties, it can be necessary to intervene for diagnostic purposes to generate information about the nature and source of the anomaly. And interventions intended to be therapeutic may turn out to have mostly diagnostic value when the response of the monitored process to the intervention is different than expected. Diagnostic value of interventions is also important to help generate possible hypotheses to be ruled out or considered as possibilities.

The early studies of operators in nuclear power emergencies provided a great deal of data about anomaly response as a generic form of cognitive work that was later buttressed by studies in the surgical operating room and in space shuttle mission control. Figure 3.2 reproduces the original figure summarizing the model of anomaly response as a form of cognitive work. The best description of the model uses NASA cases and is presented in detail in Chapter 8 of Woods and Hollnagel (2006).

In the early 1980s, cognitive modeling was just beginning to become popular in the form of software made to reason similar to people (e.g., Allen Newell's SOAR 1990 or John Anderson's ACT-R 1983). Emilie Roth and I were engaged to develop a cognitive model for nuclear control rooms during emergencies. We approached the modeling task in two new ways (Woods and Roth 1986). First, we did not try to

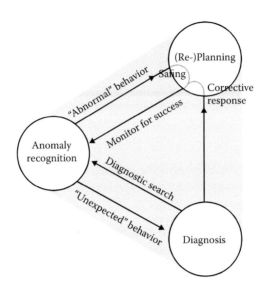

Figure 3.2   The multiple intermingled lines of reasoning in anomaly response. (From Woods, D.D. (1994). Cognitive demands and activities in dynamic fault management: Abduction and disturbance management. In N. Stanton (Ed.), *Human Factors of Alarm Design* (pp. 63–92). London: Taylor & Francis, p. 79. Reprinted by permission of author.)

model a person, a set of people, or a team. Instead, we set out to model the *cognitive demands* any person, intelligent machine, or combination of people and machines would have to be able to handle. For example, in anomaly response, there is the potential for a cascade of disturbances to occur, and all joint cognitive systems have to be able to keep pace with this flow of events, anomalies, and malfunctions.

Second, we set out to model the cognitive demands of *anomaly response* as a form of cognitive work. The model of anomaly response that we built from observations was a form of abductive reasoning (Peirce 1955). In abductive reasoning, there is a set of findings to be explained, potential explanations for these findings are generated (if hypothesis A were true, then A would account for the presence of finding X), and competing hypotheses are evaluated in a search for the "best" explanation based on criteria such as parsimony. Abduction allows for more than one hypothesis to account for the set of findings, so deciding on what subset of hypotheses best "covers" the set of findings is a treacherous process.

There were people developing AI software to reason abductively so we teamed with one developer, Harry Pople, who was working on the *Caduceus* software to perform diagnosis in one area of medicine (Pople 1985). With Pople, we tried to work through how his goals for software development could be adapted to do anomaly response in control rooms for engineered processes. The collaboration formed in part because of common interest in approaches based on abduction and in part because Pople had run into trouble as he tried to develop the *Caduceus* software to handle cases his previous software system *Internist* couldn't handle—he had run into cases where his experimental software had gotten stuck on one explanation. The way the findings to be explained presented themselves over time turns out to make a difference in what explanations appeared to be best, and his software had difficulty revising as additional information became available. Pople thought that looking at a time-dependent process would help him improve his software under development. Besides, Emilie Roth and I had lots of data on how anomaly response worked and sometimes didn't work so well (Woods 1984b, 1994, 1995a).

However, improving the software didn't go as planned. Characteristics of anomaly response kept breaking his experimental software. The first problem was determining what are the findings to be explained. The AI-ers had previously used a trick—they had determined, in advance, a fixed set of findings to be explained. The machine didn't have to figure out what is a finding or deal with a changing set of findings; the findings were handed to it on a silver (human developer) platter. By the way, when it comes to all things AI, remember that there is always a hidden trick—good CSE-ers find ways to break automata and plans by focusing on patterns of demands that challenge any agent or set of agents whatever the combination of human and machine roles.

Modeling the cognitive demands of anomaly response required hooking the machine reasoning software to multiple dynamic data channels—hundreds and even thousands for a nuclear plant at that time (and think of the advances in sensing since then that provide access to huge data streams). Even though we narrowed in on a very limited set of critical sensor feeds, the machine was quickly victimized by data overload. There were many changes going on all the time; which of these changes needed to be explained? In our model, the answer is *unexpected changes* based on the data on how people do anomaly response. But figuring out what are unexpected events is quite difficult and requires a model of what is expected or typical depending on context. Some changes could be abnormal but expected depending on the situation and context and therefore not in need of a new or modified explanation. Imposing even more difficulty, the absence of an expected change is a finding very much in need of explanation. Abductive inference engines had no way to compute expectations, but people do use expectations, and, more surprisingly, the mind generates expectations very early in the processing of external cues in order to determine which out of very many changes need more processing (Christoffersen et al. 2007; Klein et al. 2005).

Next, the software had to deal with actions that occurred before any diagnosis was reached and accepted. Abnormal changes demanded interventions or at least the consideration of intervention without waiting for more definitive assessments to occur. Many of these actions were taken by automated systems or by other human agents. These actions produced more changes and expected and unexpected responses. If there was an action to reduce pressure and pressure stayed the same, oops, that's a new unexpected finding in need of an explanation; that is, what is producing that behavior? Did the action not happen as intended or instructed? What broke down in moving from intention to effect? Or perhaps, are there some other unrecognized processes or disturbances going on whose effects offset the action to reduce pressure?

Another issue in abductive reasoning and in anomaly response is where do hypotheses come from? AI approaches snuck in a trick again—the base set of hypotheses was preloaded. Yes, in abductive reasoning computations, the software could build lots of different composites from the base set. But in actual anomaly response, generating hypotheses is part of the cognitive work, a difficult demand.

Studies showed that having diverse perspectives helps generate a wider set of possible hypotheses, as do many other factors. One thing is clear though: Asking for a single agent, human or machine, to generate the widest set of possible hypotheses all by themselves is too much. The right teaming with the right interplay will do much better.

These and other factors created an ironic twist—we didn't need a more sophisticated diagnostic evaluation in our abductive reasoner. What we needed were new modules beyond the standard AI software to track what was unexpected and to generate provisional assessments ready to revise as everything kept changing (Roth et al. 1992; Woods et al. 1987b, 1988, 1990a).

While I could go on in great depth about different aspects of anomaly response as a critical activity in joint cognitive systems at work, several things stand out. Anomaly response is the broadest reference model for diagnostic processes; some settings relax parts of the demands or function in a default mode as mere classification. But broadly speaking, all diagnostic activities are some version of anomaly response. Second, the model we developed in the mid-1980s is still the best account of the demands that have to be met by any set of agents of whatever type and in whatever collaborative configuration. And third, we still don't have software that can contribute effectively to the hard parts of anomaly response as part of a joint cognitive system. It's disappointing.

3.10   BRITTLE MACHINE EXPERTS

The early 1980s witnessed the explosion of software that promised to mechanize cognition. This is interesting in part because there have been several waves of technology since promising the same (and in 2017, we are in the middle of a new intense wave of promises about mechanized cognition and action). And remember, CSE started with the realization that either/or architectures of whatever form—some variation on either people do it or machines do it—function poorly and require people to work around the limits (Woods 1985). As I summarized years later in the first law of cooperative systems—it's not cooperation if either you do it all or I do it all (see Chapter 10 of Woods and Hollnagel 2006).

Technically, the performance of either/or architectures saturates quickly as alerting rate, exception rate, and anomaly rate go up. This ceiling is too low practically and people have to step into the breach to keep the work systems functioning when anomalies and exceptions occur—which inevitably happens more than developers expect. By the way, Norbert Wiener, one of the founders of modern computer science, warned us about this in 1950 (see Norbert's Contrast in Chapters 11 and 12 of Woods and Hollnagel 2006).

Well, I witnessed all of these issues in action, for the first of many times, in the 1980s expert systems era. When Hollnagel and I wrote the first CSE paper circa late 1981 to early 1982, we were thinking quite explicitly about steering efforts in a different direction. The either/or claims of knowledge engineers and expert system developers in the early 1980s were a prime target for a joint cognitive system analysis. Plus, it would be a good way to show the power of our ideas for CSE. My analysis revealed the basic conflict of the standard approach being advocated at the time. In the standard approach, the AI system was supposed to substitute for the unreliable human (except in those days, automating human hands and eyes was too expensive, but everyone thought that automating human higher cognitive functions was about to be quite easy). On the other hand, the AI developers attempted to escape all responsibility for their software's results by having the person act as a check and backup should the software prove unable to perform the task or deliver inaccurate results. My diagram of this situation is reproduced in Figure 3.3 (from Woods 1985).

There was conflict at the heart of this architecture:

The responsible human roles could exert control of the cognitive work and use the AI system as a tool, even though it was designed to be a substitute and no tool at all.

Or,

The people could retreat from their responsibility as a problem holder and leave the machine to function by itself, except the machine, if it really acted on its own, wasn't as good as its developers expected.

By recognizing the conflict and that it was untenable for responsible human roles, I had the starting point to develop a rich set of concepts for understanding the behavior of joint cognitive systems of machines and people. The concepts remain critical today and include the following:

1. Machines are brittle at their boundaries; events will challenge those boundaries; and the boundaries are uncertain and dynamic. Plus, developers will overestimate what the machine can handle by itself and underestimate surprises.

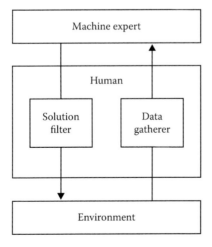

Figure 3.3 A version of an either/or human–machine architecture from the movement to build machine expert systems in AI in the early 1980s. (From Woods, D.D. (1985). Cognitive technologies: The design of joint human–machine cognitive systems. *AI Magazine, 6*, 86–92, Figure 1, p. 87, Copyright 1985 by AIAA. Reprinted with permission of AIAA.)

2. Machines are neither problem holders nor stakeholders; those are only human roles. However, some stakeholders use new machine capabilities to try to dominate other human roles from a distance.
3. The hidden role for people to filter out bad computer recommendations or actions is a poor joint system architecture—inevitably, people will under-utilize the machine or will unduly rely on it.
4. Architectures create responsibility–authority double binds, and these double binds undermine the performance of the whole system in unexpected ways.
5. There are basic regularities in how people adapt to the double bind. One is for practitioners to develop a covert work system to meet their responsibility to overcome the brittleness of the machine. A second pattern is where some practitioners work only to meet the minimum definition of their role—role retreat—to avoid responsibility when they no longer have effective authority to handle variations, difficulties, and anomalies.

I was shocked when *AI Magazine* was willing to publish my analysis (Woods 1985). What I needed next was an empirical case to demonstrate the brittleness of machine expert systems and to show the different ways people adapted to the brittleness given the pressures to succeed and the potential sanctions for failure. And the real world dropped one into our lap. An expert system was being developed in the standard AI methodology for the time (Roth et al. 1987). The technicians worked on-site to repair broken electromechanical devices while the machine expert system would be centrally located. Communication bandwidth was quite limited then, so our team was commissioned to design the codes that the machine expert would use to communicate with the technicians in the troubleshooting process. I immediately recognized the opportunity to observe the interplay between human practitioner and expert machine much more deeply, even though it was a much less demanding task than the ones we had been studying. We simply observed each troubleshooting episode the developer used to iteratively expand and tune the machine's rules to solve more and more cases. I set up a template that abstracted each interaction to see if it followed the assumptions of AI—if the expert machine worked as technologists imagined, there would be a straight path from initial symptom to correct diagnosis. But what would happen if there were deviations from the straight path? We observed and compiled events that threw the troubleshooting process off of the straight path, and what happened to get troubleshooting back on a path toward a solution (Roth et al. 1987).

The results were a confirmation of the 1985 analysis. In summary:

- Complicating factors and surprises occurred much more often than anticipated.
- The machine expert was very brittle at the boundaries; developers had little idea of where these boundaries were or how brittle the machine expert really was.
- The design of the machine expert was intended to substitute for the technician, which created a responsibility–authority double bind.

- The technicians as problem holders had to adapt to troubleshoot the device successfully; they innovated ways to use the machine expert as a tool even though nothing had been designed to make the machine expert tool-like.
- Some other technicians adapted by role retreat, and troubleshooting was least successful then.

These results of the 1985 and 1987 papers have been confirmed most notably and quite convincingly by a set of studies by Phil Smith and colleagues in other domains (Guerlain et al. 1996; Smith et al. 1997). In particular, Layton et al. (1994) demonstrated clearly that asking people to monitor the machine and filter out bad computer recommendations or actions is a poor joint system architecture.

The reaction of technologists to these findings and concepts was quite interesting—they discounted the discrepant findings (one of the signs indicating they were fixated on a hypothesis about human–machine joint systems). I was told—a reaction, that older technology was brittle, but my latest technology is better—a reaction I called "a little more technology will be enough, this time" (Woods 1987a).

Ignoring the fundamental brittleness of machines didn't make the machines less brittle, though eventually it did give rise to my call for a new field of Resilience Engineering to study, model, and design ways to overcome the epidemic of brittleness as our systems became more and more interdependent in a complex world (Hollnagel et al. 2006; Woods 2015).

Meanwhile, it is still popular to build either/or architectures where the machine does all the work and people are supposed to monitor the machine to jump in when it is doing the wrong thing. Or for more dynamic situations, the machine does all the work till it can no longer do so, and then it dumps the messy situation into some human's hands—bumpy transfers of control. With every deployment of an either/or architecture (or degrees of either/or as in levels of automation) comes new opportunities to observe the same unintended and undesirable effects as I observed in the mid-1980s. Yet, technologists and stakeholders still believe that the results will be different with the next new technology even though it is the same either/or architecture—another key sign of fixation (see Chapters 10 and 11 of Woods and Hollnagel 2006 for a synthesis that includes several iterations of what I first saw with machine expert systems). These studies highlighted the importance of another facet of the original intent behind our formulation of CSE—how to design new technological capabilities so that they functioned as tools supporting human roles as they met the demands of cognitive work.

3.11 COGNITIVE TOOLS

Using design to better support cognitive work was an explicit and important part of how Hollnagel and I envisioned the role of CSE. To paraphrase the punchline of my paper for the 1985 meeting on Intelligent Decision Support (Woods 1986): Intelligence does not lie in devices, artifacts, computations, algorithms, or automata; intelligence lies in the creation and utilization of these capabilities in the pursuit of human purposes.

There was an explicit design goal for all three of the collaborations I was involved in during the early to mid-1980s: how to intelligently use the new technological capabilities to better support joint systems to meet the cognitive demands of work. In particular, I worked on a new kind of computerized control room to support the cognitive work of emergency operations. Figure 3.4 reproduces an example of what nuclear control rooms looked like in 1980. Data from each sensor are presented in a single display; the displays are spatially distributed, but some computer-based displays have been added. In addition, note the annunciator or alarm lights at the top and the status panel below them. To the experienced operator, these indicate that the plant is in the middle of a major emergency—the plant should be shut down; emergency cooling should be running to compensate for several disturbances in several basic functions. In general, I worked on design techniques and concepts that could lead to better designs, ones "that make us smart" to use Don Norman's phrase.

As soon as I was engaged in observing emergency operations with an eye to design backfit computerized systems as part of the post-TMI fixes in 1980, I realized that my background in perception and attention was great preparation for innovating new techniques and concepts. In perception, I learned how very complex mechanisms functioned and how to uncover those functions for a system that already existed. My new task was to understand how the practitioners used tools and artifacts to meet the demands of emergency operations.

To illustrate this, consider this case of the clicking control rod position indicators. We were observing operator teams in a simulated case in the training center. Over several runs of one case, we noticed that one of the operators would move to the reactor display/control area, but we couldn't tell how he knew when to check out information about these systems. When we asked, the operators made only vague general comments that were not helpful at all. After a while, it dawned on us that he was listening to the clicking sounds made by mechanical counters that indicated the position of the control rods as they moved in and out of the reactor. It turned out that

Figure 3.4 A nuclear power plant control room during a simulated fault (training simulator). (From D.D. Woods, personal collection. Reprinted by permission of collector.)

the activity of the automatic rod control system could be monitored by listening to the sounds made by the mechanical counters as the automatic control system moved the rods to regulate reactor power and to shut down the reactor if faults occurred.

We then observed how the operators used this serendipitous auditory display. When the rod control system was active (clicking) and the operator didn't expect much activity, he would begin to investigate. When the rod control system was inactive (not clicking) and the operator did expect activity, he would begin to investigate. When changes in rod control system activity and the associated clicking sounds matched expectations, the operators went on about other tasks in the control room and there were no observable signs that the operators were monitoring or thinking about rod control. Note the earlier comments about the role of expectations in modeling anomaly response, especially how hearing no sounds can be an anomalous finding when clicking was expected. When we then asked the operators if they listened for unexpected changes in the clicking sounds, the operators responded, of course. The sounds were good clues as to what the rod control system was doing and helped them know when something funny was going on in that part of the process. Now, this was a naturally occurring example of what makes for a cognitive tool or artifact.

The auditory display example illustrates how practitioners find and make tools work for them. There is a great line of research on the difference between cognitive artifacts created or shaped by practice/practitioners and stuff designed by outsiders and delivered to practitioners (e.g., see Cook and Woods 1996; Nemeth et al. 2006; Perry and Wears 2012). I also mention this line of work with the profound sense of humility all designers should possess because what they design for others may have effects quite different from what was intended.

3.12   ALARM OVERLOAD

Another example of the value of the new joint cognitive systems perspective involved the design and use of alarms in process control systems (Woods 1995a). As I observed teams handle simulated emergencies, I could see how the alarms undermined cognitive work in some ways and assisted cognitive work in other ways. The annunciator or alarm lights are at the top of the control panel in Figure 3.4. Each annunciator lit up when a threshold was crossed or a state changed. In emergencies, there was alarm overload as faults produced a cascade of disturbances as was discussed earlier about anomaly response. Plus, many of the alarms signaled status, changes in state, or events, rather than signaled what was abnormal, and this adds volume, noise, and numbers to the overload.

The default assumption was that the alarms should be computerized and use new software technology to become more intelligent. Interacting with developers, I saw how the engineering process resulted in a proliferation of uninformative alarms. Engineers defined thresholds that set off alarms for each individual device, component, or subsystem they were responsible for developing or modifying. To be thorough and conservative, each engineer (a) assigned many threshold alarms and (b) set the threshold (the criterion) so as to minimize misses. With respect to (a), assigning alarms one at a time from the bottom up resulted in a mass of thresholds that produced a large volume of alarms quickly when faults occurred and triggered sets of disturbances. The ability of the human practitioners to process the signals was quickly overwhelmed whenever anything vaguely serious went wrong. Simply computerizing the same kinds of alarms would do nothing to reduce overload.

With respect to (b), setting thresholds to minimize misses guaranteed maximizing false alarms as a basic constraint in signal detection theory. When I pointed this out to engineers, their response then was people should just pay attention to all alarms regardless. When I heard developers say people should just be more careful, I knew that statement was an indicator of the design of a bad human–machine system. As someone who had studied and used signal detection theory, increasing false alarms was guaranteed to reduce the informativeness of any alarm, which in turn would dramatically reduce the effectiveness of the human monitor, independent of any motivational factors. In other words, the engineers were using the wrong unit of analysis for designing an effective monitoring system. They should be designing the joint cognitive system to meet the cognitive demands of monitoring in anomaly response.

The joint cognitive system for monitoring consisted of two stages: a distributed set of detectors monitoring different parameters, components, and subsystems and a second-stage human monitor who shifted focus depending on the pattern of outputs from the first stage. The standard way alarms were assigned and set up guaranteed that this two-stage cognitive system could not function well when faults triggered sets of disturbances that changed over time. This was an early case of a type of human–machine architecture whose performance has a very low ceiling as the number and rate of problems go up.

The next question was how to confirm the joint cognitive system analysis. I teamed with Robert Sorkin (Sorkin and Woods 1985) to build a simple two-stage signal detection system as a model of the general joint cognitive system for monitoring and then evaluated how it performed under various conditions. The only assumption was that the second stage was under some workload constraint (finite resources relative to the number of channels in the first stage). No matter how we set up the two stages and no matter how well the first-stage monitors performed, increasing false alarms dramatically reduced performance of the work systems as a whole. In other words, the increasing rate of false alarms reduced the informativeness of the signal so much that the signal was useless *mathematically*. The result confirmed what I had observed— alarms were (and should be) ignored because they did not provide any information and had low power to discriminate normal from abnormal conditions. The activity of operators was not the result of psychological or motivational factors of people but a basic property that had been designed into the joint cognitive systems inadvertently and with undesirable impacts.

Today, we might use the statistic positive predictive value (PPV) as an easy initial surrogate for our original analysis—increasing false alarms even slightly produces big decreases in PPV. If we compute this statistic for most alarm systems, we find an epidemic of very low PPV. Even today, most alarms carry no information and have little capability to discriminate conditions that warrant shifts in attention and work from those that do not (Rayo and Moffatt-Bruce 2015). It is particularly interesting that this work in 1985 established the path to develop better measures for use in alarm system testing and design. In other words, it provided a quantitative analysis—what human systems researchers and engineers are always criticized for not providing. Yet, the response I received from both engineers and managers then was that the results weren't the right kind of quantitative analysis because it was unfamiliar, it required significant engineering work, and the results would require significant engineering changes with associated costs. I have received the same response recently when I provided the technical basis for how to improve alarm system design.

However, all was not lost because I was part of a team using CSE to design a new kind of computerized control room. In January of 1982, under deadline pressure in a project to make diagnosis more intelligent, I conceived a new kind of alarm system stimulated in part by the functional modeling approach of Morten Lind and Jens Rasmussen, which was one of the conceptual foundations for our control room design approach. The concept I called a *disturbance board* was one way to combine digital processing with the best characteristics of the analog control room (Woods et al. 1986, 1989). The goal of the design was to represent the cascade of disturbances that flowed from any combination of faults in the plant as they occurred in real time. To do this, first I redefined what could be alarming based on how a process could malfunction; the specific criteria were based on the type of malfunction and the specific engineered process in question.

Second, we created a high-level functional map of the nuclear plant viewable from anywhere in the control room. The key was to create emergent context sensitivity. Otherwise, the computational burden would be overwhelming and too brittle. The technique I came up with was to make the set of currently active alarm conditions compete for a limited amount of display space available for each function within the functional map of the plant. It worked. As a fault or faults occurred, they set off a set of disturbances over time (Woods 1994). The disturbance board allowed operators to see this chain of disturbances start and spread through the different functions and parts of the plant. The alarms that indicated the severity, or the alarms that were the strongest evidence for the presence of a malfunction, won the competition for the limited space based on local ordering criteria so that operators always saw the best evidence characterizing the abnormal state of the plant. Designers could regulate the degree of competition for space by spreading out or collapsing the number of units in the functional map to tune the design so as to best visualize the cascade of disturbances.

The concept was the first and still the only alarm system designed based on Rasmussen's approach to functional modeling. Jim Easter helped me produce a semi-dynamic mockup for a couple of accident cases that showed the promise of the new concept. Bill Elm and I led a team to produce a complete design that was then implemented (AWARE was the product name) and built into one power plant. Alas, the human–machine architecture in our design conflicted with the architecture that the industry as a whole had chosen, which was based on strict adherence to procedures so the product was not adopted further. Yet alarm overload and data overload remain pressing problems today (Rayo and Moffatt-Bruce 2015; Woods 1995a; Woods et al. 2002).

3.13 VISUAL MOMENTUM AND REPRESENTATION DESIGN

Understanding how practitioners used tools and artifacts to meet the demands of emergency operations was only a prelude to the design challenge. Modeling how the joint system worked to meet demands was needed, and this knowledge then facilitated development of new design concepts to support cognitive work by using new computer processing and display capabilities in novel ways (Woods 1998). The alarm system case above is just one demonstration from the early days of CSE.

In another example, I introduced the concept of visual momentum (Woods 1984a) and a set of design techniques around that concept for use in our work to develop what was called the safety parameter display system in 1980. Given the computer technology of the day, each display was thought of and laid out as a separate entity. A display system was a set of these individual displays organized in a rough hierarchy from top to detail. What chunk of the plant went on an individual display? Typically, the designer chose by default a system or subsystem. Surprisingly, I still frequently find sets of displays organized this way, and this approach has left us the very misleading but common expression—"drill down."

I looked at the control room layout of displays and controls, and despite many weaknesses (such as the "one display–one indicator" philosophy; Goodstein 1981), an experienced operator could navigate the panels reasonably smoothly to find the data they were looking for, and they had some ability to notice changes or activity in other areas of the control room and shift their focus to another place (as in the rod control indicator example). When I looked at the default standard of individual displays in a hierarchy, it was obvious to me that the design created a narrow keyhole and a high navigation burden.

In the language of perception and attention, the question is—how do you know where to look or focus next? The answer in perception and attention in natural environments is we have various orienting perceptual functions that help us reorient or refocus to new events or changes (Woods 1995a; for a modern take given today's technology, see Morison et al. 2015). The default design of the set of displays undermined or eliminated all of the perception and attention mechanisms people normally use—a thing that definitely makes us dumb by changing fast, reliable perceptual process into a slow, serial, deliberate cognitive activity.

An example is the emergency cooling system, which should turn on at the beginning of an emergency and which has three subsystems. Each of the three subsystems is designed to provide emergency cooling for different ranges of pressure in the reactor, and there is a transition between the three that depends on how pressure changes. Some of the transition was manual and a good deal was automatic. The operator had to monitor that transitions across the three subsystems occurred as desired and expected and had to take the appropriate manual actions as part of this task. Plus, all of this was contingent on how the pressure was changing. The default display system designs throughout the industry divided the emergency cooling system into three separate displays with no overlap. As a result, an operator had to switch among four to five different displays in order to look at all of the parameters involved in monitoring transitions.

One computerized control room had shifted from a spatial layout of displays constituting 60 feet of control board, to selecting any one of over 32,000 displays they could call up on about eight computer displays (cathode ray tubes in those days). In one trial, we ran with this kind of conventional computerization of the traditional analog control room; the test operators knew in advance the simulated scenario and they still could not keep up with the pace of the cascading events as they struggled to navigate from one display to the next.

To reduce the navigation burden, I took advantage of my knowledge of perception to innovate a set of techniques that would empower the human capabilities for reorienting and refocusing so they could function in this new virtual data space. In school, when I was teaching a perception course, I had stumbled across a paper by the perception researcher Julian Hochberg and his colleague Virginia Brooks that looked at what made film cuts in cinematography comprehensible (Hochberg and Brooks 1978). They used the term *visual momentum* to capture their sense of what made a transition in cinematography comprehensible (Hochberg 1986). I borrowed their term to refer to the key ideas behind the set of techniques and to refer to the set as a whole. Some of the techniques I identified to build visual momentum occurred in transitions in cinematography and some were new innovations required to address the new world of virtual data spaces. An example of the former is the concept of longshots; an example of the latter is side effect views (see Woods and Watts 1997).

The set of techniques to build visual momentum has been explained in several places (see Bennett and Flach 2012; Watts-Perotti and Woods 1999; Woods and Watts 1997), but I will make a few retrospections. First, the concept has been reinvented and relabeled several times (e.g., focus plus context; Card et al. 1991), but if people are to find relevant data and avoid data overload, the techniques are necessary whatever the label (Woods et al. 2002).

Second, there is a core idea behind the use of these techniques: enhancing navigation requires a conceptual topology, and almost always, there is more than one conceptual topology. One quip I use frequently to provide some summary guidance is: "Navigation mechanisms should be a model of the topic being navigated."

The default design in 1980s process control was to cut up the system diagrams into pieces that would conveniently fit on display one full screen at a time and this guaranteed near-zero visual momentum. There was no real topology. Today, we see the default design topology most frequently used is physical space. This reference frame does have topological properties, but it is used as the only frame of reference and leads to a collection of icons scattered over the surface of a top-down map. This default ignores the capabilities of technology today to permit changes in perspective and ignores the rules from perception on coherent perspective shifts (see Morison et al. 2009, 2015). Plus, collections of icons are a very limited form of symbol, which undermines the search for meaning when used in isolation and used too much (Woods 1991, 1995b).

To provide the basis for visual momentum in the control room case (and for other reasons), we used two interconnected topologies. One was based on systems that carry out the functions needed in different situations, and the other was based on needed functions that are carried out by different systems under different conditions. These are two parallel ways to monitor the process under supervisory control, and the two frames of reference are closely interconnected. When a practitioner looked at some part or level of the system topology, the display system was smart enough to show the functions that were affected by the activities of that part of the system. When a practitioner looked at some part or level of the functional topology, the display system was smart enough to show the systems that provided or could provide that function. In this way, we had the means to visualize a *side effect view*—the first use of that visual momentum technique.

I used the visual momentum techniques in the guidelines for our team's design of the safety parameter display system (1980–1981) and then in our work on a complete computerized control room (1982–1985). I presented the concept at the cognitive modeling of nuclear power plant operators in 1982, and the first paper came out in 1984.

Visual momentum, as I argued in 1982, is just one concept for aiding the search for meaning in massive amounts of changing data. The risk in emergency operations was data overload. As technology continues to collect and transmit ever larger fields of data, data overload continues to be a dominant risk in analytics and in human–robot interaction today (Morison et al. 2015; Patterson et al. 2001; Zelik et al. 2010). As CSE started up as a design activity, escaping data overload was one major challenge. The technique and concepts that emerged then are even more valuable today (Woods et al. 2002).

3.14   THE COGNITIVE SYSTEM TRIAD

The examples above from the early days of CSE help us see another fundamental concept about what makes a joint cognitive system. Hollnagel, Roth, and I were looking at a new kind of system—an emergent system that arises in the interactions among (1) the demands the world imposes on cognitive work, (2) the interplay of multiple agents who do cognitive work (joint and distributed cognition), and (3) the properties of the artifacts, representations, and tools that enable cognitive work (Hollnagel and Woods 1983, 2005; Woods and Hollnagel 2006; Woods and Roth 1988a,b). This triad is the fundamental dynamic and adaptive unit of analysis, since all human–machine work systems involve interactions across multiple roles and levels. Focusing on any component, slice, or level of this in isolation without considering the interactions and what emerges from those interactions oversimplifies, misleads, and sets up unintended consequences.

3.15   WHY CSE THEN AND NOW?

In this chapter, I have selected a few examples from my participation in the origins of CSE to give today's practitioners a sense of what things were like during the start-up period of the field. These examples illustrate the key ideas that we thought were, or would be, the basis for the field. I want to close with two general ideas about why we focused our energies in those days on developing and on advocating for this field.

First, the name—why did Hollnagel and I use "cognitive systems engineering" in the "New Wine in New Bottles" paper. Around 1981, as we were already writing that paper, Don Norman was using cognitive engineering as the label and argued that we needed to develop this capability as an application-oriented partner to cognitive science (Norman 1987). We were not happy with this framing. The shift underway was much more than stuffing personal computers into cognitive psychology laboratories, or stuffing the mind into computations. It was also much more than repairing poor designs in specific applications. It was a shift from compensating for human limits to active expansion of people's ability to adapt in the face change and surprise. It was a shift in the base unit of analysis to the joint cognitive system, not people versus technology. It was a shift in phenomena under study—the target is observing, modeling, and supporting how people adapt in the face of different complexities.

Our addition of the "systems" word was intended to be much more than a small variation, and the subtitle followed in that spirt—something new and much more fundamental was afoot. We explicitly intended that the label could be parsed in two different ways. One intent was engineering a cognitive system, in particular, a joint cognitive system. This framing directly addressed the rise of AI and countered the new version of either/or human–machine architectures that came along with the new technology. Today, we still find people trying to design and justify work systems where the machine either handles all of the tasks or, when it can't, transfers it all to a person. We saw this as technically wrong in 1981 and everything since has only confirmed our position.

The alternative parsing is that we were proposing a new form of systems engineering—a cognitive, systems engineering—oriented toward understanding how human work systems, including technology and organizational factors at both sharp and blunt ends of the sociotechnical system, continuously adapt both to cope with complexity and to take advantage of new capabilities. Over the years since 1981, the latter meaning has gained more prominence, at least to both Hollnagel and myself (e.g., chapter 12 of Woods and Hollnagel 2006; Woods and Branlat 2010), as the themes of complexity and adaptation have come to dominate modern systems and led us to create the field of resilience engineering (Hollnagel et al. 2006).

REFERENCES

Abelson, R.F. (with 51 others) (1976). *Proposal for a Particular Program in Cognitive Sciences*. Proposal to the Alfred P. Sloan Foundation, New York.

Anderson, J.R. (1983). *The Architecture of Cognition*. Cambridge, MA: Harvard University Press.

Bennett, K.B. and Flach, J.M. (2012). Visual momentum redux. *International Journal of Human–Computer Studies,* 70, 399–414. doi:10.1016/j.ijhcs.2012.01.003.

CAIB (*Columbia* Accident Investigation Board). 2003. *Report, 6 vols*. Government Printing Office, Washington, DC. www.caib.us/news/report/default.html.

Card, S.K., Mackinlay, J.D., and Robertson, G.G. (1991). The information visualizer: An information workspace. *CHI 91 ACM Conference on Human Factors in Computing Systems*, New York: ACM Press.

Christoffersen, K., Woods, D.D., and Blike, G. (2007). Discovering the events expert practitioners extract from dynamic data streams: The mUMP Technique. *Cognition, Technology, and Work*, 9, 81–98.

Clancey, W.J. (1985). Heuristic classification. *Artificial Intelligence*, 27, 289–350.

Cook, R.I., Render M.L., and Woods, D.D. (2000). Gaps in the continuity of care and progress on patient safety. *British Medical Journal*, 320, 791–794.

Cook, R.I. and Woods, D.D. (1996). Adapting to new technology in the operating room. *Human Factors*, 38, 593–613.

De Keyser, V. and Woods, D.D. (1990). Fixation errors: Failures to revise situation assessment in dynamic and risky systems. In A.G. Colombo and A. Saiz de Bustamante (Eds.),

Systems Reliability Assessment (pp. 231–251). Dordrecht, The Netherlands: Kluwer Academic.

Gibson, J.J. (1979). *The Ecological Approach to Visual Perception*. Boston: Houghton Mifflin.

Goodstein, L. (1981). Discriminative display support for process operators. In J. Rasmussen and W. Rouse (Eds.), *Human Detection and Diagnosis of System Failures* (pp. 433–449). New York: Plenum Press.

Guerlain, S., Smith, P.J., Obradovich, J.H., Rudmann, S., Strohm, P., Smith, J., and Svirbely, J. (1996). Dealing with brittleness in the design of expert systems for immunohematology. *Immunohematology*, 12, 101–107.

Hochberg, J. and Brooks, V. (1978). Film cutting and visual momentum. In J.W. Senders, D.F. Fisher, and R.A. Monty (Eds.), *Eye Movements and the Higher Psychological Functions* (pp. 293–313). Hillsdale, NJ: Lawrence Erlbaum Associates.

Hochberg, J. (1986). Representation of motion and space in video and cinematic displays. In K.R. Boff, L. Kaufman, and J.P. Thomas, (Eds.), *Handbook of Human Perception and Performance*, Vol. I. (pp. 1–64). New York: John Wiley & Sons.

Hoffman, R.R. (Ed.) (1992). *The Psychology of Expertise: Cognitive Research and Empirical AI*. Mahwah, NJ: Erlbaum.

Hollnagel, E., Pedersen, O., and Rasmussen, J. (1981). *Notes on Human Performance Analysis*. Roskilde, Denmark: Risø National Laboratory, Electronics Department.

Hollnagel, E. and Woods, D.D. (1983). Cognitive systems engineering: New wine in new bottles. *International Journal of Man–Machine Studies*, 18, 583–600 [originally Riso Report M2330, February 1982] (Reprinted *International Journal of Human–Computer Studies*, 51(2), 339–356, 1999 as part of special 30th anniversary issue).

Hollnagel, E. and Woods, D.D. (2005). *Joint Cognitive Systems: Foundations of Cognitive Systems Engineering*. Boca Raton, FL: Taylor & Francis.

Hollnagel, E., Woods, D.D., and Leveson, N. (2006). *Resilience Engineering: Concepts and Precepts*. Aldershot, UK: Ashgate.

Hutchins, E. (1991). The social organization of distributed cognition. In L.B. Resnick, J.M. Levine, and S.D. Teasley (Eds.), *Perspectives on Socially Shared Cognition* (pp. 283–307). Washington, DC: American Psychological Association. doi:10.1037/10096-012.

Hutchins, E. (1995). *Cognition in the Wild*. Cambridge, MA: MIT Press.

Hutchins, E.L., Hollan, J.D., and Norman, D.A. (1985). Direct manipulation interfaces. *Human–Computer Interaction*, 1, 311–338. doi:10.1207/s15327051hci0104_2.

Klein, G., Woods, D.D., Bradshaw, J.D., Hoffman, R.R., and Feltovich, P.J. (2004). Ten challenges for making automation a "team player" in joint human–agent activity. *IEEE Intelligent Systems*, 19, 91–95.

Klein, G., Pliske, R., Crandall, B., and Woods, D. (2005). Problem dectection. *Cognition, Technology, and Work*, 7(1), 14–28.

Layton, C., Smith, P.J., and McCoy, C.E. (1994). Design of a cooperative problem-solving system for en-route flight planning: An empirical evaluation. *Human Factors*, 36, 94–119.

Malin, J., Schreckenghost, D., Woods, D.D., Potter, S., Johannesen, L., Holloway, M., and Forbus, K. (1991). *Making Intelligent Systems Team Players*. NASA Technical Report 104738, Johnson Space Center, Houston, TX.

Morison, A.M., Voshell, M., Roesler, A., Feil, M., Tittle, J., Tinapple, D., and Woods, D.D. (2009). Integrating diverse feeds to extend human perception into distant scenes. In P. McDermott and L. Allender (Eds.), *Advanced Decision Architectures for the Warfighter: Foundations and Technology* (pp. 177–200). Boulder, CO: Alion Science and Technology.

Morison, A., Woods, D.D., and Murphy T.B. (2015). Human–robot interaction as extending human perception to new scales. In R.R. Hoffman, P.A. Hancock, M. Scerbo, R. Parasuraman, and J.R. Szalma (Eds.), *Handbook of Applied Perception Research*, Volume 2 (pp. 848–868). New York: Cambridge University Press.

Neisser, U. (1976). *Cognition and Reality*. San Francisco: W.H. Freeman.

Nemeth, C, O'Connor, M., Klock, P.A., and Cook, R. (2006). Discovering healthcare cognition: The use of cognitive artifacts to reveal cognitive work. *Organization Studies*, 27, 1011–1035.

Newell, A. (1990). *Unified Theories of Cognition*. Cambridge MA: Harvard University Press.

Norman, D.A. (1980). Cognitive engineering and education. In D.T. Tuma and F. Reif (Eds.), *Problem Solving and Education* (pp. 81–95). Hillsdale, NJ: Erlbaum.

Norman, D.A. (1986). Cognitive engineering. In D.A. Norman and S.W. Draper (Eds.), *User Centered System Design* (pp. 31–61). Hillsdale NJ: Erlbaum.

Norman, D.A. (1987). Cognitive science—cognitive engineering. In J.M. Carroll (Ed.), *Interfacing Thought* (pp. 325–336). Cambridge MA: MIT Press.

Norman, D.A. (1993). *Things That Make Us Smart*. Reading MA: Addison-Wesley.

Patterson, E.S., Roth, E.M., and Woods, D.D. (2001). Predicting vulnerabilities in computer-supported inferential analysis under data overload. *Cognition, Technology and Work*, 3, 224–237.

Peirce, C.S. (1955). Abduction and induction. In J. Buchler (Ed.), *Philosophical Writings of Peirce* (pp. 150–156). London: Dover (original work published, 1903).

Perry, S.J. and Wears, R.L. (2012). Underground adaptations: Case studies from health care. *Cognition, Technology, and Work*, 14, 253–260.

Pople, H.E. Jr. (1985). Evolution of an expert system: From internist to caduceus. In I. De Lotto and M. Stefanelli (Eds.), *Artificial Intelligence in Medicine* (pp. 179–203). New York, NY: North-Holland.

Rasmussen, J. and Lind M. (1981). Coping with complexity. In H.G. Stassen (Ed.), *First European Annual Conference on Human Decision Making and Manual Control*. New York: Plenum (also as Risø-M-2293. Electronics Department, Risø National Laboratory, Roskilde, Denmark 1981).

Rayo, M.F. and Moffatt-Bruce, S.D. (2015). Alarm system management: Evidence-based guidance encouraging direct measurement of informativeness to improve alarm response. *BMJ Quality and Safety*, 24, 282–286. doi:10.1136/bmjqs-2014-003373.

Roth, E.M., Bennett, K., and Woods D.D. (1987). Human interaction with an "intelligent" machine. *International Journal of Man–Machine Studies*, 27:479–525, 1987 (reprinted in E. Hollnagel, G. Mancini, and D.D. Woods (Eds.), *Cognitive Engineering in Complex, Dynamic Worlds*. London: Academic Press, 1988).

Roth, E.M., Woods, D.D., and Pople, H.E. (1992). Cognitive simulation as a tool for cognitive task analysis. *Ergonomics*, 35, 1163–1198.

Smith, P.J., McCoy, C.E., and Layton, C. (1997). Brittleness in the design of cooperative problem-solving systems. *IEEE Transactions on Systems Man, and Cybernetics, Part A: Systems and Humans*, 27, 360–371.

Sorkin, R.D. and Woods, D.D. (1985). Systems with human monitors: A signal detection analysis. *Human–Computer Interaction*, 1, 49–75.

Watts-Perotti, J. and Woods, D.D. (1999). How experienced users avoid getting lost in large display networks. *International Journal of Human–Computer Interaction*, 11, 269–299.

Wiener, N. (1950). *The Human Use of Human Beings: Cybernetics and Society*. New York: Doubleday.

Woods, D.D. (1982). Visual momentum: An example of cognitive models applied to interface design. In T.B. Sheridan, J. Jenkins, and R. Kisner. (Eds.), *Proceedings of Workshop on Cognitive Modeling of Nuclear Plant Control Room Operators* (pp. 63–72). NUREG/CR-3114, August 15–18, 1982.

Woods, D.D. (1984a). Visual momentum: A concept to improve the cognitive coupling of person and computer. *International Journal of Man–Machine Studies*, 21, 229–244. doi:10.1016/S0020-7373(84)80043-7.

Woods, D.D. (1984b). Some results on operator performance in emergency events. In D. Whitfield (Ed.), *Ergonomic Problems in Process Operations*, *Inst. Chem. Eng. Symp. Ser. 90*.

Woods, D.D. (1985). Cognitive technologies: The design of joint human–machine cognitive systems. *AI Magazine*, 6, 86–92.

Woods, D.D. (1986). Paradigms for intelligent decision support. In E. Hollnagel, G. Mancini, and D.D. Woods (Eds.), *Intelligent Decision Support in Process Environments* (pp. 153–173). New York: Springer-Verlag.

Woods, D.D. (1987a). Technology alone is not enough: Reducing the potential for disaster in risky technologies. In D. Embrey (Ed.), *Human Reliability in Nuclear Power*. London: IBC Technical Services.

Woods D.D. (1987b). Cognitive engineering in complex and dynamic worlds. *International Journal of Man–Machine Studies*, 27, 479–525.

Woods, D.D. (1988). Coping with complexity: The psychology of human behavior in complex systems. In L.P. Goodstein, H.B. Andersen, and S.E. Olsen (Eds.), *Mental Models, Tasks and Errors* (pp. 128–148). London: Taylor & Francis.

Woods, D.D. (1991). The cognitive engineering of problem representations. In G.R.S. Weir and J.L. Alty (Eds.), *Human–Computer Interaction and Complex Systems* (pp. 169–188). London, England: Academic Press.

Woods, D.D. (1993). Process-tracing methods for the study of cognition outside of the experimental psychology laboratory. In G. Klein, J. Orasanu, R. Calderwood, and C.E. Zsambok (Eds.), *Decision Making in Action: Models and Methods* (pp. 228–251). Norwood, NJ: Ablex.

Woods, D.D. (1994). Cognitive demands and activities in dynamic fault management: Abduction and disturbance management. In N. Stanton (Ed.), *Human Factors of Alarm Design* (pp. 63–92). London: Taylor & Francis.

Woods, D.D. (1995a). The alarm problem and directed attention in dynamic fault management. *Ergonomics*, 38, 2371–2393.

Woods, D.D. (1995b). Towards a theoretical base for representation design in the computer medium: Ecological perception and aiding human cognition. In J. Flach, P. Hancock, J. Caird, and K. Vicente, (Eds.), *An Ecological Approach to Human Machine Systems I: A Global Perspective*. Hillsdale, NJ: Lawrence Erlbaum.

Woods, D.D. (1998). Designs are hypotheses about how artifacts shape cognition and collaboration. *Ergonomics*, 41, 168–173.

Woods, D.D. (2003). Discovering how distributed cognitive systems work. In E. Hollnagel (Ed.), *Handbook of Cognitive Task Design* (pp. 37–53). Hillsdale, NJ: Lawrence Erlbaum.

Woods, D.D. (2005). Creating foresight: Lessons for resilience from Columbia. In W.H. Starbuck and M. Farjoun (Eds.), *Organization at the Limit: NASA and the Columbia Disaster* (pp. 289–308). Malden, MA: Blackwell.

Woods, D.D. (2015). Four concepts of resilience and the implications for resilience engineering. *Reliability Engineering and Systems Safety*, 141, 5–9.

Woods, D.D. and Branlat, M. (2010). Hollnagel's test: Being "in control" of highly interdependent multi-layered networked systems. *Cognition, Technology, and Work*, 12, 95–101.

Woods, D.D., Dekker, S.W.A., Cook, R.I., Johannesen, L.L., and Sarter, N.B. (2010). *Behind Human Error*, 2nd edition. Aldershot, UK: Ashgate

Woods, D.D., Elm, W.C., and Easter, J.R. (1986). The disturbance board concept for intelligent support of fault management tasks. In *Proceedings of the International Topical Meeting on Advances in Human Factors in Nuclear Power Systems* (pp. 65–70). United States: American Nuclear Society.

Woods, D.D., Elm, W.C., Lipner, M.H., Butterworth III, G.E., and Easter, J.R. (1989). Alarm Management System. Patent 4816208. Filed: February 14, 1986. Date of Patent: March 28, 1989.

Woods, D.D. and Hollnagel, E. (2006). *Joint Cognitive Systems: Patterns in Cognitive Systems Engineering.* Boca Raton, FL: Taylor & Francis.

Woods, D.D., O'Brien, J. and Hanes, L.F. (1987a). Human factors challenges in process control: The case of nuclear power plants. In G. Salvendy (Ed.), *Handbook of Human Factors/Ergonomics* (pp. 1724–1770). New York: Wiley.

Woods D.D. and Patterson, E.S. (2000). How unexpected events produce an escalation of cognitive and coordinative demands. In P.A. Hancock and P. Desmond (Eds.), *Stress, Workload and Fatigue* (pp. 290–302). Hillsdale, NJ: Erlbaum.

Woods, D.D., Patterson, E.S., and Roth, E.M. (2002). Can we ever escape from data overload? A cognitive systems diagnosis. *Cognition, Technology, and Work, 4,* 22–36.

Woods, D.D., Pople Jr., H.E., and Roth, E.M. (1990a). *The Cognitive Environment Simulation as a Tool for Modeling Human Performance and Reliability* (2 volumes). U.S. Nuclear Regulatory Commission, Washington DC Technical Report NUREG-CR-5213.

Woods, D.D. and Roth. E.M. (1986). *Models of Cognitive Behavior in Nuclear Power Plant Personnel* (2 volumes). U.S. Nuclear Regulatory Commission, Washington DC Technical Report NUREG-CR-4532.

Woods D.D. and Roth, E.M. (1988a). Cognitive engineering: Human problem solving with tools. *Human Factors, 30,* 415–430.

Woods D.D. and Roth, E.M. (1988b). Cognitive systems engineering. In M. Helandier (Ed.), *Handbook of Human–Computer Interaction.* New York: North-Holland (reprinted in N. Moray (Ed.), *Ergonomics: Major Writings.* London: Taylor & Francis, 2004).

Woods, D.D., Roth, E.M., and Bennett, K.B. (1990b). Explorations in joint human–machine cognitive systems. In S. Robertson, W. Zachary, and J. Black (Eds.), *Cognition, Computing and Cooperation* (pp. 123–158). Norwood, NJ: Ablex Publishing.

Woods, D.D., Roth, E.M. and Pople Jr., H.E. (1987b). *Cognitive Environment Simulation: An Artificial Intelligence System for Human Performance Assessment* (3 volumes). U.S. Nuclear Regulatory Commission, Washington DC Technical Report NUREG-CR-4862.

Woods D.D., Roth, E.M., and Pople, Jr., H.E. (1988). Modeling human intention formation for human reliability assessment. *Reliability Engineering and System Safety, 22,* 169–200.

Woods, D.D. and Watts, J.C. (1997). How not to have to navigate through too many displays. In M.G. Helander, T.K. Landauer, and P. Prabhu (Eds.), *Handbook of Human–Computer Interaction,* 2nd edition. Amsterdam, The Netherlands: Elsevier Science.

Woods, D.D., Wise, J.A., and Hanes, L.F. (1981). An evaluation of nuclear power plant safety parameter display systems. In *Proceedings of the Human Factors Society Annual Meeting,* 25, 1:110–114.

Zelik, D., Patterson, E.S., and Woods, D.D. (2010). Measuring attributes of rigor in information analysis. In E.S. Patterson and J. Miller (Eds.), *Macrocognition Metrics and Scenarios: Design and Evaluation for Real-World Teams* (pp. 65–83). Aldershot, UK: Ashgate.

4 Medication Reconciliation Is a Window into "Ordinary" Work

Richard I. Cook

CONTENTS

4.1 INTRODUCTION: ORDINARY WORK IS EXTRAORDINARY

Our appreciation for expertise is driven by the high drama of critical decision making. Gary Klein's descriptions of fire incident commanders or the world of naval combat information centers (Klein 1998) are compelling, at least in part, because the stakes are palpably high. But expertise is also essential in the conduct of ordinary work. This is the sort of work that appears routine or even mundane to outsiders and is often described by practitioners as being *pro forma*, unimportant, or even clerical in character. The speed and apparent ease with which experts do ordinary work belies the sophistication required. Woods and Hollnagel (2006) call this the *Law of Fluency*:

"Well"-adapted cognitive work occurs with a facility that belies the difficulty of the demands resolved and the dilemmas balanced (p. 20).

Ordinary expert work is often repetitive. For air traffic controllers or medical practitioners, a work shift is composed of a series of encounters of airplanes or patients. A physician or a nurse, for example, might "see" (engage and interact professionally with) 20 or more patients in a single clinic day. An air traffic controller might interact with a few dozen aircraft in his or her sector during a shift.

Each encounter entails cognitive effort: the practitioner must shift attention to (or back to) a specific situation, acquire (or reacquire) the relevant context for that situation, use that context to determine what is now relevant to that situation, search for now relevant data, analyze and assess the importance of those data, and determine what needs to be done in furtherance of the goals associated with that situation. These steps are preparatory. They set the stage for the sort of decision making most often studied formally.

But these steps also involve a sort of decision making. Attending to, searching for, and comparing to discover and make sense of data require expertise. The ability of the expert to "pay attention" to some things and ignore others derives from that individual's training, experience, knowledge, and practice. Analyzing and assessing depend entirely on what the expert has identified or selected as meaningful and relevant. Laboratory experiments do not capture these preparatory steps because they deliberately exclude the distractors, noise, red-herrings, and subtle features present in the real world. Yet, it is readily apparent that expertise is, in large part, knowing what to ignore and what to attend to and being able to do so in various situations.

In some settings, experts can process large quantities of data quickly and efficiently. Medical clinics and air traffic control centers are deliberately organized to this end. The physical layout and the information displays have become optimized for this purpose. While this may make the expert's cognitive effort more efficient, it does not eliminate the cognitive work. The expert must still make sense of the patient or the air traffic pattern before anything useful can be done.

It has been clear for some time that this sensemaking is an iterative, "bootstrapping" process in which early cognitive work has the effect of making further effort productive. The result of this process has been described as a "mental model" (Gentner and Stevens 1983). Empirical validation of the mental model details is a difficult problem (Rouse and Morris 1986), mostly because cognitive processes are inaccessible to observers. Although it is indisputable that experts develop and use mental representations of the world as they work, characterizing these representations in any detail remains quite difficult. So quick, facile, and subtle is the expert's sensemaking that it remains largely out of researchers' reach, at least for the real-world situations where it plays a significant role and where cognitive systems engineering is likely to be most rewarding.

A major goal of cognitive systems engineering is to devise ways of organizing the external environment so that practitioners interact with it and with each other smoothly, powerfully, and precisely over the entire range of situations that can occur. But this goal is frustrated by our inability to characterize cognitive work itself.

Medication reconciliation (*MedRec*) is a good example of "ordinary" work that is poorly supported by existing tools and methods largely because its cognitive requirements are misunderstood. This chapter explores some of the details *MedRec*, shows how cognitive systems engineering can be applied to understand the details of the work, and shows how this approach adds to the larger understanding of what practitioners do. The chapter will show that (1) *reconciliation* is inherently the resolution of a contradiction or discrepancy, (2) the discrepancy arises from the construction of a mental model of the external world and exists only in this mental model, and (3) resolving the discrepancy requires cognitive search and revision of the mental model. Close examination of the cognitive work of *MedRec* reveals important aspects of clinician expertise and, paradoxically, shows why efforts to make it reliable have failed.

4.2   WHAT'S ALL THIS ABOUT *MEDREC*?

Although there are many people who rarely take a prescribed medicine, a growing number of patients take three medicines several times each day and some take a dozen or more. The average number of medications at admission to hospital is around eight; the average at discharge is more than nine (Kramer et al. 2014). Medications form the principal armamentarium of medical practice and the efficacious medicines relieve much of the burden of illness, including chronic conditions such as hypertension and diabetes. Virtually no useful medicine is free of toxicity, side effects, and allergic potential, and many medicines potentiate or interfere with the effects of other medicines. Medical practice focuses on optimizing the risk–benefit trade-off of medications (prescribed and "over the counter") for the patient. The scale of medication use is staggering. The U.S. Food and Drug Administration approves approximately 40 new medicines for use each year. There are approximately 4 billion prescriptions written in the United States each year, resulting in approximately US $370 billion in spending (Kleinrock 2015).

Medicines may be prescribed by different practitioners at different times. Medical conditions wax (and, thankfully, sometimes wane) and the individual's response to a particular medicine may change over time. In response, type and quantity of medicine may be altered, sometimes frequently. Patient response to medication is somewhat idiosyncratic so that a particular medicine "works" for one person but not another. Similarly, the severity of side effects or allergic reactions varies widely across patients. Beyond the prescribing, patients vary widely in their adherence to medication regimens (described as patient "compliance"). For these reasons, the current medication regimen for an individual is a unique summation of that person's medical past.

An accurate account of the individual's medications is important. For this reason, virtually every patient contact with a practitioner will involve creating, retrieving, or consulting a list of that patient's medications. A wide variety of studies have shown that medication lists are inaccurate for 30% to 60% of hospitalized patients (Barnsteiner 2005; Gleason et al. 2004; Kramer et al. 2014). Discovering and correcting these inaccuracies is called *medication reconciliation* (widely known as *MedRec*). Successful *MedRec* creates an accurate account of an individual patient's medications. *MedRec* is performed at least a million times each day in the United States. *MedRec* is now formally required in many settings (JCAHO 2006; Thompson 2005).

A variety of aids for *MedRec* exist. These are mostly means for tabulating and displaying existing lists of medications for comparison (Gleason et al. 2010;

Plaisant et al. 2013). Despite its importance, little is known about how *MedRec* is accomplished or what factors contribute to its success or failure. As of this writing, there are virtually no published research results describing how practitioners detect discrepancies in medication lists or how they resolve these discrepancies. The absence of research results severely limits our ability to design and test aids to reconciliation or to provide sound guidance on how to perform this important task.

MedRec is a field ripe for study. Research in this area is highly likely to produce measurable improvements in medical care quality and basic and applied research in this area should be a high priority.

4.3   WHY *MEDREC* MATTERS

An example of the *MedRec* problem is a case presented by Scott Silverstein (Hancock 2013). According to the court documents, this is what happened:

> *Example 1*: Silverstein's mother, an elderly woman with heart disease, was admitted to hospital with a stroke. Her routine medication, *sotalol* (a beta-blocker), was not included in the medications she was scheduled to receive while in the hospital. The absence of *sotalol* led to her developing atrial fibrillation, an irregular heart rhythm. Because the irregular rhythm predisposes a patient to develop blood clots in the heart, the medicine *heparin* was used to anticoagulate her blood. While receiving *heparin*, she developed an intracranial hemorrhage. This hemorrhage required emergency neurosurgery. Ultimately, she was left neurologically devastated and died in the hospital. In theory, *MedRec* should have led to Silverstein's mother receiving *sotalol* while in the hospital. Abrupt discontinuation of beta blockade, it is alleged, led to her death.

The example demonstrates the potential cascade of consequences arising from seemingly small details in medical information handling. The medication in this instance was unrelated to the reason for admission to the hospital; it was not the center of attention. The patient's heart rhythm at admission was not disturbed because the medication was still present and countering the medical condition she had. This balance was achieved long before the admission.

The description does not make clear the intensity associated with admission for an acute neurological condition. Time pressure, the nature of the emergency room and its inevitable stress and noise, the awkward way that medication data must be entered into the computer, the many other data that also must be recorded, all make this routine job difficult. Yet, this situation is exactly the sort in which *MedRec* is important. So too would be other *MedRecs* during the course of hospitalization, at discharge, and during follow-up visits to her physician.

More broadly, the need for *MedRec* arises from a gap in the continuity of care (Cook et al. 2000). Several factors may contribute to such gaps. Transfers of authority and responsibility across caregivers or units, or institutions, entry or departure from care locations, and changing disease processes have inherent potential for creating medication gaps (Rodehaver and Fearing 2005). Experience with these gaps encourages efforts to prevent their results from propagating through patient care (Whittington and Cohen 2004).

Gaps that generate the need for *MedRec* are common in modern medical settings. The same technological progress and organizational change that make treatments more effective also increase the complexity of care, the number of clinicians involved in care, and the number and specificity of interventions available. The large number of transitions of care is a major contributor to medication adverse events, accounting for roughly half of all such occurrences (Bates et al. 1997; Gleason et al. 2004; Pronovost et al. 2003).

Incomplete or incorrect medication information is common (Barnsteiner 2005) and widely recognized as a source of devastating consequences for patients (Landro 2005). It is not surprising that such a complicated activity is not always carried out successfully. It is, however, surprising that existing evidence indicates that it fails about half the time.

4.4 CURRENT APPROACHES AND THE PROBLEM THEY POSE

The need for reliable *MedRec* has spawned a variety of efforts to understand and improve the recording of medications (Gleason et al. 2004; Sullivan et al. 2005). The work of reconciliation can consume more than 30 minutes of clinician work time (Rozich et al. 2004 cited in Ketchum et al. 2005), suggesting that the process of reconciliation requires substantial effort. Restoration of continuity of medication information by pharmacists dedicated to that task just after hospital admission took an average of 11.4 minutes *with a range of 1 to 75 minutes* (Gleason et al. 2004). This significant range suggests that *MedRec* must sometimes be rather easily accomplished but can also be quite demanding. Although there are no published data regarding the effort required to reconcile medications at other points in the hospital stay, Pronovost et al. (2003) report that changes to medication use during hospital stay are extensive. *MedRec* at discharge is likely to require substantial cognitive work.

Medical literature on *MedRec* tends to focus on its importance rather than on its conduct. Assertions about *MedRec* align with stakeholder interests. A bone of contention is who should have the right and responsibility for the conduct of *MedRec*. Nurses (Burke et al. 2005), pharmacists (Andrus and Anderson 2015; Hart et al. 2015; Okere et al. 2015), pharmacy technicians (Cater et al. 2015), and physicians (Liang et al. 2007) have claim to be particularly well positioned, specially qualified, or uniquely capable of performing *MedRec*. It is no surprise that there is no clarity about the responsibility for *MedRec* or that this should lead to frustration (Lee et al. 2015).

Computer-based approaches to *MedRec* have been proposed and tested (e.g., Amann and Kantelhardt 2012; Batra et al. 2015; Giménez-Manzorro et al. 2015; Li et al. 2015). Computer-based approaches generally present lists for review or aid in recording information. None of these approaches have been particularly successful at making *MedRec* reliable in practice.

The "brown bag" approach to *MedRec* (Becker 2015; Colley and Lucas 1993; Nathan et al. 1999; Sarzynski et al. 2014) is a practical adaptation to *MedRec*. Here, the patient is told to collect all his or her medications in a bag and bring them to the medical encounter where a clinician examines and records them. This approach has the great advantage of capturing those medications that the patient possesses and, presumably, takes. It may offset the consequences of poor patient recall, which is a common problem (Jones et al. 2015). It cannot, of course, be applied to the hospitalized patient or the emergency room patient. It also requires coordination before the planned clinical encounter.

Even in the outpatient setting, the approach is vulnerable. Patients may bring old or unused medications that have been discontinued or replaced by other medications or doses (Martinez et al. 2012). This is one mechanism for generating "zombie" medications. Zombie medications are those that have been discontinued or are no longer used by the patient but that have been revived by a *MedRec* process gone awry. Zombie medications can be found in many electronic medical records.

Despite its limitations, using medication containers to aid cognitive work is an important instance of direct aiding of the *MedRec* process itself. The patient's physical medication containers represent a resource that can be exploited to make *MedRec* easier or more reliable (Hollnagel 2009). Significantly, this approach was developed well before the term "medication reconciliation" became common in the literature. The containers are naturally occurring "cognitive artifacts" (Hutchins 1995), available to practitioners who may incorporate their use into the practitioners' cognitive work. Recognizing and analyzing existing cognitive artifacts is one means for exploring distributed cognition in the wild (Nemeth et al. 2004).

The "brown bag" approach also makes it clear that *MedRec* output is necessarily a compromise between competing sources. Which medications the patient actually ingests, injects, or applies can be a subset or superset of the drugs prescribed by one or several clinicians over time. Despite claims to the contrary (e.g., Gleason et al. 2010), there is no "gold standard" or "single source of truth" that can be objectively known to be the "correct" list of medications. *Reconciliation* implies more than declaring a winner. The most accurate representation of a patient's medications will sometimes be found in one place, sometimes in another, and sometimes will exist nowhere and have to be constructed.

The clinician's goal in *MedRec* is to create a valid (i.e., accurate, up to date, concise, coherent, complete, comprehensive, etc.) representation of the patient's medications. This representation can then be used as a starting place for further medical care. But this goal cannot be assured because there is no objective standard. At best, the result is a momentary, conditional, highly contingent approximation.

What then is "reconciliation"? What is going on when practitioners work to assemble, correlate, understand, and summarize the multiple sources of information available into that moment's best estimate of the medicated state of an individual patient? As the question implies, the answer lies in the ways that practitioners *make sense* of the patient using available information or searching out additional information about that patient.

4.5   WHAT WE CAN INFER ABOUT THE COGNITIVE WORK OF RECONCILIATION

The plain language sense of reconciliation is of bringing together two or more disjointed or conflicting entities and altering them so that the conflict between them is reduced or eliminated. The reconciliation of "medication reconciliation" is prompted by a conflict or discrepancy in a clinician's understanding of a patient and their medications.

There are many ways that a single clinician's understanding can represent a conflict or discrepancy. An obvious case is where the clinician recognizes that the patient appears to be taking a medication for a condition that the patient is not known to have.

> *Example 2*: A clinician in the orthopedic clinic is reviewing the medical record for a 55-year-old patient, a former college football player, who has come to discuss a possible knee replacement. The medical record indicates that, along with pain medications and a cholesterol-lowering drug, the patient has been prescribed an antihypertensive, that is, a medicine that reduces blood pressure. High blood pressure is common and patients commonly receive medication for its control. The clinician infers from this information that this patient has hypertension (high blood pressure). Like all medical records, this patient's chart (actually a series of windows appearing on a computer display) contains a problem list showing the active medical conditions the patient is known to have. Most of these are not significantly related to the purpose of the clinic visit but the clinician notes that hypertension is not listed there. Reviewing the problem list further, the clinician finds the initials "BPH." This is a common abbreviation for benign prostatic hyperplasia, a prostate enlargement condition that is common in men over age 50. BPH can make urination difficult. Returning to the medication list, the clinician notes that the antihypertensive medication is listed as "Hytrin." Hytrin is a registered trademark name for the generic medicine *terazosin*. Terazosin is an antihypertensive medication but it also is used in the treatment of BPH. The clinician concludes that the patient is taking terazosin for benign prostatic hyperplasia rather than for hypertension.

Such situations are common. Clinicians spend much of their time meeting and assessing new patients. Physicians, nurses, pharmacists, respiratory therapists, and others are educated, trained, and experienced with this work and they do it quickly and efficiently.

We can understand the example as reflecting the construction and modification of a specific mental *representation* of the patient, that is, a semantic entity composed of the clinician's understanding of the patient. Clinicians develop, revise, prune, add to, and correct these representations over time, using medical records, conversations with the patients and families, examinations, and other sources of information.

Completeness and coherence are important signals that the representation reflects the important aspects of the world correctly. The presence of conflicting information within the representation or—as in this case—the absence of information that should be present (i.e., incompleteness) signals that the representation contains a discrepancy. Discrepancy generates the need for *reconciliation*.

In Example 2, the clinician discovers and then resolves a discrepancy by updating the representation. Figure 4.1a is a schematic showing a small, relevant portion of the practitioner's mental model. We can infer the sequence of cognitive work to be something like this: First, the clinician visually processes and recognizes a word in the

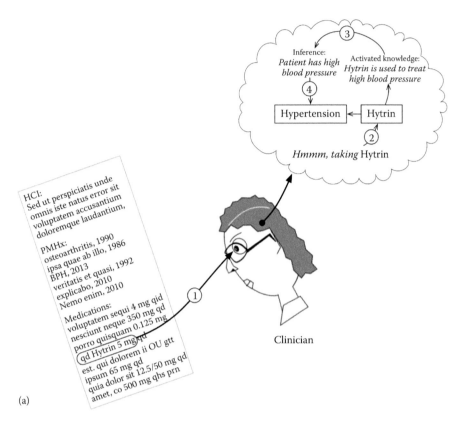

(a)

FIGURE 4.1 (a) An orthopedic surgeon is reviewing the medical record of a patient who is a candidate for a total knee replacement for relief of severe osteoarthritis. The inferred sequence includes (1) recognizing the word *Hytrin*; (2) incorporating the name *Hytrin* into the mental representation of this patient; (3) activation of related, relevant knowledge that *Hytrin* is a medicine useful for treating high blood pressure (hypertension); (4) inferring that the patient has hypertension and incorporating this medical condition into the mental representation.   (*Continued*)

record ("H-y-t-r-i-n"). In this specific context, the clinician understands the presence of this word (i.e., a character sequence appearing on a display) to mean that the patient is regularly taking the associated medicine. *Hytrin* is added to the mental representation. Based on deep knowledge, the clinician understands that *Hytrin* is used for treatment of the medical condition *hypertension*. This understanding is incorporated into the representation as the presence, in this patient, of the medical condition *hypertension*. As shown in Figure 4.1b, the mental model prompts the clinician to search for *hypertension* in the condition list (under "PMHx," which stands for *previous medical history*). But this search fails, creating a discrepancy: the clinician's representation includes *hypertension* but that condition is not included in the place where such conditions should appear. In Figure 4.1c, we see that the discrepancy prompts the clinician to further review the record and to discover "BPH" in the problem list. The clinician associates "BPH" with the medical condition *benign prostatic hyperplasia*. Again,

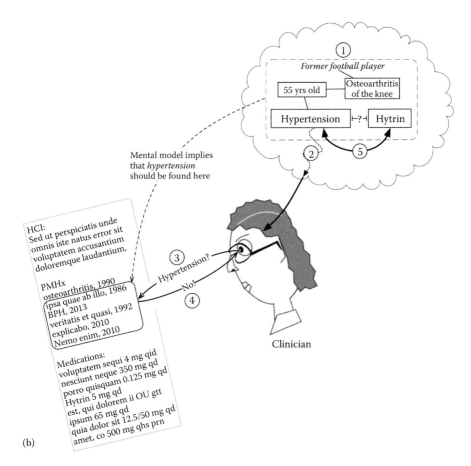

FIGURE 4.1 (CONTINUED)   (b) (1) The mental model of the patient, which previously included his age, history of playing *football*, and the medical condition *osteoarthritis* is now linked with the condition *hypertension* and the medication *Hytrin*; (2) the presence of a new (to this representation) condition hypertension prompts the surgeon to look for that condition in the patient's medical record; (3) and (4) the surgeon scans the medical record looking for the medical condition hypertension in an unsuccessful visual search that (5) causes a discrepancy to exist between the expectations flowing from the model and the data available in the world.   (*Continued*)

based on deep knowledge, the clinician understands that *Hytrin* is a treatment for this condition. The clinician revises the representation, removing *hypertension* in favor of BPH. Figure 4.1d shows the result. The clinician's mental model now contains *Hytrin* associated with BPH and BPH associated with advancing age. This portion of the clinician's mental model is a man who has *BPH* that is treated with *Hytrin*.

The example shows some cognitive features of reconciliation:

1. The impetus for reconciliation is discovery of a discrepancy. Without a discrepancy, there is nothing to reconcile. The presence of a discrepancy is a necessary but not sufficient precondition for reconciliation.

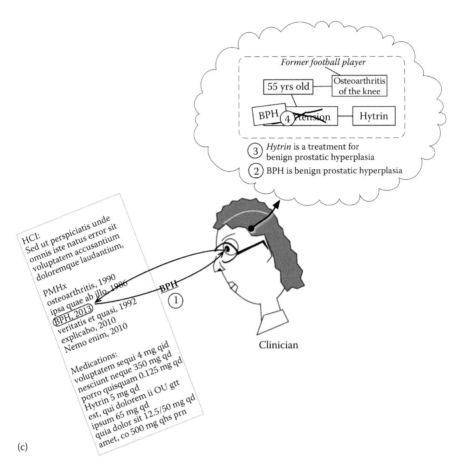

FIGURE 4.1 (CONTINUED) (c) (1) The surgeon finds "BPH" under conditions; (2) *BPH* stands for benign prostatic hyperplasia, a common condition in older men; (3) activation of related, relevant knowledge about *BPH* is that it is sometimes treated with *Hytrin*; (4) the preceding inference that hypertension was present in this patient is rejected in favor of (5) the representation that the patient has BPH.   (*Continued*)

2. The discrepancy is created while constructing a representation of the world. It arises when new data are at odds with a specific expectation generated by the existing representation. The expectation flows from the representation structure. An important consequence is that discrepancies exist in the representation, not in the external world. The factors that govern generation of representations determine whether a discrepancy will arise. If different individuals identify the same discrepancy, it is *a priori* likely that their approaches to generating representations are similar.

3. Resolving the discrepancy requires search. The discrepancy cannot be resolved by interrogating its contributing components. Instead, resolution requires additional data and knowledge. Getting the data or knowledge

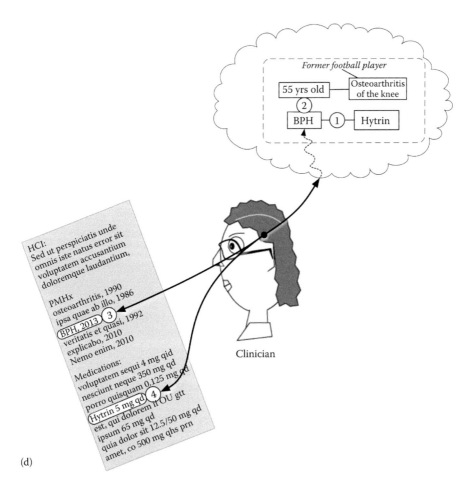

(d)

FIGURE 4.1 (CONTINUED) (d) The mental model is revised with *BPH* connected (1) to *Hytrin* and (2) to the patient age. The coherence between the data indicating (3) BPH and (4) Hytrin and the mental model has resolved the discrepancy.

requires search for data in the world (e.g., evidence that the discrepancy is arising from trying to incorporate stale data into the representation), search for knowledge in the head (e.g., the representation is flawed because it was too simple to accommodate the new data and that increasing the sophistication of the representation accommodates the elements into a representation without a discrepancy), or some combination. Search is an integral element of reconciliation. Factors that influence the success or efficiency of search will influence the quality of reconciliation.

4. Reconciliation resolves the discrepancy. Search leads to a revision of the representation, updating or correcting it such that the discrepancy no longer exists. In the example, revising the representation from "hypertension" to "BPH" restored its internal consistency and eliminated the discrepancy.

5. The reconciliation process terminates when the discrepancy is resolved. When the need for reconciliation is gone, the cognitive work that constitutes reconciliation ends.

6. The process is quick and smooth. The entire activity takes place in less than a few seconds. The activity is not sharply differentiated from the surrounding stream of cognition. The clinician is either unaware or only vaguely aware that it has taken place.

The most significant features of this sequence, other than its speed, are the requirement for search and the central role played by a specific, malleable representation.

4.6 SEARCH

Although reconciliation is triggered by discrepancy, the process of reconciliation is intimately connected with search. The clinician searches for candidates that may allow resolution of the discrepancy. This process involves exploring the world in order to discover candidates. This exploration is guided by the clinician's knowledge. The search in this instance is successful because (1) the information that resolves the discrepancy is close at hand and (2) the clinician's knowledge and experience identify that information and its relevance to resolving the discrepancy.

In non-human animals, where it is most studied, search is highly optimized for specific conditions and purposes, for example, pigeons finding food. Animals have a rich repertoire of approaches to search (sometimes called "strategies") and they vary ("choose") their approach based on the conditions in the world and the success they experience in striving for their goal. An approach that yields a satisfactory reward quickly and with low effort is preferred. If such an approach fails, more effortful approaches are used (see Davelarr and Raaijmakers 2012).

The decision to invest effort in search is inherently probabilistic. Search for a particular target may be successful or unsuccessful. Although unsuccessful search may lead to further search using a different approach, if the target's value is low, it may be better to abandon the effort entirely. Even for important targets, repeated failure suggests that the target cannot be acquired. Experience plays an important role here. If experience shows that persistence is likely to be rewarded, search may continue. If experience shows that initial failure presages unsuccessful search, early failures may truncate search more quickly.

For reconciliation, the value of search targets may vary enormously. In Example 2, the target (resolution of a discrepancy between the implication of a medicine and an expected medical condition) has low value *relative to the purpose of the clinician* who is an orthopedic surgeon considering a patient for a knee replacement. The search is immediately rewarding: the physician quickly finds a suitable explanation for the discrepancy. If the early search effort were unrewarding, the search itself might be abandoned, leaving the discrepancy unresolved. The world is, after all, awash in discrepant data and the job of the surgeon is to determine if the patient is a suitable candidate for a specific operation, not repair the world!

This same target might be of far greater significance to a different practitioner. If the situation were slightly different, say the clinician was a urologic surgeon, the inference drawn from the presence of *Hytrin* would be that the patient is being treated for BPH, a urological condition that would be at the center of any clinical interaction between doctor and patient.

Knowing that the target is certainly present in the field of search encourages continuing if the field itself is crisply bounded and not too large. Conversely, a "needle in the haystack" problem encourages early abandonment. The clinician performing *MedRec* may encounter a range of situations between these extremes. Available information is sometimes rich, well organized, and accessible, sometimes sparsely scattered and difficult to obtain.

4.7 SEARCH SUCCESS DEPENDS ON KNOWLEDGE ACTIVATION

In Example 2, the content and organization of the medical record make the search immediately productive. The search stops because the discrepancy is resolved by the retrieval from memory of a relationship between two dissimilar items, the medicine and the medical condition. This requires activation of rather specialized knowledge, here the relationship between *benign prostatic hyperplasia* and the medication *terazosin*. Although we do not know the details of that cognitive activity—for example, we cannot tell how the knowledge is encoded or what specific features of the data presentation cause it to be activated—there is no doubt that it is highly specialized. The example shows that it is also rapid.

Experimental approaches to cognitive search do not begin to approach the sophistication of the knowledge activation in the example. Such experiments often have subjects search memory repeatedly for words related to a concept ("List as many animals as you can") or to match items from a previously provided list. The surgeon here is not searching for a well-defined target but for data that can be incorporated into an explanation of a discrepancy in the mental model being constructed. This is a crucial element of reconciliation: if the clinician does not recognize at the moment that the presence of "Hytrin" on one page is related to the presence of "BPH" on another, the search will continue without success and eventually be abandoned. Although we have no direct access to the knowledge structures and patterns of activation that comprise the effort devoted to reconciliation, they are certainly formidable and remarkably efficient.

4.8 QUICK SUCCESS LIMITS AWARENESS OF SEARCH

The discrepancy in the first example is quickly resolved. A discrepancy arises from work of constructing a mental representation of the patient. The discrepancy prompts a search. The clinician combines data readily available in the world and knowledge readily available in the mind to update her mental model of the patient. In a sense, the discrepancy is not so much corrected as erased by updating the mental representation; the representation's internal consistency eliminates the discrepancy.

It is likely that most reconciliation is like this if only because, were it not so, the cognitive work of clinicians would be so difficult that nothing would get done! Medical practice requires continuous attention to and revision of clinicians' mental models. The smooth, untroubled fashion in which this plays out derives from expertise. This capacity for quickly creating a sophisticated, relevant, conservative mental model for a specific instance or case is an essential element of expertise.

The efficiency of reconciliation also comes from the symbiotic relationship between the clinician cognitive processes and the structure of the external world. Constructing representations is what clinicians *do* and the cognitive artifacts available in the world are purposefully suited to this. Artifacts and expertise have evolved together and are in an ecological relationship with each other. The ecological relationship has made the process so smooth that this sort of cognitive work is mostly invisible to outsiders and also to the practitioners themselves. Far from being a staged performance called out at specific moments, reconciliation is a cognitive activity that is more or less always happening. It is a core feature expertise.

4.9   PRYING OPEN RECONCILIATION

Reconciliation requires cognitive work. This work includes making, comparing, and revising mental representations of the external world. There is almost no published research on how experts produce, test, and recast these representations. There is a body of literature on how experts can be tricked into creating faulty or inconsistent representations. But this literature is devoted to demonstrating the fragility of expertise, usually by some manipulation of one or another detail of the problem presentation. It is safe to say that the world does not routinely manipulate details in this way and that these demonstrations have little to tell us about how experts routinely achieve success.

Here lies a conundrum: although we recognize the importance and power of expertise, the Law of Fluency means that we are likely to find it difficult to devise means for exposing the experts' cognitive processes that do not also hobble them in ways that distort their expertise. In effect, expertise hides its sources from both researcher and expert.

We have developed an experimental approach that provides some insight into reconciliation (Vashitz et al. 2011, 2013). Experienced clinicians are given a set of index cards representing the medical conditions and medications of a single patient. They are told to consider these cards the description of a patient to be placed under their care. The subjects are then asked to arrange the cards any way they like and describe the patient.

The clinicians organize the cards and describe the patient in quite similar ways. Although the task involves a moderately complex patient, clinicians can complete the task very quickly, often within a minute. After organizing the cards, the experts described the medically relevant situation for the "patient" in similar ways.

The clinicians appeared to use a rather specific strategy for selecting and arranging cards. They initially chose the medical conditions closely related to their own specialty and medicines related to those conditions. They did this very quickly, usually in less than 10–20 seconds. The related cards were grouped together, away from the remaining cards. They then organized the cards for conditions and medicines less relevant to their specializations, loosely grouping subcollections by a physiological system. The process had the advantage of stepwise reducing the complexity of the card pattern, moving the familiar, organized cards away from the less relevant (and, presumably, less familiar) cards (Figure 4.2.)

In some of the trials, we discretely removed a few entries from the card deck, making the resulting collection incomplete. When cards were missing, the subjects began the task in much the same way. The activity slowed markedly, however, as the arrangement neared completion. The experts uniformly detected that there was something missing from the collection they were arranging. When the "missing" card

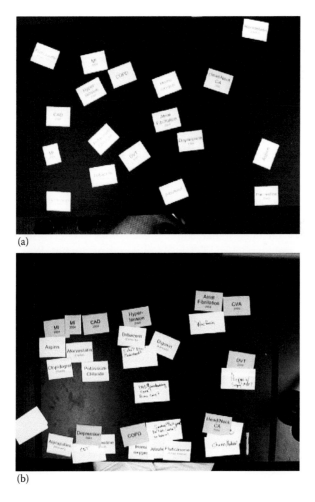

(a)

(b)

FIGURE 4.2   (a) Before and (b) after card layouts. The photos are screen captures from a video camera mounted above the table on which the cards were placed. The entire process of rearrangement and verbal commentary was video recorded for each subject.

was provided, they quickly placed it with the associated medical condition. Their card organization and the ability to detect something missing are strong evidence that the clinicians are doing more than simply matching "things that go together." They are constructing a mental representation that specifies meaningful, high-order symbolic and coherent relationships that characterize a specific individual, that is, a mental *model* of the patient.

During its construction, the discrepancy between the mental model and the field of data prompts a search for and failure to find the "missing" cards. The mental model construction creates expectations about what should be found in the data field. The search for the expected data fails. The slowing of performance at this stage, we believe, comes from a sort of reverberating search that cyclically interrogates the external world and the mental model, seeking a resolution. The absence of expected data in the external world might signal that the mental model is flawed (this was the case in Example 2). Alternatively, the external data might be missing a key datum. In the real world, a clinician might look to other data sources but our experimental protocol constrains the field of available data so that the search fails quickly. Real-world searches can fail too, of course, or be so unproductive that they are abandoned.

Clinician experts know from experience that the real world is often ill-behaved with respect to their cognitive work. They know that their mental models of patients are critical to their clinical work. They also know that their mental models can be flawed and that internal inconsistencies in the models may signal the presence of flaws. To function successfully in the real world of work, the expert must make good judgments about whether a particular discrepancy signals missing data or a flawed model. For example, a well-ordered list of things implies internal coherence and completeness that scattered data do not. Some relationships in mental models are strong while others are weak. Some discrepancies are likely to be important (e.g., the presence of two medicines that amplify each other's effects such as coumadin [Warfarin] and acetaminophen [Tylenol]) while others are not. Whether to pursue a discrepancy must depend on the balance between its significance in the moment, the projected likelihood that further search will be rewarding, and the pressure of competing demands for attention. The sources of success and failure in *MedRec* are to be found in this balance.

Of course, the experimental task bears only limited resemblance to the real work of *MedRec*. There are no external distractors, no images of the patient, and no sensory input except a few symbols. The work is not even presented as a *medication* reconciliation problem. Arguably, however, the task is representative of the kind of cognitive work that clinicians must do in order to accomplish *MedRec*, and it offers some insight into why *MedRec* goes astray half the time.

Far from being some form of list processing, reconciliation is an aspect of the deep structure of expert cognition. Reconciling is inherent in representation construction. Representation construction and revision is the core skill required for "making sense." Experts do not simply assemble all the available data into an undifferentiated montage. Instead, they process data in an orderly, critical way that employs a mental model to support (i.e., test and explore) their apprehension of the external world and its significance.

4.10 WHAT COGNITIVE SYSTEMS ENGINEERING CAN TELL US ABOUT *MEDREC*

A cognitive system has three elements: "the world to be acted on, the agent or agents who act on the world, and the [external] representations through which the agent experiences the world" (Woods and Roth 1990). There is an ecological relationship between these elements. Changes in one have effects on the others and change is frequent in all three. The system is necessarily complex and adaptive. Engineering a cognitive system is the deliberate configuration of one or more elements to achieve better adaptive performance.

In cognitive systems engineering terms, *MedRec* is a process accomplished by a cognitive agent in order to create an external representation of the world. Cognitive systems engineering principles lead us to look for the kinds of data that can be used, how uncertainty is managed, what practical limitations on effort exist, and how the trade-off point between completeness, quality, and effort is struck (Hollnagel 2009).

The cognitive work of reconciliation is truly launched by the occurrence of a discrepancy. Discrepancy exists in the mental representation, not in the world, and it promotes a search for data (in an external representation) or relationships (in the agent's mental model) that can be used to resolve the discrepancy.

Aiding discrepancy detection has received some attention. Methods have been developed to aid in the construction and comparison of lists. The automated presentation of lists optimized for cross-list comparison has been proposed and tested (Plaisant et al. 2013). Improvements, albeit small ones, have been reported with some forms of computer support (Giménez-Manzorro et al. 2015).

Aiding discrepancy *resolution* has received much less attention. The experiments described indicate that clinician experts build their mental models by cross-referencing different kinds of data. Rather than internally representing a patient as lists of items, clinicians create representations that link medications to the conditions for which they are applied (and also to those conditions with which they might interfere, including other medications!). It is likely that *MedRec* can be improved by enhancing the ability of experts to construct and test their models, and exploiting the relationship between medical condition and medication is an obvious target.

A strategy of aiding reconciliation by linking medications to medical conditions is certainly feasible. All medications are directed at specific medical conditions. Most medical conditions for which patients seek care are treated with medicines. Although the relationship is not necessarily one to one (a single medicine may have multiple effects and a single condition may be treated with multiple medicines), the relationships are explicit, durable, and well understood. The obvious way to accomplish this is to present the medication together with its associated medical condition and with its provenance. The experiments also suggest a preferred order for presentation of these linked entities. Clinicians begin the ordering process with the conditions and medications most directly associated with their own specialties, which, presumably, correspond as well to the specific issue for which the patient's data are being reviewed. A cognitive aid would present the entities in this order and grouped together by physiological system.

The sensitivity of clinicians to missing or aberrant data is likely derived from living with a great deal of imperfect information. Practitioners are inundated with data from every side. They routinely confront fields of data that can include incomplete, misleading, or even incorrect data. Evaluating data credibility is an important element of expertise. Strongly attested data are, other things being equal, more likely to be correct than weakly attested ones. The pedigree or provenance of information is useful in evaluating competing or ambiguous data. For medications, provenance includes (among other things) the prescriber's and renewer's names and roles, the date originally prescribed, the nature of modifications to dosage or frequency over time, the noted reactions to the medication, the inferences about their causes, the date the prescription was discontinued, the reason for discontinuing, the reported frequency of use (especially for "use as needed" medicines), and so forth. Ironically, in order to aid discrepancy detection, some *MedRec* aids actually strip provenance away from data. Stripping away the provenance does not eliminate the processes that use that information, it just makes those processes less reliable. Instead, a useful strategy for aiding *MedRec* is to actively display the provenance and the relationships.

4.11   WHAT *MEDREC* TELLS US ABOUT COGNITIVE SYSTEMS ENGINEERING

Exploring *MedRec* as reconciliation rather than list comparison opens up a much larger space of inquiry. What was initially a nagging technical issue is, with careful study, recognized as a symptom of a larger and more complicated set of problems that are closely related to the workplace. The need for *MedRec* derives directly from the current, fractionated approach to medical care as well as from the proliferation of medicine. Instead of being a clerical task, *MedRec* is a complex, technically demanding activity that draws on the deep structure of the domain of practice. Looking closely into *MedRec* sheds light on both the cognitive work and the obstacles to developing effective support. Finally, the inquiry circles back to the sharp-end practitioners and their contribution to rationalizing and making sense of the fragmented, disjoint field of data that represents the patient. The word "reconciliation" is perfectly appropriate for this work. The word also applies to our own efforts to understand why the problem of *MedRec* is obstinate despite the investments made in improving it.

This pattern is common in cognitive systems engineering. Apparently, simple tasks associated with ordinary (but important) work resist solutions predicated on the appearance and associations. The simplicity mirage creates the impression that it is the practitioners who are at fault. This is the source of many common attributions of failure to "human error" (Woods et al. 2010) including inattention, willful disregard of policies, and so on. When a few (perhaps many) cycles of solutions and beatings fail to produce the desired effect, closer examination of the work shows it to be complicated and difficult. Gradually, it becomes clear that expertise is important and that the representations are poorly tailored for the real tasks. What formerly appeared to be solutions are now seen as clumsy attempts to replace expertise. This sets the stage for substantive work that reveals features of the underlying domain—often intractable features linked to hard endpoints such as cost pressure or time pressure.

Other forays into cognitive systems engineering have produced important results by concentrating on aspects of "ordinary." The programming of infusion pumps (Cook et al. 1992; Henriksen et al. 2005a; Nunnally et al. 2004), the assignment of intensive care unit beds (Cook 2006), and voice loops in space mission control (Patterson et al. 2008; Watts et al. 1996) are examples. In each case, inquiry into a prosaic, apparently simple part of ordinary work leads to broad appreciation and even broader application. These investigations are not one-offs but prolonged engagements that demonstrate the deeply productive and durably useful nature of cognitive systems engineering. In many cases, these investments pay off handsomely for decades and have reverberations far outside of the original study domains.

Cognitive systems engineering requires understanding the prosaic details of cognitive work. Ironically, the Law of Fluency frustrates attempts to map the expert's problem space. The development of cognitive engineering closely tracks the development of tools and methods that help to map that space. Practitioners' use of cognitive artifacts can map out substantial parts of the work (Henriksen et al. 2005b). Deliberately looking for "what makes work hard" can focus attention on critical factors (Nemeth et al. 2008). Practitioners adapt information technology to their specific needs and the cycles of adaptation provide useful information about the underlying work and introduction of new technology is a naturally occurring experiment (Cook 2002; Cook and Woods 1996). Mapping the uncertainties and conflicting goals that confront experts and the way trade-offs across this multidimensional space are negotiated is especially useful (Woods and Cook 2002). Together, these tools and methods are often sufficient to "get in" to a new area.

Significantly, virtually all of the interesting areas of practice are in turmoil. New technology and new forms of organization appear continuously. The pace and scope of change tax the ingenuity and adaptive capacity of individuals, teams, and larger groups.

This is especially true for information technology itself where the pace of change is so great that it is spawning new paradigms. The shift from waterfall development to continuous delivery has reduced the time between software upgrades for very large systems to, in some cases, minutes. Such systems have almost organic complexity and physiological behaviors. Developing for, managing, and troubleshooting such systems must occur during change. Because these systems are highly distributed and extensively self-regulating, they are themselves cognitive agents. This raises the stakes for cognitive systems engineering and promises to open up new, productive lines of inquiry (Allspaw 2015). Cognitive systems engineering inquiry into problems like *MedRec* offers a means to leverage understanding in order to make real progress in building joint cognitive systems. This leverage is essential if explosive growth systems such as new information technology are to be made resilient.

ACKNOWLEDGMENTS

Mark Nunnally, Christopher Nemeth, Yuval Bitan, Geva Greenfield (nee Vashitz), Christine Jette, Michael O'Connor, Kiku Härenstam, and others have made important contributions to this work. A number of clinicians, both physicians and nurses, participated as subjects in the experiments. I gratefully acknowledge all of them. Both through his writing and in person, David Woods has inspired and exasperated the author for so long and in so many ways that it is no longer possible to discern his discrete contribution. The author is solely responsible for any errors [sic].

REFERENCES

Allspaw, J. (2015). *Trade-Offs under Pressure: Heuristics and Observations of Teams Resolving Internet Service Outages.* Unpublished Masters Thesis, Lund University, Lund, Sweden.

Amann, S., and Kantelhardt, P. (2012). Medication errors and medication reconciliation from a hospital pharmacist's perspective. *Zeitschrift für Evidenz, Fortbildung und Qualität im Gesundheitswesen,* 106: 717–722.

Andrus, M.R., and Anderson, A.D. (2015). A retrospective review of student pharmacist medication reconciliation activities in an outpatient family medicine center. *Pharmacy Practice (Granada),* 13: 518.

Barnsteiner, J.H. (2005). Medication reconciliation: Transfer of medication information across settings-keeping it free from error. *American Journal of Nursing,* 105: 31–36.

Bates, D.W., Spell, N., Cullen, D.J., Burdick, E., Laird, N., Petersen, L.A. et al. (1997). The costs of adverse drug events in hospitalized patients. Adverse Drug Events Prevention Study Group. *JAMA,* 277: 307–311.

Batra, R., Wolbach-Lowes, J., Swindells, S., Scarsi, K.K., Podany, A.T., Sayles, H. et al. (2015). Impact of an electronic medical record on the incidence of antiretroviral prescription errors and HIV pharmacist reconciliation on error correction among hospitalized HIV-infected patients. *Antiviral Therapy,* 20: 555–559.

Becker, D. (2015). Implementation of a bag medication reconciliation initiative to decrease posthospitalization medication discrepancies. *Journal of Nursing Care Quality,* 30: 220–225.

Burke, K.G., Mason, D.J., Alexander, M., Barnsteiner, J.H., and Rich, V.L. (2005). Making medication administration safe: Report challenges nurses to lead the way. *American Journal of Nursing,* 105: 2–3.

Cater, S.W., Luzum, M., Serra, A.E., Arasaratnam, M.H., Travers, D., Martin, I.B. et al. (2015). A prospective cohort study of medication reconciliation using pharmacy technicians in the emergency department to reduce medication errors among admitted patients. *Journal of Emergency Medicine,* 48: 230–238.

Colley, C.A., and Lucas, L.M. (1993). Polypharmacy. *Journal of General Internal Medicine,* 8: 278–283.

Cook, R.I. (2002). Safety technology: Solutions or experiments? *Nursing Economics,* 20: 80–82.

Cook, R.I. (2006). Being bumpable: Consequences of resource saturation and near-saturation for cognitive demands on ICU practitioners. In D.D. Woods and E. Hollnagel (Eds.), *Joint Cognitive Systems: Patterns in Cognitive Systems Engineering* (pp. 23–35). New York: CRC Press.

Cook, R.I., Render, M., and Woods, D.D. (2000). Gaps in the continuity of care and progress on patient safety. *BMJ: British Medical Journal,* 320: 791.

Cook, R.I., and Woods, D.D. (1996). Adapting to new technology in the operating room. *Human Factors: The Journal of the Human Factors and Ergonomics Society,* 38: 593–613.

Cook, R.I., Woods, D.D., Howie, M.B., Horrow, J.C., and Gaba, D.M. (1992). Case 2-1992. Unintentional delivery of vasoactive drugs with an electromechanical infusion device. *Journal of Cardiothoracic and Vascular Anesthesiology,* 6: 238–244.

Davelarr, G.J., and Raaijmakers, J.G.W. Human memory search. In P.M. Todd, T.T. Hills, and T.W. Robins (Eds.), *Cognitive Search: Evolution, Algorithms, and the Brain* (pp. 178–193). Cambridge MA: MIT Press.

Gentner, D., and Stevens, A.L. (1983). *Mental Models*. Hillsdale, NJ: Erlbaum.

Giménez-Manzorro, Á., Romero-Jiménez, R.M., Calleja-Hernández, M.Á., Pla-Mestre, R., Muñoz-Calero, A., and Sanjurjo-Sáez, M. (2015). Effectiveness of an electronic tool for medication reconciliation in a general surgery department. *International Journal of Clinical Pharmacology*, 37: 159–167.

Gleason, K.M., Groszek, J.M., Sullivan, C., Rooney, D., Barnard, C., and Noskin, G.A. (2004). Reconciliation of discrepancies in medication histories and admission orders of newly hospitalized patients. *American Journal of Health-System Pharmacy*, 61: 1689–1695.

Gleason, K.M., McDaniel, M.R., Feinglass, J., Baker, D.W., Lindquist, L., Liss, D. et al. (2010). Results of the Medications at Transitions and Clinical Handoffs (MATCH) study: An analysis of medication reconciliation errors and risk factors at hospital admission. *Journal of General Internal Medicine*, 25: 441–447.

Hancock, J. (2013). *Health Technology's "Essential Critic" Warns of Medical Mistakes*. *Kaiser Health News*. Retrieved June 15, 2015, from http://khn.org/news/scot-silverstein -health-information-technology/.

Hart, C., Price, C., Graziose, G., and Grey, J. (2015). A program using pharmacy technicians to collect medication histories in the emergency department. *Pharmacy and Therapeutics*, 40: 56–61.

Henriksen, K., Battles, J.B., Marks, E.S., and Lewin, D.I. (Eds.). (2005a). *Making Information Technology a Team Player in Safety: The Case of Infusion Devices (Vol. Advances in Patient Safety: From Research to Implementation (Volume 1: Research Findings))*. Rockville, MD: Agency for Healthcare Research and Quality.

Henriksen, K., Battles, J.B., Marks, E.S., and Lewin, D.I. (Eds.). (2005b). *Cognitive Artifacts' Implications for Health Care Information Technology: Revealing How Practitioners Create and Share Their Understanding of Daily Work (Vol. Advances in Patient Safety: From Research to Implementation (Volume 2: Concepts and Methodology))*. Rockville, MD: Agency for Healthcare Research and Quality.

Hollnagel, E. (2009). *The ETTO Principle: Efficiency–Thoroughness Trade-Off: Why Things That Go Right Sometimes Go Wrong*. Burlington, VT: Ashgate.

Hutchins, E. (1995). *Cognition in the Wild*. Cambridge, MA: MIT Press.

JACHO. (2006). Using medication reconciliation to prevent errors. *Sentinel Event Alert*, 35: 1–4.

Jones, G., Tabassum, V., Zarow, G.J., and Ala, T.A. (2015). The inability of older adults to recall their drugs and medical conditions. *Drugs and Aging*, 32: 329–336.

Ketchum, K., Grass, C.A., and Padwojski, A. (2005). Medication reconciliation: Verifying medication orders and clarifying discrepancies should be standard practice. *American Journal of Nursing*, 105: 78–85.

Klein, G.A. (1998). *Sources of Power: How People Make Decisions*. Cambridge, MA: MIT Press.

Kleinrock, M. (2015). *Medicine Use and Spending Shifts: A Review of the Use of Medicines in the U.S. in 2014*. Retrieved October 1, 2015, http://www.imshealth.com/.

Kramer, J.S., Stewart, M.R., Fogg, S.M., Schminke, B.C., Zackula, R.E., Nester, T.M. et al. (2014). A quantitative evaluation of medication histories and reconciliation by discipline. *Hospital Pharmacist*, 49: 826–838.

Landro, L. (2005). Teaching doctors to be nicer. *Wall Street Journal (East Edition)*, pp. D1, D4.

Lee, K.P., Hartridge, C., Corbett, K., Vittinghoff, E., and Auerbach, A.D. (2015). Whose job is it, really? Physicians', nurses', and pharmacists' perspectives on completing inpatient medication reconciliation. *Journal of Hospital Medicine*, 10: 184–186.

Li, Q., Spooner, S.A., Kaiser, M., Lingren, N., Robbins, J., Lingren, T. et al. (2015). An end-to-end hybrid algorithm for automated medication discrepancy detection. *BMC Medical Informatics and Decision Making*, 15: 12.

Liang, B.A., Alper, E., Hickner, J., Schiff, G., Lambert, B.L., Gleason, K. et al. (2007). *The Physician's Role in Medication Reconciliation: Issues, Strategies and Safety Principles*. Chicago, IL: American Medical Association.

Martinez, M.L., Vande Griend, J.P., and Linnebur, S.A. (2012). Medication management: A case of brown bag-identified medication hoarding. *The Consultant Pharmacist*, 27: 729–736.

Nathan, A., Goodyer, L., Lovejoy, A., and Rashid, A. (1999). Brown bag' medication reviews as a means of optimizing patients' use of medication and of identifying potential clinical problems. *Family Practice*, 16: 278–282.

Nemeth, C.P., Cook, R.I., O'Connor, M., and Klock, P.A. (2004). Using cognitive artifacts to understand distributed cognition. *IEEE Transactions on Systems, Man and Cybernetics, Part A: Systems and Humans*, 34: 726–735.

Nemeth, C.P., Kowalsky, J., Brandwijk, M., Kahana, M., Klock, P.A., and Cook, R.I. (2008). Between shifts: Healthcare communication in the PICU. In C.P. Nemeth (Ed.), *Improving Healthcare Team Communication: Building on Lessons from Aviation and Aerospace* (pp. 135–153). Aldershot, UK: Ashgate.

Nunnally, M., Nemeth, C.P., Brunetti, V., and Cook, R.I. (2004). Lost in menuspace: User interactions with complex medical devices. *IEEE Transactions on Systems, Man and Cybernetics, Part A: Systems and Humans*, 34: 736–742.

Okere, A.N., Renier, C.M., and Tomsche, J.J. (2015). Evaluation of the influence of a pharmacist-led patient-centered medication therapy management and reconciliation service in collaboration with emergency department physicians. *Journal of Managed Care and Specialty Pharmacy*, 21: 298–306.

Patterson, E.S., Watts-Perotti, J., and Woods, D.D. (2008). Voice loops: Engineering overhearing to aid coordination. In C.P. Nemeth (Ed.), *Improving Healthcare Team Communication: Building on Lessons from Aviation and Aerospace* (pp. 79–95). Aldershot, UK: Ashgate.

Plaisant, C., Chao, T., Wu, J., Hettinger, A.Z., Herskovic, J.R., Johnson, T.R. et al. (2013). Twinlist: Novel user interface designs for medication reconciliation. *AMIA Annual Symposium Proceedings*, 2013: 1150–1159.

Pronovost, P., Weast, B., Schwarz, M., Wyskiel, R.M., Prow, D., Milanovich, S.N. et al. (2003). Medication reconciliation: A practical tool to reduce the risk of medication errors. *Journal of Critical Care*, 18: 201–205.

Rodehaver, C., and Fearing, D. (2005). Medication reconciliation in acute care: Ensuring an accurate drug regimen on admission and discharge. *Joint Commission Journal on Quality and Patient Safety*, 31: 406–413.

Rouse, W.B., and Morris, N.M. (1986). On looking into the black-box—Prospects and limits in the search for mental models. *Psychological Bulletin*, 100: 349–363.

Rozich, J.D., Howard, R.J., Justeson, J.M., Macken, P.D., Lindsay, M.E., and Resar, R.K. (2004). Standardization as a mechanism to improve safety in health care. *Joint Commission Journal on Quality and Patient Safety*, 30: 5–14.

Sarzynski, E.M., Luz, C.C., Rios-Bedoya, C.F., and Zhou, S. (2014). Considerations for using the 'brown bag' strategy to reconcile medications during routine outpatient office visits. *Quality in Primary Care*, 22: 177–187.

Sullivan, C., Gleason, K.M., Rooney, D., Groszek, J.M., and Barnard, C. (2005). Medication reconciliation in the acute care setting: Opportunity and challenge for nursing. *Journal of Nursing Care Quality*, 20: 95–98.

Thompson, C.A. (2005). JCAHO views medication reconciliation as adverse-event prevention. *American Journal of Health-System Pharmacy*, 62: 1528, 1530, 1532.

Vashitz, G., Nunnally, M.E., Bitan, Y., Parmet, Y., O'Connor, M.F., and Cook, R.I. (2011). Making sense of diseases in medication reconciliation. *Cognition, Technology and Work*, 13: 151–158.

Vashitz, G., Nunnally, M.E., Parmet, Y., Bitan, Y., O'Connor, M.F., and Cook, R.I. (2013). How do clinicians reconcile conditions and medications? The cognitive context of medication reconciliation. *Cognition, Technology and Work*, 15: 109–116.

Watts, J.C., Woods, D.D., Corban, J.M., Patterson, E.S., Kerr, R.L., and Hicks, L.C. (1996). Voice loops as cooperative aids in space shuttle mission control. Proceedings of the 1996 ACM Conference on Computer Supported Cooperative Work. November 16–20, 1996 Boston, MA: USA.

Whittington, J., and Cohen, H. (2004). OSF healthcare's journey in patient safety. *Quality Management in Health Care*, 13: 53–59.

Woods, D.D., and Cook, R.I. (2002). Nine steps to move forward from error. *Cognition, Technology and Work*, 4: 137–144.

Woods, D.D., Dekker, S., Cook, R., Johannesen, L., and Sarter, N.B. (2010). *Behind Human Error* (2nd ed.). Burlington, VT: Ashgate.

Woods, D.D., and Hollnagel, E. (2006). *Joint Cognitive Systems: Patterns in Cognitive Systems Engineering*. Abingdon, UK: CRC Press.

Woods, D.D., and Roth, E.M. (1990). Cognitive systems engineering. In M.G. Helander (Ed.), *Handbook of Human–Computer Interaction* (pp. 3–43). Amsterdam, Holland: Elsevier.

Part II

Understanding Complex Human–Machine Systems

5 Engineering the Morality of Cognitive Systems and the Morality of Cognitive Systems Engineering

P. A. Hancock

CONTENTS

5.1 INTRODUCTION

At first blush, cognitive systems engineering and morality would seem to be very strange bedfellows. Formally, they appear to occupy different worlds of discourse. That is, they are apparently so divided in their essence and nature that they can well be dissociated in people's minds, rather in the same way that we parse the underlying academic disciplines within our nominal "universities." Dividing the human enterprise in this Aristotelian fashion has, *pro tem*, proved to be a rather useful and effective strategy, as our advancing understanding of the world demonstrates. However, every generative act is a moral act and so whether those pursuing CSE are aware of it or not, a basic morality underpins all of their actions and all of their aspirations. What I propose, assert, and argue for in this chapter is that CSE and morality are neither mutually exclusive domains nor indeed inherently divorced or separable in any meaningful way. In fact, I believe that they are essentially and necessarily intertwined one with the other and, more importantly, interlinked in ways that can substantially advance the understanding of each. My purpose here is to establish and subsequently promote such a unified and synergistic perspective.

Initially then, it might appear to the reader that I have undertaken a rather singular enterprise and that few, if any, have previously trod this path and sought to explore this connection. However, this would be an incorrect assumption, as indeed there are a number of individuals who have begun to exploit this bridge, now in rather extensive levels of detail and sophistication (e.g., see Dekker 2007a; Dekker et al. 2013a). Such insights also subserve a larger endeavor, which is to establish the greater philosophical foundation of all of human–technology science (e.g., Dekker and Nyce 2012; Dekker et al. 2013b; Hancock 2012b; Hancock and Diaz 2001; Hancock and Drury 2011; Reed 1996). As a foundation for my present discourse, however, I use two immediate areas of concern to CSE in order to carry the thread of the argument forward; these being (1) human error and (2) function allocation (FA), respectively. I hasten to add that these are only two examples of what I believe to be much more general and even ubiquitous principles.

5.2 HUMAN ERROR

Human error has long been considered perhaps the central issue in safety, and most especially in the etiology of failures in the operations of large-scale technical systems that characterize the genesis of CSE (see Perrow 1984; Reason 1990). It is my protestation that error is sin, clothed in a modern guise (and see also the argument in Dekker 2007b). Across the recent centuries, theological transgressions have evolved into secular mistakes. Perhaps error and sin can consequently be perceived as twin brothers who differ only in the name of their Father. Human error also subserves a surrogate purpose in contemporary morality, which is primarily founded upon our instantiation of the still evolving Judeo-Christian ethos. Sin and error each consists of nonconformity to a standard set by an external authority. Much depends on what one believes about the nature of that authority as the essential arbiter of justice under such circumstances. Error is a more acceptable modern term than sin, especially in an increasingly secular society that considers itself underpinned by "science." If one truly believes in a totally and absolutely powerful deity, then, in Christian terms, the determination of and punishment meted out must be left to that deity (… vengeance *is mine sayeth the Lord*). However, since the first codifications of Hammurabi, we as human societies seem collectively unpersuaded by the absolute power of punishment by any god. Consequently, we look to wreak our own form of social vengeance in the temporal world—or more adroitly and acceptably *to enforce the law*. Often used to impose the will of the relatively powerful on the relatively powerless, the ascendancy of human law in the determination of sin, then error, and then transgression, then unintended action, may actually be a metric of secular success over expressions of certain fanaticisms of faith. As we shall see, neither the framework of faith nor that of legal precedent is particularly revealing or appealing for future progress in understanding either safety or the human condition (and see Figure 5.1). Nor are such percepts well founded, especially as we look to its more modern, "rational" scientific underpinnings.

FIGURE 5.1 One of the first ever recorded customer complaints provides us much insight into the relative progress of our attitudes toward failure in contractual relationships. That is, Tell Ea-nasir: Nanni sends the following message. "When you came, you said to me as follows: 'I will give Gimil-Sin (when he comes) fine quality copper ingots.' You left then, but you did not do what you promised me. You put ingots that were not good before my messenger (Sit-Sin) and said: 'If you want to take them, take them; if you do not want to take them, go away!' What do you take me for, that you treat somebody like me with such contempt? I have sent as messengers gentlemen like ourselves to collect the bag with my money (deposited with you) but you have treated me with contempt by sending them back to me empty-handed several times, and that through enemy territory. Is there anyone among the merchants who trade with Telmun who has treated me in this way? You alone treat my messenger with contempt! On account of that one (trifling) mina of silver which I owe(?) you, you feel free to speak in such a way, while I have given to the palace on your behalf 1080 pounds of copper, and umi-abum has likewise given 1080 pounds of copper, apart from what we both have had written on a sealed tablet to be kept in the temple of Samas. How have you treated me for that copper? You have withheld my money bag from me in enemy territory; it is now up to you to restore (my money) to me in full. Take cognizance that (from now on) I will not accept here any copper from you that is not of fine quality. I shall (from now on) select and take the ingots individually in my own yard, and I shall exercise against you my right of rejection because you have treated me with contempt." (See Mödlinger, B.M. (2015, October 21.) *Ordering Copper.* Retrieved October 17, 2016, from https://arsenicloss.com/author/moedimani1/page/2/.)

5.3 FUNCTION ALLOCATION

The second thread that I explore (and I believe a second thread is necessary to show that my ideas are not confined to just one specific area) is that of function allocation (FA). FA is another point of divergence between CSE and what might be termed the more "traditional" human factors and ergonomics (HF/E) community (e.g., see De Winter and Hancock 2015; Hancock et al. 2013; Sheridan et al. 1998). Here, the apparently innocent questions as to *who* does *what, when* and *how* are, I would aver, necessarily embedded in the subtext of *why* such task distributions are undertaken in the first place. Allocation decisions are largely the purview of some external arbiter (nominally in the expression of a deliberate designer[s]) who actually decides on such questions. Dictating the nature of a person's work activity is very much a moral issue and one that, as we shall see, is largely superseded in the modern world by questions of profit, disguised as the unending striving for "efficiency" and "growth." Moral judgments, in such cases, extend their reach well beyond the proximal physical, technical work system to hand into the very heart of what any society seeks to be.

While I will argue that morality is a human-centric and thus relativistic term, it is nevertheless the case that in a world of social pragmatism, some goals, aims, and objectives can be identified as being for a collective "good." Other actions, which create physical damage, injury, misery, and death, are predominately viewed as "bad" and residually as expressions of "evil" by a society still riven with moralistic myopia (Mill 1863). Despite the absence of any ratio scale of moral certainty, from the pragmatic perspective, CSE remains a subjectively laudable moral endeavor *pro tem.* Precisely to whose eventual "good" the innovations generated by CSE actually contribute represents a minefield of contemporary contention (Hancock 2014). As can be seen in Figure 5.2, it is evident that the increased levels of productivity (the antithesis of sloth and inefficiency) that have accompanied the growth of CSE have disproportionately benefited progressively fewer and fewer individuals. At the same time, as is shown by the inset in Figure 5.2, we have been generating this putative profit very much at the expense of global sustenance. This is not even a Faustian bargain but the act of a pathologic and viral species that even contemplates enshrining individuality above all other forms of "good" (Rand 1943). However, with the notion of such eventual "good" in mind, I seek to conclude this chapter by espousing some tenets for CSE's immediate future in both the scientific and moral dimension. But first, to advance the present discussion, I have to consider the nature of cause that underpins all of the arguments promulgated in this chapter. Thus, FA is a surface expression of certain underlying moral imperatives that have, in our modern world, been liberally prostituted to the great god of "profit." As G. Gordon Liddy is reputed to have observed, "*If one wants to understand what's going on, follow the money.*"

5.4 NATURE AND CAUSE

If we are to begin at the beginning, as it were, we have to penetrate the philosophical foundations upon which many of our assumptions are based (and see Dekker 2007b). Thus, I first consider Aristotle's four orders of cause as my initial point of departure; this despite their own foundation in the etiology of existence per se. These four

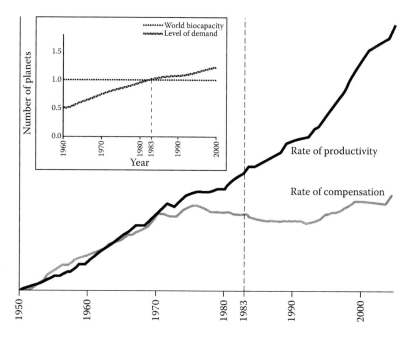

FIGURE 5.2 Disbursement of the fruits of increasing productivity enabled by technology. While advances in CSE may well create safer and more efficient complex technical systems, CSE cannot ignore the distribution of the subsequent benefit of those advantages that are rendered.

causes are, respectively, (1) material cause, (2) formal cause, (3) efficient cause, and (4) final cause (and see Hill 1965).

At the risk of distorting the original purpose of Aristotle's conception, I want to propose here that each type of cause is necessarily embedded, that is, situated, within its forebear. By this, I mean that they are intrinsically ordered and hierarchical in nature. From the start, it would seem that CSE (like almost all of engineering) is primarily associated with the goal of specifying efficient cause. Thus, CSE seeks to answer the "how" questions of life by defining local, proximal, and systematic causal relationships among the technological elements within its particular realm of study. Indeed, in its earliest incarnations, CSE was an elaboration of simple causal chains that individually are more rightly characterized by micro-systems analyses; this itself being one category of general systems theory (Bertalanffy 1972). From the latter, and more traditional HF/ E approach, the individual effects of a small, limited set of variables have typically been examined in terms of their influence on another highly restricted set of outcome variables. As far as possible, extraneous sources of experimental "error" are isolated and excised to the degree that any experimenter is able to do so. Hence, limited but fairly strong inferences could and can be drawn from such structured studies (De Winter 2014). As the science of CSE has evolved, now over a period of several decades, the level of focus has been elevated from such micro-level concerns to a more macro-level perspective (e.g., Dul et al. 2012; Klein et al. 2006; Marras and Hancock 2013).

Notwithstanding the above observations, one central goal of CSE remains the specification of causal relationships even though researchers in CSE have become very aware of, and sensitive to, the fact that the circumstances they are studying have proved to be evidently more complex and resistant to such "linear" understanding than might be revealed by the simple addition of one or two further incremental variables to any prior study (Miller 1986). Indeed, in its very essence, complexity is what the word *systemic* connotes (and see Hoffman et al. 2009; Kauffman 1993; Paries 2006).

In contrast to CSE, morality apparently belongs to the realm of ultimate purpose, or Aristotle's "final" cause. This is not to say that CSE and its study of systems does not have a purpose—indeed it does, at least an *a priori* declared purpose. But rather it is the *primary* focus of morality that is concerned with "why" (Aristotle's final cause) as opposed to "how" (Aristotle's efficient cause) questions. An example of the persistent dissociation between domains that feature "how" and those that emphasize "why" can be found in numerous influential and popular scientific communications. For example, founded in his own studies of biological systems, the naturalist Stephen Jay Gould (1997) was sufficiently persuaded of the viability of this division that he proposed that these types of endeavor occupy different parts of understanding, which he called "*non-overlapping magisteria*." I reject this division absolutely. In previous work (Hancock 2009), and to a degree in opposing Moray (1993), I have asserted that "*purpose predicates process*." That is, the way in which any goal, intention, or purpose is to be enacted in the first place (i.e., the enabling process) is critically dependent on the nature of that goal (i.e., its fundamental purpose).

The above assertion can appear to be a polemical one since the necessary relationship between purpose and process can well appear to be remote in time and space. Our modern interlocking society creates the appearance of a causal fabric in which such linkages between nominal cause and putative effect must be traced across the contingencies of historical development (Moray and Hancock 2009). In the same way that "*purpose predicates process*," so, reciprocally, "*process potentiates purpose*." That is, the goals that we can envisage for our future are in actuality predicated upon the techniques, technology, and teleology of the present time. This is perhaps easier to comprehend if we understand that capacities such as language, logic, mathematics, and time are each themselves only cognitive tools that serve human goals; whether such goals prove to be adaptive or maladaptive in respect of the whole species or not. So "why" and "how" are locked together in the same way that perception and action are symbiotic and bonded both theoretically and practically in an inextricable manner (Gibson 1979; Mace 1977). Axiomatically, purpose and process stand in this same mutual symbiotic relationship. We can highlight this mutuality on a conceptual level by examining one formal antithesis here. This antithesis would contest that purpose without process is by fiat ineffectual, while process without purpose is by definition aimless.

In the traditional view, Aristotle's four causes are seen as individual and largely clastic forms of understanding. That is, one can discuss material cause independent of final cause, or efficient cause separately from formal cause, and so on, for each pairwise combination. However, here I claim that these types of causes are necessarily linked together and further that this necessary relationship is a nested one as I have illustrated in Figure 5.3. Thus, material cause—what things are made of—necessarily provides a required basis for formal cause—how things are shaped. So,

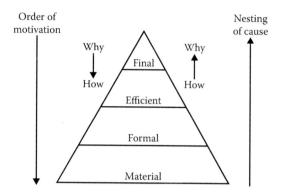

FIGURE 5.3 Nesting of the four forms of Aristotelian causes and the directionality of the motivating force in human activities.

the form that any item, object, entity, or organism can assume is always contingent upon its material being—*ab initio*. (It is, of course, possible for humans to conceive of multiple forms that never have had any actual material basis. However, even these very notions of an extracted concept are contingent on an *a priori* material basis.) Thus, I see formal cause as an embedded subset of material cause that logically precedes it; there being no form before materiality (contra Plato 360 BCE, or Berkeley 1710). Pursuing this nesting principle further, it is then logical that efficient cause—the "how" of existence—is itself nested within formal cause. That is, how things happen are necessarily contingent upon their form. By extension, it is then evident that final cause—the "why" of life in which purpose is expressed as human goals and aspirations—must be embedded in efficient cause. More formally, purposes being expressed in terms of what goal is aspired to (eg., its moral and ethical foundations) are necessarily nested in the how (e.g., CSE systems) such goals are to be achieved.

The emergent and interesting paradox here is that while the sequential nesting is founded first on material cause, that is, a bottom-up relationship, from material to formal to efficient and to final cause, human discourse, in contrast, derives initially from final cause and flows in a top-down manner. I have shown this bidirectionality explicitly in Figure 5.2. I make no comment here about the processes of neonate cognitive development of these concepts or indeed about their emergence across all of human evolution (but in respect of this, see Moray 2014). This bidirectionality of cause nested on a material foundation, as opposed to human motivations being found first in final cause, is the essential and fundamental foundation for my assertion that morality is embedded in the way in which CSE or technology is generally expressed in the world (Rasmussen 1983).

The above postulate can certainly take on the appearance of a highly disputable one since phenomenally we "know' that human beings can represent goals and purposes as cognitively abstract constructs. These constructs might then readily appear as being independent of the way that we conceive of technology or indeed any sorts of physical infrastructure or material form that might support such goals. However, this only *appears* to be the case as a result of the aforementioned dissociation in space and time between diverse purposes, the way they have been achieved in the past as well as the remnants that such prior actions have left upon the human landscape. Indeed, this represents one especially interesting and involving example of the apparent notion that formal cause can precede material cause, although as I have indicated, this assertion is a fallacious one. The notion that the abstract can precede the concrete also accrues from the breadth of what actually connotes human tools, that is, both its material (e.g., a computer) and nonmaterial (e.g., language) forms. Hence, limited individual human apperception then may well introduce the illusion of a separation between purpose and process but, as with free will, illusion is all that it remains.

Furthermore, it is often highly convenient and even lucrative for certain segments of our modern society (e.g., the legal community) to persist in the pretense of such a viable dissociation (i.e., the status quo). This form of traditional narrative is one that humans have persuaded themselves of, now for a number of millennia (Dekker 2007b). It is from this rational impasse that the unhelpful, but persistent culture of "blame" and "compensation" emerges and is sustained (Evans 2014). Unfortunately, such division leads to compartmentalized and ultimately impoverished thinking, which the CSE community has laudably questioned (i.e., the partial rejection of the status quo; and see Klein et al. 2014; Reason 2013). So much then for the formal foundation and premise for my present argument. My task now is to clothe such assertions with the previously identified concrete exemplars, intimately familiar to those in CSE.

5.5   REDUCTION AND ERROR

To explore the specific links between morality and CSE cited above, I look further at the two noted areas that have formed some of the central concerns in the historical development of CSE. These respectively are the issues of human error (e.g., Reason 1990) and FA (e.g., Dekker and Woods 2002; Hancock and Scallen 1998). Further consideration of these exemplar concepts requires us to examine and explain the role of, the value of, as well as the potential flaws in reductionistic approaches and strategies that characterized the respective early explorations of both error and FA (e.g., Chapanis 1965; Heinrich 1931). Reductionism has been an important, indeed essential, weapon in science's arsenal. As an approach to understanding, it has been tremendously successful over a period of decades and even centuries (De Winter 2013). However, as I like to note, reductionism is a good servant but a poor master. Disassembling things into their elemental parts, looking to understand those parts, and then reassembling them to seek greater comprehension of the revitalized and revivified whole has provided humankind with some of its most fruitful and fundamental insights. However, as I have asserted and now reiterate, reductionism is a useful servant but a very poor master. Laudable though the accrued insights have been, reductionism is not without its drawbacks and intrinsic limitations. These are expressed primarily when, as an investigative strategy, it overdominates any investigator's mode of thinking (Miller 1986). Such concerns have been aired and articulated extensively by those at the founding base of CSE and have even been extensively represented in Concept-Map form (Hoffman and Militello 2008). Some of these critics have made valid and trenchant points in their respective observations, while other criticisms stand in great danger of devolving into *ad hominem* attacks, as each of the respective constituencies willfully or unintentionally ignores the motivating philosophical bases of their respective interlocutors.

One evident problem of reductionism derives from the very first act of "parsing" of any issue, problem, entity, and/or conception of concern into its component parts. This "parsing" is a direct result of personal deliberation and is a process that can never be without opinion-based and conceptual conflict (Hancock 2012a). What is to one individual a "clear," "natural," and "obvious" line of division (predicated perhaps upon dimensions such as the timescale of operations or internal and/or external physical divisions of the formal structure of any system at hand) can, to another researcher, be neither natural nor obvious at all. Rather, someone else's divisions can and often do appear to be arbitrary, artificial, and obscure (Hoffman and Hancock 2014). Thus, the nominal parts (objects of the study) that accrue from this first act of division are intrinsically contentious in their very genesis.

In the physical sciences, dissent as to the initial act of division has traditionally been somewhat more muted than in those realms involving more complex social entities. For example, splitting the atom was a long-cherished but highly debated goal in physics, but few at that time disputed that atoms and their (assumed) constituent parts were the appropriate entity(ies) of study. Of course, this is not to say that the parsing in the physical sciences has not been without contention or indeed that such contentions have had important value when certain basic assumptions have been questioned. However, the "fracture" lines of parsed physical systems, especially in early incarnations (e.g., matter), were more readily agreed (i.e., directly perceived?) by the community of involved scientists. For investigators of living systems, such division lines are arguably never so evident or so straightforward (Hancock and Hancock 2009; Kirlik 2012; Schrodinger 1944). In HF/E, for example, some researchers have advocated that the traditional focus on the human (part one) and the machine (part two) does not necessarily provide either a "natural" or indeed advantageous division (see, e.g., Flach and Dominguez 1995). The latter authors thus emphasized "use-centered design" over "user-centered design," a difference of only one letter but one of rather considerable theoretical import. Other scientists, such as David Woods himself, have similarly suggested different ways to frame and establish the respective "unit of analysis" (see Hoffman et al. 2002).

Parenthetically, the parsing (framing) problem is one that proves to be much more general than CSE's concerns and indeed precedes even the first formal expressions of CSE as an area of study. However, changing one's perspective as to what is the relevant "unit" of analysis requires the ability to suspend one's learned expectations as to what connotes the appropriate parsing of any particular problem. The fundamentally empathic step away from one's comfortable "perspective" is an act that can be difficult and threatening, especially for those embedded in certain accepted and traditional paradigms. Regardless of disparities among different scientific communities as to what connotes any appropriate parsing strategy, other limitations to reductionism persist.

The understanding that is derived from any parsing process is necessarily founded upon the primary act of separation. The next, and often unstated subtext, concerns the issue of interactions among these now divorced parts. In applied psychological research, as we have seen, these concerns are actually often framed as potential sources of "error variance." In engineering models of human performance, for example in the famous crossover model of tracking, this element was referred to as "remnant" (McRuer and Jex 1967; and see MacCallum 2003 for a general discussion of models;

and see also Jagacinski and Flach 2003). Regardless of the specific discipline-based label employed, the underlying issue is that parsed units, when subsequently "run" as models or reconnected modules, do not behave exactly like the integrated systems they purport to represent in the real world (Miller 1986). This division and isolation strategy remains a persistent and problematic issue in modeling and simulation, which is frequently offered as a form of pabulum for full-scale testing (and see also Flach 1998). Whatever the empirical insights that are derived from this scientific "divide and conquer" strategy, they are inevitably limited by this shortfall. In contrast, pure ratiocination is not contaminated in the same way but is itself limited by the imagination of the individual involved and the manner in which such visions can be instantiated (Kant 1787). Here, CSE might be conceived as an intermediary "station of the cross" between these different approaches to understanding.

This parsing problem raises its head even more evidently than when the nominal process of reintegration occurs. I say nominal since, in much of the literature in a wide variety of human sciences, such reintegration is never actually attempted. Thus, memory specialists, attention researchers, those in artificial intelligence, and motor skills scientists each respectively and independently develop their own understanding of issues (such as error). But often this happens in isolation, one area from another. These discipline-based divisions tend to create situations in which the componential elements of the complex system to hand (such as the human operator) are never actually *put back together again* to see whether microscopic predictions work together at more macroscopic levels (and see Johnson 2001; Miller 1986). Here, error provides a most relevant case study since memory researchers report on their observed errors of memory (e.g., Schacter 2001) and, similarly, attention researchers report nominal failures of "situation awareness' (Endsley 1995), while motor skills researchers report on motor selection and execution errors (and see Newell 1991). Unfortunately, these discrete efforts occur in relative isolation from each other as well as from the large-scale practical concerns exhibited in CSE. Sadly, for important and overall encompassing issues such as error, other than rather puerile descriptive taxonomies, no comprehensive integrative understanding has really accrued from such reductionistic and discipline-driven efforts. The broad area of human error, like certain equivalent concerns that transcend the macro-level/micro-level comparison, remains a critical and pragmatic aspect of large-scale technical systems operation and, of course, their spectacular failure (Perrow 1984). However, it is arguable that discipline-based insights have still not fully penetrated the macro-level discussion. In this way, reductionism can act to dissociate scientific insight with its appropriate application in critical real-world circumstances, postponed by the reductionists' insistence that genuine science involves discrete studies. The prospect of such a never-ending enterprise generates significant frustration, which, in turn, initiates calls for paradigmatic evolution (e.g., Hollnagel 2014; Hollnagel and Woods 1983; Mitropoulos et al. 2005) and it is to this foundation in frustration that I now turn.

5.6 CSE'S STIMULUS IN FRUSTRATION

As I have noted, and perhaps need to reinforce here, reductionism has contributed tremendously to our overall understanding of the world. However, it is the case that reductionism's progressively more evident shortfalls, founded in some of the basic constraints that I have discussed, eventually generates what might be considered an inevitably growing level of frustration. This seems especially true with respect to the prediction of future states in actual complex systems. In respect of this general frustration, CSE itself derived initially from concerns for predictions of safety and operational efficiency in such systems (see, e.g., Rasmussen et al. 1994). Pragmatically then, CSE originated with frustrations, including frustrations over prediction, frustrations centered on poor technology designs, and even a more general frustration with controlled factorial experimentation divorced from notions of ecological relevance, validity, and utility. Over time, those frustrations became evident in the fact that while those in traditional HF/E were assumedly making some degree of iterative progress by examining the way in which the individual operators were affected by, and could affect, a restricted segment of a particular one human–one machine work system, predictions as to what would actually happen in large-scale work systems in the "real world" were falling woefully short of the sorts of levels of explanatory utility required for actual implementation (Klein 1998). This particular shortfall has raised its head in the HF/E community most especially around issues such as FA.

As I noted earlier, there are fewer more important places in which morality, ethical judgment, and CSE come together more closely, and thus more publicly, than in the decision to say *who* or *what*, does *what*, *when*, and *how*. Disputes surrounding this decision have historically been founded ostensibly upon what the machine could perform that the human could not perform (e.g., see De Winter and Hancock 2015; Fitts et al. 1951). Here, the argument was couched in negative terms, as exemplified by the MABA–MABA structure; that is, what machines could still *not* do. The "Fitts' List" framework was one in which the important features of machines (e.g., high memory capacity, logic) exactly compensated for the human's limitations (e.g., poor memory, emotional bias). As computational capacities have increased, the residual functions remaining to the human operator have been largely a litany of those tasks that computers had yet to be engineered so as to be able to accomplish. This left human beings, in Kantowitz's most evocative phrase, as the *sub-systems of last resort*. The moral concern with respect to a human operator's choice in terms of their own personal expectations, aspirations, dignity, and enjoyment of work was largely neglected, ignored, or deferred to the technical discussion of evolving automation capacities (Hancock et al. 2005; Hoffman et al. 2008). Human hopes, desires, or more generally the hedonomic dimensions of work have very rarely been emphasized as imperatives, where the evolving technical capacities of the computer system have almost without exception served to drive the debate (Feigh and Prichett 2014; Hancock 2014; Pritchett et al. 2014; Sheridan et al. 1998).

Barriers to practical implementation of theoretical FA policies have included major issues such as acontextuality (trying to develop viable generic FA policies) and underspecificity (not knowing enough about the whole work environment). Interactive effects that had been intentionally excluded from the reductionist-oriented laboratory (i.e., relegated to the context) then become the very pith and substance of activity in actual implementations. Questions of operator long-term motivation, job security, expectancy, de-skilling, boredom, fatigue, expertise, anxiety, stress, and illness—all of which are critical in the world outside the laboratory—tend to be absent or at best de-emphasized in the short-term, task-naïve niche of the sterile laboratory environment (De Winter and Dodou 2014). Even the most sophisticated of empiricists, however, cannot envisage all possible operational circumstances, and so the laboratory exercise, however sophisticated the model, the simulation, or surrogate "world" employed, is necessarily underspecified (i.e., inevitably composed of regions of obligatory ignorance). These practical concerns (frustrations) were voiced rather cogently by Fuld (1993) who decried the failure of descriptive models such as MABA–MABA as design heuristics for actual FA policies (cf., Dekker and Woods 2002; Hancock and Scallen 1996; Parasuraman et al. 2008). Fuld (1993) was rather adamant that such allocation policies, especially the largely static allocation procedures he had tried to apply, were simply inadequate to deal with actual dynamic operations. Such shortfalls have been experienced and documented in many dimensions of HF/E applications and not just FA or error alone. This is where David Woods has not only stood out as a scientist but has acted to lead the CSE community toward a much deeper and more profound understanding of the cognitive dimensions of all forms of work (see, e.g., Hollnagel et al. 2007; Woods et al. 2010; and indeed the entire Woods oeuvre).

Thus, even as CSE grew historically in trail of more and more spectacular technical disasters that compose some of our science's doxology, the stimulus was not just public outcry alone. The juxtaposition of scientific innovation linked to large-scale public failures was at once, and paradoxically, both enervating and energizing. Variously derived predictive thrusts founded upon reductionistic insights—be they pragmatic, synthetic, or theoretical—that should have helped to account for how these events were happening were at best limited and at worst vacuous. Frustration in science, like all other human pursuits, leads to innovation. In response, CSE leaders expressed their unhappiness and rightly broached a more complex and embracing form of science (e.g., Hollnagel 2012; Rasmussen 1983; Reason 1990; Woods et al. 2010).

As these respected leaders are fond of noting, an apparently simple iterative change in technology is not just a matter of componential replacement but rather such changes act to alter the very nature of the work to be accomplished. However, and by extension, CSE as an area is itself an agent of change that alters perspectives and also subsequently changes the very way we approach the moral foundation of work systems operation. The fact that we still cannot adequately predict or anticipate the spatiotemporal occurrence or frequency of failure in large-scale systems, and thus cannot specify when, where, how, and why they will occur, is a representation of the shortfall of the traditional approach. Of course, it is also a shortfall of nascent CSE, although the inability to predict might be much more profound, being a necessary but inescapable part of the human condition (see Hancock 2013). Both traditional HF/E and CSE have each understandably and frequently hoped and advertised far more than they were able to actually deliver at that moment in time. The effectiveness of such advertising has, in part, been contingent upon the persuasiveness of individual interlocutors, of which CSE has several outstanding orators, and perhaps none more so than David Woods. But why, in the end, should we be concerned about morality at all when it appears that we have not yet "solved" the efficient causal aspects of work systems that are surely the proximal problems at hand? I submit that it is in morality, and its basis in natural philosophy, which is where we need to search for our inherent source of predictive failure. It is to this endeavor that I now propose to turn.

5.7   THE CSE–MORALITY AXIS

If my premises are viable concerning the respective embedding of Aristotelian causes, it would appear, from a bottom-up perspective, that we must first understand efficient cause (the source of change) in order then to comprehend final cause (the goal of change). However, I believe that human intention, as well as its embedded morality, flows downward (see Figure 5.3). That is, any causal narrative requires that we first state or imply our human purposes from which process flows, rather than vice versa. In consequence, I would argue that the shortfall in CSE's capacity to predict emanates from its insufficient philosophical comprehension rather than a lacuna of systematic and even expansive empirical studies.

This shortfall is now coming to the fore, where the pursuit of daunting and apparently intractable problems leads us to begin to question our basic underlying philosophical assumptions. It is only by this arduous enquiry, as compared, say, to a potentially infinite series of ethnographic case studies, that our next substantive step toward progress can be achieved. Recent views on human error (e.g., Dekker 2001; Woods et al. 2010), for example, have indicated insightfully that error's expression, error's conceptual genesis, and indeed error's philosophical illusion are necessarily contingent upon the stated purpose of the work system at hand. However, to make even more concrete the linkage between morality and technology, I want to present two extreme but illustrative cases of technicians who initially focused on efficient cause only to have final cause subsequently swamp and overwhelm them as both scientist/engineer and person.

5.7.1   CASE STUDY #1 IN MORALITY AND TECHNOLOGY: THE KALASHNIKOV RIFLE

The first is a recently "resolved" case and concerns the soldier and inventor Mikhail Kalashnikov. Credited with the creation of the AK-47 rifle during the Second World War, it has been estimated that there are approximately 100 million Kalashnikov rifles in the world today, of which interestingly some 90% have been produced illegally. Close to his own death, Kalashnikov was wracked with both doubt and guilt enquiring whether "*If my rifle took lives, does it mean that I, Mikhail Kalashnikov, aged 93, a peasant woman's son … is guilty of those people's deaths, even if they were enemies.*" For those who can divorce purpose and process, the answer here is the traditional, simple, and antiseptic—no—and the comforting denial of responsibility. This stance protests that the fabricator of a technology is not responsible for the way in which his or her creation is subsequently employed when let loose in the world. Sadly, after a lifetime of contemplation, Kalashnikov was unable to sustain such a convenient and comforting dissociation. And it is further interesting to note that any such separationist principle is in fact in direct contradiction to the original code of Hammurabi, which is perhaps the first such social contract that we possess between artisan and consumer.

5.7.2 CASE STUDY #2 IN MORALITY AND TECHNOLOGY: THE ATOMIC BOMB

A similar, and perhaps even more tragic case, is that of J. Robert Oppenheimer. Driven under the similar impetus of war and nationalistic fervor, Oppenheimer was the pivotal figure in the development of the atomic bomb. Oppenheimer might have been able to salve his conscience with respect to the pure numbers of dead he had helped generate (as compared, e.g., to Kalashnikov), and also by the notion that he was engaged in a nominal "life or death" struggle between his own country and those of the Axis who had similar programs in development. However, he was tragically insightful enough to fully understand what he had actually created and what would be its destructive power both then and in the future. That knowledge helped to destroy him, almost as effectively as the victims of the actual detonations. It is, however, possible to argue that because these weapons can generate such a prohibitive vision of Armageddon, they might actually serve to inhibit wider-scale conflict. It is, however, a position for which I have little empathy and which has been rightly rejected (see Bronowski 1956, 1973).

These two cases, although perhaps seemingly extreme, illustrate that unless the individual is a sociopath or distinctly mentally disturbed in some fashion, standards of contemporary morality can exert a devastating psychological impact upon a technology's creator. This self-same principle is surely applicable to the modern and contemporary case of the designers, fabricators, and even operators of modern armed drone systems (and see Hoffman et al. 2014). Here, death is rained from the sky rather in the manner of the God of the Old Testament, but without the indemnification that presumably omniscience brings. Thus, any sustainable narrative embraced by those involved in the creation and operation of effective technical systems must present the required and ever publicly laudable set of aims and goals. Sans ethical principles, the process can well appear to be one solely of an arid efficiency that can, for example, promote the throughput of gas chambers, or the creation of autonomous killing weapons with an apparently dispassionate equanimity.

But now, we must take the ultimately difficult step of understanding that ethical principles and moral precepts are themselves neither fixed, immutable, nor "natural" constraints but rather are contextually and culturally driven "norms" that fluctuate rather wildly across time, being contingent upon history and location. Thus, all of the issues I have raised have no single and unequivocal resolution. However, like CSE itself, morality must co-evolve with the contexts that surround it.

5.8 THE RELATIVITY OF MORALITY

Up to this point, I have been talking about conditions that largely prevail at present. To extend beyond the present to a consideration of the future, we must delve more deeply into the relativity of morality. There is a current "rent" or "tear" in the philosophical fabric of what is often represented to be the bedrock of "traditional" Western thought. Through the power of technology, such "Western" thought has come to dominate the world stage (see, e.g., Friedman 2005; Fukuyama 1992). Historically, this schism between purpose and process can be traced somewhat facilely to the death of Francis Bacon and the noncompletion of his *Great Instauration*, his great renewal of thought. Science, as we know it, emerged from his completed componential work, the *Novum Organum* (and see also Descartes 1637) but unfortunately Bacon left incomplete the necessary bonding of his new science with morality. So, at the base of science itself in the very early 1600s, when the initial path of progress was set, there was an inadvertent schism, a fracture that persists to the present day. This developed in western philosophy into the division between natural philosophy and moral philosophy, which having come apart have never been united again. This is one way in which western philosophy stands in contrast with a number of eastern-emanating philosophies. Such a fissure has been reinforced and entrenched in the centuries since Bacon, and indeed the present chapter is one small appeal to understand how we might begin to anneal this division.

Up until the emergence of CSE, many "systems-based" errors were thought of as the proximal problem of the closest human operator who was in some fashion error-prone or in other ways limited, biased, or even worse somehow "deserving" of such a fate (Arbous and Kerrich 1951; Dekker 2007b; Flach and Hoffman 2003; Wilder 1927). These have been couched in psychological terms such as lacks of attention, failures of situation awareness, and so on. It might be argued that these are somewhat more benign labels than one tellingly derived from the English military for pilot error: that is, a *lack of moral fiber*. These ways of stating the problem directly entail a stance that it is necessarily the advancements in technology that will circumvent such failures (Hancock 2000). Just occasionally, there arise assertions that collective social governances also bear some responsibility for such mishaps.

Blame, the moral pronouncement and disapprobation of the powerful collective on the behavior of the poor or poor individual or indeed any other less powerful collective, remains an intrinsic result of this divide. Here, error (sin, and its correlate—evil) becomes an inherent property or characteristic of that identified individual or minority. The excision of that one deviant operator or small group of operators quickly follows and thus the standard narrative is sustained. This notion echoes the disease model of harm in getting rid of the defect or one individual source. One great contribution of CSE has always been to look to expose this myth and to seek an alternative perspective from which to frame our future progress away from this relative myopia. However, technical application and moral thinking have not really evolved commensurately (but see Scheutz and Malle 2014). Blame remains the status quo, even to a degree in CSE where "blame" raises another of its hydra-like heads in the identification that nominally illuminating methods such as task analyses are often insufficiently complete due to time and money constraints. Indeed, they often are insufficient, but cognitive task analysis is not the intrinsic path toward affirming the validity of the blame game.

Sadly, and even some would contend tragically, CSE has had, as yet, a disproportionately small impact on the way in which western society still deals with these issues of complex systems failure. Perhaps, as I have contended, this is a question of the need for a fundamental change in the foundational philosophical stance but perhaps it might be due more pragmatically to the fact that only relatively few scientists have yet embraced CSE. Morality is primarily codified in the idea of law, which persists in its entrenched and highly leveraged position, predicated upon the nominal power of precedent. The status quo here provides manifest benefits in retaining the concept of blame and the idea of error as individual human failure versus, for example, the larger systems-based perspective (Casey 1998). This intransigence is allied to, and perhaps underwritten by, the general public's perception of causation and in its associated morality of personal responsibility.

Our western morality here emanates from an essential schizophrenia that derived from the concept of a munificent but omnipotent deity. A degree of free will, the degree of which of course is never fully specified, is permitted. However, when that individual expresses that free will, the deity sustains no culpable responsibility for any unfortunate turn of events. In contrast, the deity garners great "praise" for any of the beneficial aspects of that event. More colloquially, heads God wins, tales humans lose. Of course, this notion of deistic causality has not remained constant, even across the relatively brief history of western thought (Calasso 1994). When one possesses a combination of both religion and the law (in essence, the Lord's spiritual and the Lord's temporal), which are in accord on this issue, one has a very powerful lobby indeed for inertia. Such a constituency has a vested interest in the status quo and so no great concern for changing it. Thus, CSE may be embraced by science and some limited segments of the business community but has yet to exert any proportionate social and moral impact because of the intransigence of the identified inertial institutions. Whether we can change our basic foundations of morality, in this case by use of efficient cause, such that morality can be made an affordance of the technical system to hand, remains an open empirical question.

5.9 CONCLUSIONS

As a result of the historical antecedents, which in the western world have divided process from purpose, our present circumstances generate complex work systems of great technological sophistication but erected upon ethical bases that are implicit, inconsistent, incomplete, and ultimately inoperable. Utilitarian, patchwork moral principles, derived from a dissolving base of religious dogma, are now morphing into ad hoc, secular legal arcana whose appeal to various poorly defined and developed human delusions, such as the notion of error, free will, and the omnivorous drive for greater efficiency, continues to subserve the dominant segment of profit-based capitalism (and polemically, see Dawkins 2009). However, a true search for enlightenment burns off such excesses, and CSE must, in aspiration at least, be intimately involved in such evanescence.

Moral relatively has to be recognized, yet contemporary pragmatism employs and emphasizes an "as if" policy such that the search for moral certitude continues. Like error itself, morality is a convenient illusion that helps sustain societies in their necessary collective delusion. Morality is an intimate part of the narrative that we have always told ourselves (Homer 735 BC). However, as Richard Feynman observed, *nature will not be fooled*. Morality expresses social approbation and disapprobation and CSE subserves these moral goals (as well as other forms of purpose). What is good or bad, correct or incorrect are collective interpretations (most often after the fact) that support the individual and collective narrative that we have chosen as our conduit to collective reality (and, of course, see Nietzsche 1886).

To survive as a species, we will need to disabuse ourselves of many of these illusions and CSE appears to be a hopeful avenue to begin to achieve such a necessary re-enlightenment.

ACKNOWLEDGMENTS

I am, as ever, grateful for the wide, penetrative, and enlightening insights and comments of Dr. Robert Hoffman on an earlier version of this chapter. I am further indebted to the most important observations of Dr. Joost de Winter that have helped to refine my own thinking on such matters. I am most especially grateful to Dr. Sidney Dekker, our manifest "thought leader" in this area, for his comments and directions on the present work. Remaining errors, philosophical, theoretical, or practical, are my own and should in no way be attributed to those who have been kind enough to help me.

REFERENCES

Arbous, A.G., and Kerrich, J.E. (1951). Accident statistics and the concept of accident-proneness. *Biometrics*, 7(4), 340–432.

Berkeley, G. (1710). *A Treatise Concerning the Principles of Human Knowledge*. Dublin: Rhames.

Bertalanffy, L. (1972). The history and status of general systems theory. *Academy of Management Journal*, 15(4), 407–426.

Bronowski, J. (1956). *Science and Human Values*. London: Faber & Faber.

Bronowski, J. (1973). *The Ascent of Man*. London: BBC Books.

Calasso, R. (1994). *The Marriage of Cadius and Harmany*. New York: Vintage.

Casey, S.M. (1998). *Set Phasers on Stun: and Other True Tales of Design, Technology, and Human Error*. Santa Barbara: Aegean Press.

Chapanis, A. (1965). On the allocation of functions between men and machines. *Journal of Occupational Psychology*, 39, 1–11.

Dawkins, R. (2009). *The God Delusion*. New York: Random House.

Dekker, S.W.A. (2001). The reinvention of human error. *Human Factors and Aerospace Safety*, 1, 247–266.

Dekker, S. (2007a). What is rational about killing a patient with an overdose? Enlightenment, continental philosophy and the role of the human subject in system failure. *Ergonomics*, 54(8), 679–683.

Dekker, S. (2007b). Eve and the serpent: A rational choice to err. *Journal of Religion and Health*, 46(4), 571–579.

Dekker, S.W.A., Hancock, P.A., and Wilkin, P. (2013a). Ergonomics and the humanities: Ethically engineering sustainable systems. *Ergonomics*, 56(3), 357–364.

Dekker, S.W.A., and Nyce, J.M. (2012). Cognitive engineering and the moral theology and witchcraft of cause. *Cognition Technology and Work*, 14, 207–212.

Dekker, S., Nyce, J., and Myers, D. (2013b). The little engine who could not: "Rehabilitating" the individual in safety research. *Cognition, Technology and Work*, 15(3), 277–282.

Dekker, S.W.A., and Woods, D.D. (2002). MABA-MABA or abracadabra: Progress on human-automation coordination. *Cognition, Technology and Work*, 4, 240–244.

Descartes, R. (1637). *Discourse on Method and Meditations*. Laurence J. Lafleur (trans.) (1960). New York: The Liberal Arts Press.

De Winter, J.C.F. (2014). Why person models are important for human factors science. *Theoretical Issues in Ergonomic Science*, 15(6), 595–614.

De Winter, J.C.F., and Dodou, D. (2014). Why the Fitts list has persisted throughout the history of function allocation. *Cognition, Technology and Work*, 16, 1–11.

De Winter, J.C.F., and Hancock, P.A. (2015). Reflections on the 1951 Fitts list: Do humans believe now that machines surpass them? *Procedia Manufacturing*, 3, 5334–5431.

Dul, J., Bruder, R., Buckle, P., Carayon, P., Falzon, P., Marras, W.S., Wilson, J.R., and van der Doelen, B. (2012). A strategy for human factors/ergonomics: Developing the discipline ad profession. *Ergonomics*, 55(4), 377–395.

Endsley, M.R. (1995). Toward a theory of situation awareness in dynamic systems. *Human Factors*, 37(1), 32–64.

Evans, L. (2014). Twenty thousand more Americans killed annually because U.S traffic safety policy rejects science. *American Journal of Public Health*, 104(8), 1349–1351.

Feigh, K.M., and Pritchett, A.R. (2014). Requirements for effective function allocation. A critical review. *Journal of Cognitive Engineering and Decision Making*, 8(1), 23–32.

Fitts, P.M. (Ed.). (1951). (A. Chapanis, F.C. Frick, W.R. Garner, J.W. Gebhard, W.F. Grether, R.H. Henneman, W.E. Kappauf, E.B. Newman, and A.C. Williams, Jr.), *Human Engineering for an Effective Air-Navigation and Traffic-Control System*. Washington, DC: National Research Council.

Flach, J.M. (1998). Cognitive systems engineering: Putting things in context. *Ergonomics*, 41(2), 163–167.

Flach, J.M. and Dominguez, C.O. (1995). Use-centered design: Integrating the user, instrument, and goal. *Ergonomics in Design*, 3(3), 19–24.

Flach, J.M., and Hoffman, R.R. (January–February 2003). The limitations of limitations. *IEEE Intelligent Systems*, 94–97.

Friedman, T.L. (2005). *The World Is Flat*. New York: Farrar, Straus and Giroux.

Fukuyama, F. (1992). *The End of History and the Last Man*. New York: Free Press.

Fuld, R.B. (1993). The fiction of function allocation. *Ergonomics in Design*, 1(1), 20–24.

Gibson, J.J. (1966). *The Senses Considered as Perceptual Systems*. Boston: Houghton-Mifflin.

Gibson, J.J. (1979). *The Ecological Approach to Visual Perception*. Boston: Houghton Mifflin.

Gould, S.J. (1997). Nonoverlapping magisteria. *Natural History*, 106 (March), 16–22.

Hancock, P.A. (2000). Can technology cure stupidity? *Human Factors and Ergonomics Society Bulletin*, 43(1), 1–4.

Hancock, P.A. (2009). *Mind, Machine, and Morality*. Aldershot, UK: Ashgate.

Hancock, P.A. (2012a). Notre trahison des clercs: Implicit aspiration, explicit exploitation. In: R.W. Proctor, and E.J. Capaldi (Eds.), *Psychology of Science: Implicit and Explicit Reasoning* (pp. 479–495), New York: Oxford University Press.

Hancock, P.A. (2012b). Ergaianomics: The moral obligation and global application of our science. *The Ergonomist*, 503, 12–14.

Hancock, P.A. (2013). *On the symmetricality of formal knowability in space–time*. Paper presented at the 15th Triennial Conference of the International Society for the Study of Time, Orthodox Academy of Crete, Kolymbari, Crete, July.

Hancock, P.A. (2014). Automation: How much is too much? *Ergonomics*, 57(3), 449–454.

Hancock, P.A., and Diaz, D. (2001). Ergonomics as a foundation for a science of purpose. *Theoretical Issues in Ergonomic Science*, 3(2), 115–123.

Hancock P.A., and Drury, C.G. (2011). Does Human Factors/Ergonomics contribute to the quality of life? *Theoretical Issues in Ergonomic Science*, 12, 1–11.

Hancock, P.A., and Hancock, G.M. (2009). The moulding and melding of mind and machine. *The Ergonomist*, 464, 12–13.

Hancock, P.A., Jagacinski, R., Parasuraman, R., Wickens, C.D., Wilson, G., and Kaber, D. (2013). Human–automation interaction research: Past, present and future. *Ergonomics in Design,* 21(3), 9–14.

Hancock, P.A., Pepe, A., and Murphy, L.L. (2005). Hedonomics: The power of positive and pleasurable ergonomics. *Ergonomics in Design*, 13(1), 8–14.

Hancock, P.A., and Scallen, S.F. (1996). The future of function allocation. *Ergonomics in Design*, 4(4), 24–29.

Hancock, P.A., and Scallen, S.F. (1998). Allocating functions in human–machine systems. In: R.R. Hoffman, M.F. Sherrick, and J.S. Warm (Eds.), *Viewing Psychology as a Whole: The Integrative Science of William N. Dember* (pp. 509–539), Washington, DC: American Psychological Association.

Heinrich, H.W. (1931). *Industrial Accident Prevention: A Scientific Approach*. New York: McGraw-Hill.

Hill, A.B. (1965). The environment and disease: Association or causation? *Proceedings of the Royal Society of Medicine*, 58(5), 295–300.

Hoffman, R.R., Feltovich, P.J., Ford, K.M., Woods, D.D., Klein, G., and Feltovich, A. (2002). A rose by any other name … would probably be given an acronym. *IEEE Intelligent Systems*, July/August, 72–80.

Hoffman, R.R., and Hancock, P.A. (2014). Words matter. *Human Factors and Ergonomics Society Bulletin*, 57(8), 3–7

Hoffman, R.R., Hawley, J.K., and Bradshaw, J.M. (2014). Myths of automation, Part 2: Some very human consequences. *IEEE Intelligent Systems*, March/April, 82–85.

Hoffman, R.R., Marx, M., and Hancock, P.A. (2008). Metrics, metrics, metrics: Negative hedonicity. *IEEE Intelligent Systems*, 23(2), 69–73.

Hoffman, R.R., and Militello, L.G. (2008). *Perspectives on Cognitive Task Analysis: Historical Origins and Modern Communities of Practice*. Boca Raton, FL: CRC Press/Taylor & Francis.

Hoffman, R.R., Norman, D.O., and Vagners, J. (2009). Complex sociotechnical joint cognitive work systems? *IEEE Intelligent Systems*, May/June, 82–89.

Hollnagel, E. (2012). *The ETTO Principle: Efficiency–Thoroughness Trade-off: Why Things That Go Right Sometimes Go Wrong*. Chichester: Ashgate.

Hollnagel, E. (2014). *Safety-I and Safety-II*. Aldershot, UK: Ashgate.

Hollnagel, E., and Woods, D.D. (1983). Cognitive systems engineering: New wine in new bottles. *International Journal of Man–Machine Studies*, 18(6), 583–600.

Hollnagel, E., Woods, D.D., and Leveson, N. (2007). *Resilience Engineering: Concepts and Precepts*. Aldershot, UK: Ashgate.

Homer (735 BC/1951). *The Iliad* (Richmond Lattimore, Translator). Chicago: University of Chicago Press.

Jagacinski, R., and Flach, J. (2003). *Control Theory for Humans: Quantitative Approaches to Performance Modelling*. Mahwah, NJ: Erlbaum.

Johnson, S. (2001). *Emergence*. New York: Scribner.

Kant, I. (1787). *Critique of Pure Reason*. London: Macmillan (1929).

Kauffman, S.A. (1993). *The Origins of Order: Self-Organization and Selection in Evolution*. New York: Oxford University Press.

Kirlik, A. (2012). Relevance versus generalization in cognitive engineering. *Cognition, Technology and Work*, 14, 213–220.

Klein, G. (1998). *Sources of Power: How People Make Decisions*. Cambridge: MIT Press.

Klein, G., Moon, B., and Hoffman, R.R. (2006). Making sense of sensemaking 2: A macrocognitive model. *IEEE Intelligent Systems*, 21(5), 88–92.

Klein, G., Rasmusen, L., Lin, M.-H., Hoffman, R.R., and Case, J. (2014). Influencing preference for different types of causal explanation of complex systems. *Human Factors*, 56(8), 1380–1400.

MacCallum, R.C. (2003). Working with imperfect models. *Multivariate Behavioral Research*, 38(1), 113–139.

Mace, W.M. (1977). James J. Gibson's strategy for perceiving: Ask not what's inside your head, but what your head's inside of. In: R. Shaw and J. Bransford (Eds.), *Perceiving, Acting and Knowing: Toward an Ecological Psychology*. New York: Lawrence Erlbaum.

Marras, W., and Hancock, P.A. (2013). Putting mind and body back together: A human-systems approach to the integration of the physical and cognitive dimensions of task design and operations. *Applied Ergonomics*, 45(1), 55–60.

McRuer, D.T., and Jex, H.R. (1967). A review of quasi-linear pilot models. *IEEE Transactions on Human Factors in Electronics*, (3), 231–249.

Mill, J.S. (1863/1998). *Utilitarianism*. (R. Crisp, Ed.), Oxford: Oxford University Press.

Miller, G.A. (1986). Dismembering cognition. In: S.H. Hulse and B.F. Green, Jr. (Eds.), *One Hundred Years of Psychological Research in America* (pp. 277–298). Baltimore, MD: Johns Hopkins University Press.

Mitropoulos, P., Abdelhamid, T.S., and Howell, G. (2005). Systems model of construction accident causation. *Journal of Construction Engineering and Management*, 131(7), 816–825.

Moray, N.P. (1993). Technosophy and humane factors. *Ergonomics in Design*, 1, 33–37, 39.

Moray, N.P. (2014). *Science, Cells and Soul*. Bloomington, IN: Authorhouse.

Moray, N.P., and Hancock, P.A. (2009). Minkowski spaces as models of human–machine communication. *Theoretical Issues in Ergonomic Science*, 10(4), 315–334.

Nietzsche, F. (1886). *Beyond Good and Evil*. Leipzig: Verlag.

Newell, K.M. (1991). Motor skill acquisition. *Annual Review of Psychology*, 42, 213–237.

Parasuraman, R., Sheridan, T.B., and Wickens, C.D. (2008). Situation awareness, mental workload and trust in automation: Viable, empirically supported cognitive engineering constructs. *Journal of Cognitive Engineering and Decision Making*, 2(2), 140–160.

Paries, J. (2006). Complexity, emergence, resilience. In: E. Hollnagell, D.D. Woods, and N. Leveson (Eds.), *Resilience Engineering: Concepts and Precepts* (pp. 43–53). Burlington, VT: Ashgate.

Perrow, C.A. (1984). *Normal Accidents: Living with High-Risk Technologies*. New York: Basic Books.

Pritchett, A.R., Kim, S.Y., and Feigh, K.M. (2014). Modeling human–automation function allocation. *Journal of Cognitive Engineering and Decision Making*, 8(1), 33–51.

Rand, A. (1943). *The Fountainhead*. New York: Bobbs Merrill.

Rasmussen, J. (1983). Skills, rules, and knowledge: Signals, signs, and symbols, and other distinctions in human performance models. *IEEE Transactions on Systems, Man, and Cybernetics*, 13(3), 257–266.

Rasmussen, J., Pejtersen, A.M., and Goodstein, L.P. (1994). *Cognitive Systems Engineering*. New York: Wiley.

Reed, E.S. (1996). *The Necessity of Experience*. Yale: Yale University Press.

Reason, J. (1990). *Human Error*. Cambridge: Cambridge University Press.

Reason, J. (2013). *A Life in Error: From Little Slips to Big Disasters*. Chichester: Ashgate.

Reason, J., Manstead, A., Stradling, S., Baxter, J., and Campbell, K. (1990). Errors and violations on the roads: A real distinction? *Ergonomics*, 33(10–11), 1315–1332.

Schacter, D.L. (2001). *The Seven Sins of Memory: How the Mind Forgets and Remembers*. Boston: Houghton Mifflin.

Scheutz, M. (2014). The need for moral competency in autonomous agent architectures. In: V. Muelles (Ed.), *Fundamental Issues in Artificial Intelligence*. Berlin: Springer.

Scheutz, M., and Malle, B.F. (2014). "Think and do the right thing"—A plea for morally competent autonomous robots. *Proceedings of the IEEE International Symposium on Ethics in Engineering, Science, and Technology*. IEEE.

Schrodinger, E. (1944). *What Is Life? The Physical Aspects of the Living Cell*. Cambridge: Cambridge University Press.

Sheridan, T., Hancock, P.A., Pew, R., Van Cott, H., and Woods, D. (1998). Can the allocation of function between humans and machines ever be done on a rational basis? *Ergonomics in Design*, 6(3), 20–25.

Wilder, T. (1927). *The Bridge of San Luis Rey*. New York: Albert and Charles Boni.

Woods, D.D., Dekker, S., Cook, R., Johannesen, L., and Sarter, N. (2010). *Behind Human Error*. Farnham, United Kingdom.

6 Understanding Cognitive Work

Penelope Sanderson

CONTENTS

6.1 INTRODUCTION

The purpose of this chapter is to provide some insight into how cognitive work is conceptualized and investigated in the tradition of cognitive systems engineering advocated by David Woods and his colleagues, including in particular Erik Hollnagel and Emilie Roth. First, I survey recent treatments of cognitive work analysis (CWA) and cognitive task analysis (CTA). Then, I introduce the idea of joint cognitive systems and the cognitive systems triad—these are concepts that have been fundamental in Woods' work for decades. After that, I describe a model for human performance analysis that for some time has guided investigations into how people cope with complexity and has lain at the heart of the way Woods and his colleagues approach analysis and design of cognitive work. I then describe an investigative context dubbed "staged worlds" that Woods and colleagues use to preserve authenticity while enhancing the efficiency of investigations and I illustrate the use of staged worlds with a recently published emergency medical response example (Smith et al. 2013). Finally, the model of human performance analysis and the above methods lead to "laws" that describe joint cognitive systems at work.

6.2 ANALYZING COGNITIVE WORK THROUGH COGNITIVE WORK ANALYSIS AND COGNITIVE TASK ANALYSIS

Many approaches have emerged in CSE and cognate fields of inquiry for investigating cognitive work. Two widely used approaches are cognitive task analysis (CTA) and cognitive work analysis (CWA). Given the many excellent recent reviews of CTA and CWA, there is little need for a further detailed review of their principles and methods. However, it is worth pointing to the origins of the approaches and to the recent treatments of them.

The principles that underlie both approaches to understanding cognitive work reach back over 30 years, to the genesis of CSE (Hollnagel and Woods 1983) and to European studies of human–machine systems conducted in the decade before the Three Mile Island accident in 1979 (see treatments in Rasmussen and Rouse 1981; Sheridan and Johannsen 1976 as well as overviews in Flach 2016; Le Coze 2015, 2016, and others). Since the early 1980s, Woods has been a leader and key contributor in the development and expression of those principles, and in the development and expression of methods for understanding cognitive work.

Within CSE, CTA and CWA are often compared and contrasted, given that they are core methods for analyzing cognitive work. CSE has been defined as the analysis, modeling, design, and evaluation of complex sociotechnical systems so that workers can do their work and carry out tasks more safely, and with greater efficiency. In the context of CSE, the term *cognitive work* is usually used to represent the individual and collective sense-making activities of workers and other agents in complex sociotechnical systems. The phrase "analysis of cognitive work" usually covers activities that CSE researchers carry out when performing CWA or CTA. The main focus of this chapter is Woods' contributions not only to the *analysis* of cognitive work but also to the *design* of cognitive work. As will be seen, Woods and colleagues refer to their analytic activities as CTA, and their design activities are guided by *laws that govern cognitive work*, including laws that govern joint cognitive systems at work.

The term *CWA* is by convention reserved for the systematic approach to analyzing the constraints operating on cognitive work that emerged from the work of Rasmussen and his colleagues at Riso National Laboratories in Denmark (Le Coze 2015, 2016; Rasmussen 1986; Rasmussen et al. 1994; Vicente 1999). CWA focuses on analyzing the constraints that shape cognitive work, using a family of analytic templates that guide the identification of those constraints and their interactions. The term *CTA* is best reserved for approaches to analyzing cognitive work other than CWA, but those approaches may nonetheless share some of the theoretical commitments of CWA because they emerge from the same history. A similar distinction between CWA and CTA is respected in Lee and Kirlik's (2013) handbook of cognitive engineering, with its separate chapters on CWA (Roth and Bisantz 2013) and CTA (Crandall and Hoffman 2013). Again, however, the fact that a distinction can be made should not obscure similarities between the two approaches to the analysis of cognitive work and the intertwined history of their development.

As noted, CTA and CWA have been the subject of many recent authoritative reviews, making a repetition of their fundamentals unnecessary. CWA has received several thorough treatments over the 30 years or so of its existence. The foundational work of Rasmussen is available through the Rasmussen (1986) and Rasmussen et al. (1994) monographs. The work received a subsequent pedagogical interpretation in Vicente (1999), and many aspects of Rasmussen's way of analyzing cognitive work were presented in a recent special issue of *Applied Ergonomics* in 2017. Further monographs and edited books on CWA include Bisantz and Burns (2008), Jenkins et al. (2009), Lintern (2013), and Naikar (2013). Review chapters focusing on CWA include Sanderson (2003) and Roth and Bisantz (2013). Many recent treatments of CWA offer novel templates intended to help analysts apply the principles of CWA more effectively at each phase, or to link analyses more effectively with analyses at other phases (Ashoori and Burns 2010; Cornelissen et al. 2013; Hassall and Sanderson 2014; Naikar 2013).

Since 2000, CTA has also been thoroughly reviewed in monographs or edited books such as those by Schraagen et al. (2000) and Crandall et al. (2006), and in Hoffman and Militello's excellent monograph on CTA methods (2009). Recent reviews of CTA also include Crandall and Hoffman (2013). Treatments of CTA often describe different methods of eliciting information about cognitive work. The most helpful treatments also describe the process of understanding the phenomenology of work in a lawful way. For example, as we will see, Woods has proposed laws that express important generalities about how people and technology interact (Woods and Hollnagel 2006). Those laws have been inferred from investigating and analyzing the successes and failures of human–system integration in many domains—in other words, from different CTAs conducted in widely differing industries.

Reviews that cover both CTA and CWA include Bisantz and Roth (2008), Roth (2008), and Hoffman and Militello (2009). CTA and CWA were also covered in a major review of methodological challenges for a science of sociotechnical systems and safety (Waterson et al. 2015).

6.3 JOINT COGNITIVE SYSTEMS AND THE COGNITIVE SYSTEMS TRIAD

From the CSE perspective, the entity performing cognitive work—and therefore the entity to be investigated—is not just the individual human actor. Since their formulation of CSE, Woods and colleagues have brought a "joint cognitive systems" perspective to the development of useful theoretical frameworks for understanding cognitive work (Hollnagel and Woods 1983; Woods 1985). Hollnagel and Woods (2005) define a *cognitive system* as "a simple system capable of anti-entropic behavior" (p. 78), which could be a human or an intelligent device of some kind. They then define a *joint cognitive system* as a combination of a cognitive system plus (a) one or more further cognitive systems and/or (b) one or more objects (physical artifacts) or rules (social artifacts) that are used in the joint cognitive system's work. Clearly, there are many forms that a joint cognitive system can take that extend far beyond the individual.

A key feature of Woods' approach to the analysis of cognitive work has been the so-called *cognitive systems triad* (Woods 1988; Woods and Roth 1988), which reflects the fact that a joint cognitive system carries out its functions in a context or environment. The cognitive systems triad shows the interplay of *agents* (people, cognitive systems, and joint cognitive systems), *artifacts* (technology and representations), and the external *world* (demands, constraints, and dynamics). An early version of the cognitive systems triad (Woods 1988) is shown on Figure 6.1a, and a later version, annotated with relevant elements (Woods et al. 2002), is shown in Figure 6.1b.

As Woods notes, the cognitive systems triad is not analytically decomposable because when cognitive work takes place in complex sociotechnical systems, its three

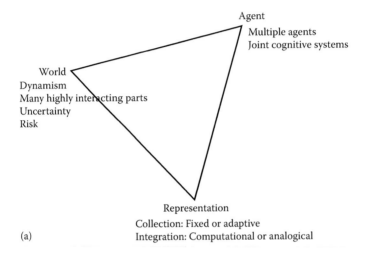

(a)

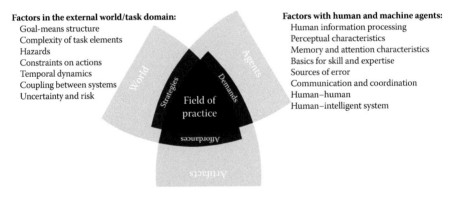

(b)

FIGURE 6.1   Two versions of the cognitive systems triad showing factors shaping cognitive work. (a) The early version (Woods 1988) and (b) the more recent version (Woods et al. 2002). ([a] Republished with permission from Taylor & Francis (CRC), from L. Goodstein, H. B. Andersen, and S. E. Olsen (Eds.), *Tasks, Errors, and Mental Models*. Bristol, PA: Taylor & Francis, Copyright 1988; permission conveyed through Copyright Clearance Center, Inc. [b] Republished with permission from the author.)

elements are inextricably linked to each other. In other words, one cannot understand cognitive work by examining each of the three elements independently of the others—by studying agents alone, artifacts alone, or the external world alone—and then trying to combine them. This fact imposes constraints on how CTA should be performed: the connections between agents, artifacts, and the external world cannot be ruptured. Accordingly, data supporting a CTA must come from qualified agents addressing authentic demands from their domain of work, using representative work artifacts and tools, as we will see in the Smith et al. (2013) example presented later in this chapter.

6.4 THEORY-DRIVEN ANALYSIS AND DESIGN

Many treatments of CTA and CWA emphasize the process rather than the purpose of performing analyses of cognitive work. A model that has appeared in different forms for over 35 years, and that reappears in the writings of Woods and colleagues, provides one way of thinking about the purpose of CTA. The model is most succinctly described diagrammatically, and an example is shown in Figure 6.2. The diagram distinguishes data-driven and concept-driven forms of analysis, and helps us focus on *why* analyses of cognitive work are done, as much as on how analyses are done.

The history of the diagram points to the intentions behind it. Early research at Riso National Laboratories about how best to support the cognitive work of nuclear power plant (NPP) operators produced powerful ways of analyzing records of cognitive

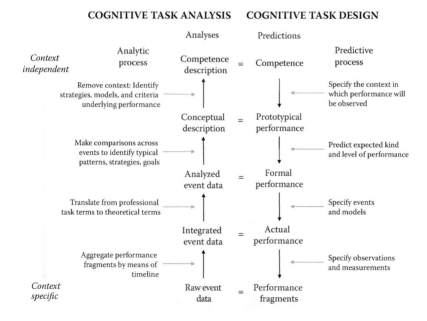

FIGURE 6.2 Levels of analysis of human performance, adapted from Hollnagel et al. (1981) and Woods and Hollnagel (2006). (Adapted with permission from D. D. Woods and E. Hollnagel, *Joint Cognitive Systems: Patterns in Cognitive Systems Engineering*. Boca Raton, FL: CRC Press, Copyright 2006; permission conveyed through Copyright Clearance Center.)

work, some of which are summarized in a Riso technical report by Hollnagel et al. (1981). The approach is also described briefly for the North American audience in one of the earliest joint papers of Woods and Hollnagel (1982).

During their analysis of incidents, events, and accidents in NPPs, Hollnagel et al. (1981) were confronted with many different sources of human performance data: event reports, post-incident reviews and interviews, recordings of performance in training simulators specific to a particular plant, and recordings of performance in more generalized research simulators. The challenge was to find a "common analytical framework" that would help researchers arrive at a conceptually coherent account of operator cognitive work, capable of both providing insight into the particular events examined and providing theoretical constructs that could be generalized and tested in other contexts. The analyses were being performed in the aftermath of the Three Mile Island accident in 1979, and the framework was intended to be a practical tool for engineers rather than a formal model for academics (Erik Hollnagel, personal communication, January 24, 2016).

Hollnagel et al. (1981) describe their process of converting raw data into forms that are useful for various purposes—initially, for deciding on training programs for NPP operators. The inputs and outputs of the process are shown in Figure 6.2, which is a reworking for present purposes of earlier versions of the diagram that appear in Hollnagel et al. (1981), Hollnagel (1986), Xiao and Vicente (2000), Woods and Hollnagel (2006), Hollnagel (2015), and other locations. The reworking in Figure 6.2 rearranges the analysis and prediction columns for better flow, rewords some elements for better understanding and, incidentally, corrects a small error that crept into prior reproductions, for example in Woods and Hollnagel (2006).

6.4.1 Left Side of Human Performance Analysis Diagram

The boxes in the CTA (left) column of the diagram show that analysts build formal, context-free accounts of cognitive work through a series of steps that combine data and impose theoretical interpretations until, at the top, a context-free theoretical description is reached that can, in principle, be applied to other contexts. We can view this as a process of abstraction from a context-specific account to a context-independent account (Xiao and Vicente 2000).

First, at the start of the process, which is seen at the bottom of the diagram, event reports, reviews, interviews, observations, and simulator-based human performance data are the *raw event data*. Second, once the raw event data are aggregated and rearranged into a whole (e.g., by being placed on a timeline) and different forms of data are integrated (e.g., performance logs and verbalization are aligned), they become *integrated event data* providing a coherent account of an individual's work processes, in professional or domain terms, but without interpretation. Third, the records of actual performance are then redescribed in a more formal language to become *analyzed event data*, where the elements of actual performance are classified into categories that may emerge from preexisting theories, ontologies, or templates relating to cognitive work, such as general information-gathering strategies or as problem-solving steps the operator uses to handle specific problems in the domain. The analyzed event data start to offer an interpretation of why the data are as they are.

Fourth, from here, data from multiple individual cases of performance are aggregated to arrive at *conceptual descriptions* for the specific context under examination. Recurring categories are noted, and the patterns of similarities and differences across the multiple cases are noted, still using the formal language introduced for analyzed event data, along with factors that might account for the similarities and differences.

Fifth, what Hollnagel et al. (1981) called a *competence description* removes reference to the specific context or contexts in which the raw data were collected. It provides a description of the "behavioral repertoire" of the operator for the general class of situations examined, and not for any particular situation or even for any particular domain. The conceptual description is placed within a broader theoretical framework that can be generalized to other contexts.

6.4.2 Right Side of Human Performance Analysis Diagram

The boxes on the cognitive task design (CTD) (right) side of Figure 6.2 handle performance prediction; they show how an evaluation can be performed of factors likely to change cognitive work, such as new work tools or new work processes. As Hollnagel et al. (1981) stated, "the competence description is … essentially the basis for performance prediction during system design" (p. 12). We can consider the process of moving from the competence description to raw data as a process of instantiation (Xiao and Vicente 2000). However, the theme of instantiation is not pursued further in the 1981 paper. The theme of instantiation reappears in Hollnagel's (1986) discussion of cognitive performance analysis, where a more explicit discussion of a top-down process of instantiation is offered. The Hollnagel (1986) diagram also specifies a top-down process of prediction, as do the more fully worked diagrams in Woods and Hollnagel (2006) and Woods (2003)—and the version in Figure 6.2.

In the CTD (right) or prediction side of Figure 6.2, the analyst starts with theories of competence, and can conjecture what the required *competence* will be for the desired cognitive work within the system. The means of support for that competence then needs to be engineered or implemented, and tested. Design specifications that produce an account of the desired *prototypical performance* can be identified, the presence of which can be tested by trying out the design with one or more workers in authentic professional contexts to produce instantiations of the prototypical performance, or *formal performance*. Formal performance is inferred from *performance fragments* combined into accounts of *actual performance* and interpreted in the appropriate theoretical frame.

The CTD or prediction side of Figure 6.2 is where abstract concepts can be put "into empirical jeopardy as explanatory or anticipatory tools in specific situations" (Woods and Hollnagel 2006). A fuller description is given in Woods and Hollnagel (2006) that includes the potential for interplay between processes on the two sides of the diagram.

The processes [in the diagram] capture the heart of Neisser's perceptual cycle as a model of robust conceptualization and revision … moving up abstracts particulars into patterns; moving down puts abstract concepts into empirical jeopardy as explanatory or anticipatory tools in specific situations. This view of re-conceptualization points out that what is critical is not one or the other of these processes; rather, the value comes from engaging in both in parallel. When focused on abstract patterns, shift and consider how the abstract plays out in varying particular situations; when focused on the particular, shift and consider how the particular instantiates more abstract patterns. ... This is a basic heuristic for functional synthesis, and the interplay of moving up and down helps balance the trade-off between the risk of being trapped in the details of specific situations, people, and events, and the risk of being trapped in dependence on a set of concepts which can prove incomplete or wrong. (Woods and Hollnagel 2006, p. 49)

The above description makes it clear that the cognitive work of the analyst who is investigating cognitive work requires tools that support rapid shifts between raw data and interpretation, and between different perspectives. Software tools for exploratory sequential data analysis (Sanderson and Fisher 1994) such as MacSHAPA (Sanderson et al. 1994) were an attempt to provide such support. The CWATool (Jenkins et al. 2007) and software supporting aspects of the Applied CWA methodology (Elm et al. 2003) have also been notable. It is not clear that the current generation of readily available software tools presents any great improvement in supporting the analyst's shifts between raw data and theoretical interpretations and between perspectives or analytic outcomes—robust tools for supporting the work of analysts analyzing cognitive work have still to be created.

6.4.3 Uses of the Human Performance Analysis Diagram

The model of human performance analysis outlined in Hollnagel et al. (1981) was quickly absorbed into Woods' thinking. For Hollnagel, Woods, and their students and close colleagues, it has guided methods for performing field investigations ever since. For example, Roth et al. (2004) and Roth and Patterson (2005) show that field investigations guided by the model can benefit from existing conceptual frameworks but also provide a means to develop new conceptual frameworks and new insights. Saleem et al. (2005) used the model to analyze VA providers' interactions with computerized clinical reminders. As will be seen in the succeeding sections, Woods (2003) extended the lessons of the model to "staged world" studies, showing how process tracing of performance in such studies can lead to high-level functional accounts of competence. Most recently, the model was revisited by Hollnagel (2015) in a resilience engineering chapter on finding patterns in everyday health care work.

Two widely cited examples of uses of the Hollnagel et al. (1981) framework are provided in Xiao and Vicente (2000) who used variants of the framework to describe (1) the process of discovering the nature of anesthesiologists' peri-operative preparations in Xiao (1994), which reflects a process of abstraction, and (2) the evaluation of ecological interface design (EID) principles for visual display design, which reflects a process of instantiation (Vicente 1991).

In the first case, involving the process of abstraction, Xiao (1994) investigated how anesthesiologists prepare their management of a patient, given that each patient is different and given that patient monitoring technology provides only partial views of the true state of the patient. Xiao aggregated field notes and recordings to arrive at an integrated description of individual anesthesiologists' planning performance. Specific preparatory strategies were then noted in the language of the domain itself, such as prefilling and systematizing the layout of syringes. Further abstraction was achieved by describing the purpose of the anesthesiologists' strategies in more general terms, such as providing reminders, offloading workload, and so on. Finally, a competence description was achieved through broad generalizations transcending specific situations, individuals, and contexts, such as statements about how expert practitioners in complex worlds manage complexity: "Experienced practitioners reduce response complexity through anticipating future situations, mental preparation, and reorganizing the physical workspace" (Xiao and Vicente 2000, p. 98). Knowledge of expert competencies could, in turn, guide the search for further instances or counterinstances in a data set, or could guide tests of generalizability to other domains.

In the second case, involving instantiation, Vicente (1991) used a theory of competent management of system disturbances to create a visual display design that would produce performance data that would confirm or refute the theory. As Woods (1998) has noted in the title of one of his papers, "designs are hypotheses about how artifacts shape cognition and collaboration" (p. 168). Starting at the top level, and based on principles of EID (Vicente and Rasmussen 1990, 1992), Vicente's theoretical claims were that (1) operators are better able to handle unanticipated variability if they can engage in knowledge-based behavior, and (2) knowledge-based behavior is best served by a display that provides a representation of the work domain that is based on an abstraction hierarchy. At the second level, two interfaces instantiated the theory in a concrete context—control of a thermodynamic process—with one interface embodying EID principles and the other not. At the third level, the objective was to construct a task that would provide "formal performance data"—aggregated experimental data capable of clearly reflecting changes in performance related to the presence or absence of EID principles. At the fourth and fifth levels, the concern was with identifying and collating the most appropriate information from participants to build the formal account.

So, on the one hand, Figure 6.2 describes a CTA process that helps us abstract lawful relationships about the interaction of people and technology. On the other hand, Figure 6.2 describes a CTD process of applying and testing those laws through changing one aspect of the cognitive systems triad, and for much of Woods' work that has involved changing the artifacts—the technology. Woods' work on CTA has been tightly linked to the extraction of regularities about sociotechnical systems that can be applied to different domains, and the generation of designs likely to support successful joint cognitive work. Too often, CTA is described in isolation from CTD, but Figure 6.2 makes it clear why CTA must be considered hand in hand with CTD. Further information about CTD from this perspective is available in Hollnagel (2003) and Woods (2003).

6.5 SHAPING THE CONDITIONS OF OBSERVATION WITH STAGED WORLDS

There are many ways in which analysts can investigate authentic professional work, each of which has advantages and disadvantages. In very early work, Woods (1985) identified an "observation problem" in psychology and particularly in the study of complex cognitive work within joint cognitive systems. Woods (1985) distinguished three mutual constraints on the ability to observe cognitive work—(1) *specificity*, which is the degree of control exercised by the observer and the repeatability of the analysis; (2) *apparent realism or face validity*, which is the fidelity of the observed work context with respect to the actual work context of interest; and (3) *meaningfulness*, which is the theoretical richness of the resulting account and its ability to be applied in other contexts.

One of the most important concerns in understanding cognitive work remains how to maximize the leverage gained from interactions with professionals in authentic work contexts, particularly with respect to the above meaningfulness dimension. Over the last 35 or more years, CSE methods have included naturalistic observation, think-aloud protocols, structured interview techniques such as the critical decision methodology, and behavioral or performance logs of professionals in their work contexts. Among these methods is Woods' idea of shaping the conditions of observation through *staged worlds* as a method for studying cognitive work.

6.5.1 STAGED WORLDS

Staged worlds are not often discussed in reviews of CWA and CTA, although they were noted in the reviews by Bisantz and Roth (2008) and Hoffman and Militello (2009). Despite this, staged worlds are an important tool in Woods' approach to understanding cognitive work.

In order to illustrate whether the conditions of observation preserve the interlinked relationships in the cognitive triad, Woods (1993; 2003; and elsewhere) contrasts staged worlds with *natural history methods* and *spartan laboratory experiments*. Natural history methods are effectively field studies, where the operation of the cognitive systems triad is undisturbed. Spartan laboratory experiments, in contrast, usually remove most of the properties of the external world and sometimes also of the agents and artifacts, in the interest of "control."

Staged worlds are simulations of work contexts that focus on specific situations or problems that practitioners may encounter and that preserve key interrelationships in the cognitive systems triad. The effectiveness of a staged world rests in how effectively the essential properties of the cognitive systems triad are preserved in the experiences created—experiences that emerge from the relationship between people, technology, and work. A staged world can create situations that might arise only very seldom in naturalistic observation, while still preserving key properties of the work domain that create an authentic, immersive experience for practitioners. As a result, a staged world is an effective and efficient means of investigating cognitive work. A staged world can be used to probe strategies, trace cognitive processes, explore the impact of new work systems, and so on. It is therefore a powerful tool for analyzing cognitive work and for understanding how practitioners cope with complexity.

The above description of natural history methods, staged worlds, and spartan laboratory experiments might suggest that only three distinct categories of observation exist, which is patently not the case. Instead, the three methods exist on a continuum. For example, a natural history method may introduce contrasts by sampling situations or contexts, or field experiments may be possible by inserting probes or prototypes into the full operational work environment. As a further example, a staged world may reproduce the work environment and its demands with different levels of breadth and depth. Some of the possibilities are discussed in Sanderson and Grundgeiger (2015) in the context of how workplace interruptions in health care have been studied, using Woods' (1985) tension among specificity, realism, and meaningfulness.

6.5.2 A STAGED WORLDS EXAMPLE IN EMERGENCY MEDICINE

A recent paper in *Annals of Emergency Medicine* co-authored by Woods (Smith et al. 2013) illustrates the use of staged worlds to support a CTA. In the paper, the principles and methods of CTA are exposed for the benefit of an audience of emergency medical and paramedical professionals. It is therefore worth describing this example in greater detail.

Smith et al. (2013) reported a CTA of the performance of experienced and less experienced paramedics as they handled simulated emergency response scenarios. The purpose of the research was to understand the cognitive strategies used by paramedics—and by the emergency medical system more generally—to adapt to novel challenges.

Participating paramedics in the Smith et al. (2013) study handled two emergency scenarios. The scenarios were based on actual cases and were developed with the help of subject matter experts and reviewed by further experts before being presented to the participants. In the first scenario, a middle-aged man presented with chest pain, suggesting an initial diagnosis of a heart attack, but the eventual diagnosis was a pulmonary embolism (blockage in an artery in the lung) rather than a heart attack. Each participant had to detect the cues for the pulmonary embolism and revise their initial diagnosis accordingly. In the second scenario, two shooting victims had to be monitored and treated simultaneously. One patient had a head wound, was unresponsive, and slowly deteriorating, whereas the other patient had a chest wound, was responsive, but indications were that he might suddenly deteriorate with a tension pneumothorax (introduction of air into the pleural space that impedes return of blood to the heart). Each participant had to detect the more immediate risk presented by the second patient and arrange an appropriate delegation of care between himself and a less-qualified emergency medical technician (EMT)-basic level partner, given the balance of risks.

The methods that Smith et al. (2013) used to elicit the paramedics' problem solving exemplify the approach to CTA advocated and practiced by Woods and colleagues since the early 1980s. First, the paramedics' cognitive strategies were investigated by observing domain practitioners handling professionally authentic situations. Second, rather than using open-ended field observation, where complex situations may not happen often enough and predictably enough to be analyzed efficiently, the researchers used "mixed-fidelity simulation" or "staged worlds" in which carefully selected complex situations were partially reconstructed and presented to practitioners. In the Smith et al. example, patients were simulated computationally, whereas the participant's EMT-basic level partner was acted by a member of the research team. Third, the researchers investigated situations that were complex and that involved cognitive challenges for the participants, rather than situations that were routine for the participants.

Similarly, the methods that Smith et al. (2013) used to analyze the records of paramedics' problem solving are typical of CTA at its best. Smith et al. sought evidence for activities that might distinguish the problem-solving processes of the experienced versus less experienced paramedics. They therefore used process tracing, "a technique that uses iterative passes through the data to capture domain-specific and progressively more abstract patterns of cognitive performance" (p. 372). The audiovisual records were transcribed and analyzed in a series of passes that moved from constructing a coherent account of the basic activities as they unfolded over time, to identifying high-level patterns of reasoning and decision making that typify different levels of expertise. The analyses involved a process of abstraction similar to that used in the human performance model of Hollnagel et al. (1981).

What the Smith et al. (2013) example does not show is the intimate connection between cognitive task *analysis* and cognitive task *design* that is also a core feature of CSE and the work of Woods and his colleagues. As noted earlier, "designs are hypotheses about how artifacts shape cognition and collaboration" (Woods 1988, p. 168).

In addition, although it presents generalizations about expertise, the Smith et al. example has a practical purpose and does not proceed to infer or invoke laws. In more recent work, Woods and colleagues have encapsulated regularities in how joint cognitive systems work into a series of laws, described below.

6.6   THEORETICAL DESCRIPTIONS OF JOINT COGNITIVE SYSTEMS AT WORK

A key question for those analyzing cognitive work is where the more formal or theoretical language comes from that is the result of the CTA or the motivation for the CTD. Specifically, what is the source of the competence description at the top of Figure 6.2?

In the original Hollnagel et al. (1981) report of the human performance model, the question driving the investigation was how best to train human operators to control NPPs. Summaries reflecting analyses at different levels of the performance analysis diagram supported different kinds of training activity. For example, aggregated performance data that preserved details of individual or team cognitive work in context—including data representations informed by formal concepts such as switches between strategies—supported training in the form of direct operator debriefing. In contrast, tools and concepts that Hollnagel et al. used to move from domain-specific to domain-independent descriptions included analytic templates such as the "human malfunction" taxonomy or the skills–rules–knowledge framework, the decision ladder, and variants of them adapted to the needs of the research. Summaries using the latter tools and concepts supported evaluation of the overall effectiveness of training programs, rather than the specification of training content.

The specific formal language or theoretical framework that might occupy the competence description at the top of Figure 6.2 will depend of course on the specific question that the analyst is investigating. The theoretical framework could have many origins and could be based on theories of expertise, learning, diagnosis, stability and control, adaptation, or decision making, among many others.

Over the years of observing how cognitive work is managed in complex sociotechnical systems undergoing change, and the challenges that people face as partners in joint cognitive systems, Woods and colleagues have developed "laws" that describe how joint cognitive systems function and that account for successes or failures in the interaction between people, technology, and work (Hollnagel and Woods 2005; Woods and Hollnagel 2006). Decades of research into the impact of new technologies in domains such as power generation, aviation, critical care, and other domains makes it abundantly clear that joint cognitive systems are not always designed in a way that avoids the pitfalls captured in some of the above laws. Therefore, an efficient way to investigate the impact of change on cognitive work is to be guided by a search for instances where these laws have been respected or violated. In other words, the analyst should be prepared to find instances where the laws are in operation but also prepared to find instances where the relationships described by the laws are present in new, surprising ways, or are absent.

Hollnagel and Woods (2005) and Woods and Hollnagel (2006) called the above universals *laws that govern joint cognitive systems at work*. They are laws in the sense of being general truths proposed about how joint cognitive systems function that have been found to hold over a wide variety of domains. As the authors note, however, the laws unfortunately appear to be "optional" in terms of whether designers respect them, yet the consequences of not respecting them are inevitable. Specifically, evidence suggests that when the laws are not respected when new technology is introduced into a work system, new complexities are introduced and operators do not have the tools to cope with those complexities.

Woods and Hollnagel (2006) proposed five general categories of the laws that govern joint cognitive systems at work. The *Laws of Adaptation* cover phenomena associated with "how cognitive systems adapt to the potential for surprise in the world of work." The *Laws of Models* cover phenomena associated with how, through models (mental or otherwise) based on the past, people project into the future. The *Laws of Collaboration* cover phenomena associated with the fact that cognitive work is distributed over multiple agents and artifacts, and so is inherently social and distributed in nature. The *Laws of Responsibility* cover phenomena associated with the fact that people modify artifacts to better achieve their own goals. Finally, *Norbert's Contrast of People and Computers* (named for Norbert Weiner) expresses the fundamental truth that "artificial agents are literal minded and disconnected from the world while human agents are context sensitive and have a stake in outcomes" (Woods and Hollnagel 2006, p. 158).

Each category of the laws that govern joint cognitive systems at work contains several more specific laws that are also generalizations about the effective or ineffective functioning of joint cognitive systems. There are too many specific laws to detail here, and they are described in more detail in Woods and Hollnagel (2006). However, some examples of the Laws of Adaptation should provide the flavor of the more specific laws and give an idea of how they might be used "top down" during an analysis of cognitive work, either as hypotheses about factors shaping cognitive work (left side of Figure 6.2) or as principles that must be respected when designing new cognitive work tasks or tools (right side of Figure 6.2).

One of the Laws of Adaptation is *context-conditioned variability*, or "the ability to adapt behavior in changing circumstances to pursue goals" (Woods and Hollnagel 2006, p. 171). When studying people's response to disturbances or changes in their work, an analyst's awareness of this law would focus their attention on changes or constancies in the kind of behavioral routines in evidence, constraints being respected, and apparent goals being pursued. Experienced operators might be quicker to recognize the change in circumstances, and quicker to find new behavioral routines that will nonetheless respect constraints and satisfy the original goals. If an analyst is aware of the regularity expressed in the concept of context-conditioned variability, then they may be quicker to recognize its absence or presence in the behavior of the operators being observed.

A further Law of Adaptation is the *Law of Stretched Systems*, which is the idea that "every system is stretched to operate at its capacity ... as soon as there is some improvement, some new technology, we exploit it to achieve a new intensity and a new tempo of activity" (Woods and Hollnagel 2006, p. 171). Awareness of this law would focus the analyst's attention not only to anticipated uses of a new technology but also to the emergence of unanticipated uses of it, potentially directed at goals other than those for which the technology was developed, and it would focus the analyst's attention on investigating the consequences of those unanticipated uses more broadly.

The benefit of laws of course is that they provide a basis for interpretation, generalization, and prediction. They are therefore an integral part of CTA and CTD. It is clear from the above high-level description that the laws all refer to some aspect of joint cognitive systems *at work*, in a work domain or environment.

6.7   CONCLUSIONS

In this chapter, I have provided a brief sketch of how cognitive work is conceptualized and analyzed in the CSE work of Woods and his colleagues. I have also briefly related Woods' approach to other communities of practice and other approaches, such as CWA and other forms of CTA, while noting that they all spring from a similar history and set of motivations. Despite this, I have only skimmed the surface of the approach that Woods and colleagues take to the study of cognitive work.

The performance analysis framework that covers both CTA and CTD is important and it deserves to sit at the core of many future investigations of cognitive work and many reviews of its methods. At the core of the framework is the role of theory—theory development, theory testing, and theory use—and it can be seen how theory that is developed in one domain of work or for one set of problems may become a powerful tool for the analyst when starting to understand cognitive work in a novel domain, or starting to investigate a novel set of problems. The set of generalizations represented in Woods' laws that govern joint cognitive work provides such theoretical leverage.

Finally, an understanding of the cognitive systems triad is essential to understanding the value of understanding cognitive work through staged worlds, alongside other methods. The cognitive systems triad emphasizes that cognitive work in a complex domain must be studied in the process of engaging with that domain, rather than separately from it. From this follows the importance of naturalistic field studies and, particularly, of staged worlds that have been constructed to provide a more efficient way of exposing authentic cognitive work.

Readers should refer to Woods and Hollnagel (2006) for an integrated description and further development of many of the themes touched on in this chapter. Many of the more informative examples and expositions are in book chapters, some of which are referenced in this chapter.

REFERENCES

Ashoori, M., and Burns, C. (2010, September). Reinventing the wheel: Control task analysis for collaboration. In *Proceedings of the Human Factors and Ergonomics Society Annual Meeting* (Vol. 54, No. 4, pp. 274–278). Thousand Oaks, CA: Sage Publications.

Bisantz, A.M., and Burns, C.M. (Eds.). (2008). *Applications of Cognitive Work Analysis*. Boca Raton, FL: CRC Press.

Bisantz, A., and Roth, E. (2008). Analysis of cognitive work. In D. Boehm-Davis (Ed.), *Reviews of Human Factors and Ergonomics: Volume 3*. Santa Monica, CA: Human Factors and Ergonomics Society (pp. 1–43).

Cornelissen, M., Salmon, P.M., Jenkins, D.P., and Lenné, M.G. (2013). A structured approach to the strategies analysis phase of cognitive work analysis. *Theoretical Issues in Ergonomics Science*, 14(6), 546–564.

Crandall, B.W., and Hoffman, R.R. (2013). Cognitive task analysis. In J. Lee and A. Kirlik (Eds.), *The Oxford Handbook of Cognitive Engineering*. Oxford, UK: Oxford University Press (pp. 229–239).

Crandall, B., Klein, G.A., and Hoffman, R.R. (2006). *Working Minds: A Practitioner's Guide to Cognitive Task Analysis*. Cambridge, MA: MIT Press.

Elm, W.C., Potter, S.S., Gualtieri, J.W., Easter, J.R., and Roth, E.M. (2003). Applied Cognitive Work Analysis: A pragmatic methodology for designing revolutionary cognitive affordances. In E. Hollnagel (Eds.), *Handbook of Cognitive Task Design*. Boca Raton, FL: CRC Press.

Flach, J. (2016). Supporting productive thinking: The semiotic context for Cognitive Systems Engineering (CSE). *Applied Ergonomics*, 59, 612–624.

Hassall, M.E., and Sanderson, P.M. (2014). A formative approach to the strategies analysis phase of cognitive work analysis. *Theoretical Issues in Ergonomics Science*, 15(3), 215–261.

Hoffman, R.R., and Militello, L.G. (2009). *Perspectives on Cognitive Task Analysis: Historical Originals and Modern Communities of Practice*. New York: Psychology Press.

Hollnagel, E. (1986). Cognitive system performance analysis. In E. Hollnagel., G. Mancini, and D. Woods (Eds.), *Intelligent Decision Support in Process Environments. NATO ASI Series*, Vol. F21. Berlin Heidelberg: Springer-Verlag (pp. 211–226).

Hollnagel, E. (2003). *Handbook of Cognitive Task Design*. Boca Raton, FL: CRC Press.

Hollnagel, E. (2015). Looking for patterns in everyday clinical work. In R.L. Wears, E. Hollnagel, and J. Braithwaite (Eds.), *Resilient Health Care, Volume 2: The Resilience of Everyday Clinical Work*. Aldershot, UK: Ashgate.

Hollnagel, E., Pedersen, O.M., and Rasmussen, J. (1981). *Notes on Human Performance Analysis*. Technical Report, Riso-M-2285. Roskilde, Denmark: Riso National Laboratory.

Hollnagel, E., and Woods, D.D. (1983). Cognitive systems engineering: New wine in new bottles. *International Journal of Man–Machine Studies*, 18(6), 583–600.

Hollnagel, E., and Woods, D.D. (2005). *Joint Cognitive Systems: Foundations of Cognitive Systems Engineering*. Boca Raton, FL: CRC Press.

Jenkins, D., Farmilo, A., Stanton, N.A., Whitworth, I., Salmon, P.M., Hone, G., Bessell, K., and Walker, G.H. (2007). *The CWA Tool V0. 95. Human Factors Integration Defence Technology Centre (HFI DTC)*, Yeovil, Somerset, UK.

Jenkins, D.P., Stanton, N., Salmon, P., and Walker, G. (2009). *Cognitive Work Analysis: Coping with Complexity*. Aldershot, UK: Ashgate.

Le Coze, J.-C. (2015). Reflecting on Jens Rasmussen's legacy (1): A strong program for a hard problem. *Safety Science*, 71, 123–141.

Le Coze, J.-C. (2016). Reflecting on Jens Rasmussen's legacy (2): Behind and beyond, a "constructivist turn." *Applied Ergonomics*, 59, 558–569.

Lee, J.D., and Kirlik, A. (2013). *The Oxford Handbook of Cognitive Engineering*. Oxford, UK: Oxford University Press.

Lintern, G. (2013). *The Foundations and Pragmatics of Cognitive Work Analysis: A Systematic Approach to Design of Large-Scale Information Systems*. Melbourne, Australia: http://www.CognitiveSystemsDesign.net.

Naikar, N. (2013). *Work Domain Analysis: Concepts, Guidelines, and Cases*. Boca Raton, FL: CRC Press.

Rasmussen, J. (1986). *Information Processing and Human–Machine Interaction. An Approach to Cognitive Engineering*. New York: North-Holland.

Rasmussen, J., and Rouse, W. B. (1981). *Human Detection and Diagnosis of System Failures*. New York: Plenum Press.

Rasmussen, J., Pejtersen, A.M., and Goodstein, L.P. (1994). *Cognitive Systems Engineering*. New York: John Wiley & Sons.

Roth, E.M. (2008). Uncovering the requirements of cognitive work. *Human Factors: The Journal of the Human Factors and Ergonomics Society*, 50(3), 475–480.

Roth, E.M., and Bisantz, A.M. (2013). Cognitive work analysis. In J. Lee and A. Kirlik (Eds.), *The Oxford Handbook of Cognitive Engineering*. Oxford, UK: Oxford University Press (pp. 240–260).

Roth, E.M., Christian, C.K., Gustafson, M., Sheridan, T.B., Dwyer, K., Gandhi, T., Zinner, M., and Dierks, M. (2004). Using field observations as a tool for discovery: Analysing cognitive and collaborator demands in the operating room. *Cognition, Technology, and Work*, 6, 148–157.

Roth, E.M., and Patterson, E.S. (2005). Using observation study as a tool for discovery: Uncovering cognitive and collaborative demands and adaptive strategies. In H. Montgomery, R. Lipshitz, and B. Brehmer (Eds.), *How Professionals Make Decisions*. Mahwah, NJ: Lawrence Erlbaum Associates (pp. 379–393).

Saleem, J.J., Patterson, E.S., Militello, L., Render, M.L., Orshansky, G., and Asch, S.M. (2005). Exploring barriers and facilitators to the use of computerized clinical reminders. *Journal of the American Medical Informatics Association*, 12(4), 438–447.

Sanderson, P.M. (2003). Cognitive work analysis. In J.M. Carroll (Ed.), *HCI Models, Theories, and Frameworks: Toward a Multi-Disciplinary Science*. San Francisco, CA: Morgan Kaufmann (pp. 225–264).

Sanderson, P.M., and Fisher, C. (1994). Exploratory sequential data analysis: Foundations. *Human–Computer Interaction*, 9(3–4), 251–317.

Sanderson, P.M., and Grundgeiger, T. (2015). How do interruptions affect clinician performance in healthcare? Negotiating fidelity, control, and potential generalizability in the search for answers. *International Journal of Human–Computer Studies*, 79, 85–96.

Sanderson, P., Scott, J., Johnston, T., Mainzer, J., Watanabe, L., and James, J. (1994). MacSHAPA and the enterprise of exploratory sequential data analysis (ESDA). *International Journal of Human–Computer Studies*, 41(5), 633–681.

Schraagen, J.M., Chipman, S.F., and Shalin, V.L. (Eds.). (2000). *Cognitive Task Analysis*. Mahwah: NJ: Lawrence Erlbaum Associates.

Sheridan, T.B., and Johannsen, G. (1976). *Monitoring Behaviour and Supervisory Control*. New York: Plenum Press.

Smith, M.W., Bentley, M.A., Fernandez, A.R., Gibson, G., Schweikhart, S.B., and Woods, D.D. (2013). Performance of experienced versus less experienced paramedics in managing challenging scenarios: A cognitive task analysis study. *Annals of Emergency Medicine*, 62(4), 367–379.

Vicente, K.J. (1991). *Supporting Knowledge-Based Behavior through Ecological Interface Design*. Unpublished PhD thesis, Department of Mechanical and Industrial Engineering, University of Illinois at Urbana–Champaign, Urbana, IL.

Vicente, K.J. (1999). *Cognitive Work Analysis: Toward Safe, Productive, and Healthy Computer-Based Work*. Boca Raton, FL: CRC Press.

Vicente, K.J., and Rasmussen, J. (1990). The ecology of human–machine systems II: Mediating 'direct perception' in complex work domains. *Ecological Psychology*, 2(3), 207–249.

Vicente, K.J., and Rasmussen, J. (1992). Ecological interface design: Theoretical foundations. *IEEE Transactions on Systems, Man, and Cybernetics*, 22(4), 589–606.

Waterson, P., Robertson, M.M., Cooke, N.J., Militello, L., Roth, E., and Stanton, N.A. (2015). Defining the methodological challenges and opportunities for an effective science of sociotechnical systems and safety. *Ergonomics*, 58(4), 565–599.

Woods, D.D. (1985). *The Observation Problem in Psychology (Westinghouse Technical Report)*. Pittsburgh, PA: Westinghouse Corporation.

Woods, D.D. (1988). Coping with complexity: The psychology of human behaviour in complex systems. In L. Goodstein, H.B. Andersen, and S.E. Olsen (Eds.), *Tasks, Errors, and Mental Models*. Bristol, PA: Taylor & Francis (pp. 128–148).

Woods, D.D. (1993). Process tracing methods for the study of cognition outside of the experimental psychology laboratory. In G. Klein, J. Orasanu, R. Calderwood, and C. Zsambok (Eds.), *Decision Making in Action: Models and Methods*. Norwood, NJ: Ablex Publishing (pp. 228–251).

Woods, D.D. (2003). Discovering how distributed cognitive systems work. In E. Hollnagel, (Ed.), *Handbook of Cognitive Task Design*. Boca Raton, FL: CRC Press (pp. 37–53).

Woods, D.D., and Hollnagel, E. (1982, October). A technique to analyze human performance in training simulators. In *Proceedings of the Human Factors and Ergonomics Society Annual Meeting* (Vol. 26, No. 7, pp. 674–675). Thousand Oaks, CA: Sage Publications.

Woods, D.D., and Hollnagel, E. (2006). *Joint Cognitive Systems: Patterns in Cognitive Systems Engineering*. Boca Raton, FL: CRC Press.

Woods, D.D., and Roth, E.M. (1988). Cognitive systems engineering. In M. Helander (Ed.), *Handbook of Human–Computer Interaction*. North-Holland: New York. (Reprinted in N. Moray (Ed.) *Ergonomics: Major Writings*. London: Taylor & Francis, 2004.)

Woods, D.D., Tinapple, D., Roesler, A. and Feil, M. (2002). *Studying Cognitive Work in Context: Facilitating Insight at the Intersection of People, Technology and Work*. Cognitive Systems Engineering Laboratory, The Ohio State University, Columbus, at: http://csel.org.ohio-state.edu/productions/woodscta/

Xiao, Y. (1994). *Interacting with Complex Work Environments: A Field Study and a Planning Model*. Unpublished PhD thesis, Department of Industrial Engineering, University of Toronto, Canada.

Xiao, Y., and Vicente, K.J. (2000). A framework for epistemological analysis in empirical (laboratory and field) studies. *Human Factors*, 42(1), 87–101.

Part III

*Designing Effective Joint
Cognitive Systems*

7 Adaptation in Sociotechnical Systems

Shawn A. Weil, Jean MacMillan, and Daniel Serfaty

CONTENTS

7.1 INTRODUCTION

In today's networked world, the nature of work has significantly changed from the way it was before the explosion in connectivity. Individuals, teams, and even organizations do not work in isolation. They work in dynamic networks, organizational contexts, and cultures that impose rules, constraints, and structures—some of which are relatively constant (e.g., the need for communication) and some of which are functionally driven (e.g., hierarchy or heterarchy). Decisions are shaped by missions, tasks, and work processes that can evolve rapidly with time and circumstances. Technology capabilities mediate and amplify these behaviors. One of the fundamental tenets of cognitive systems engineering is the shift of the unit of analysis from the individual to layers or echelons within networked sociotechnical organizations in a "coupled aggregate."

Research on organizational structures and their associated performance is intensely challenging for a multitude of reasons. Organizations are complex entities, and many variables affect their performance (Hollnagel et al. 2006). Further, the sheer size and scale of organizations makes it very difficult and costly to study organizational dynamics empirically using the methods of conventional experimentation and analysis. This chapter describes a program of CSE research that focused on a key component of sociotechnical systems—organizational structure—in the context of how well it fits with all of the other factors that affect organizational performance. The purpose of the research was to determine how organizational structure can best support effective performance and adapt to a changing world.

In this long-term, multi-institutional, interdisciplinary program of research, scientists and technologists asked questions, formulated hypotheses, and proposed theories and models at each level of aggregation in Rasmussen's framework (Rasmussen 1986). For instance, given organizational size constraints, what structure will best enable an organization to achieve its goals? What processes should be defined given the functional relationships among teams and tasks? What is the most effective mapping between human agents and technology ownership, given constraints in communication? The underlying question, as shown in Figure 7.1, is this: How can organizational-level structures be optimized for adaptability and resilience with respect to their major functional responsibilities and goals?

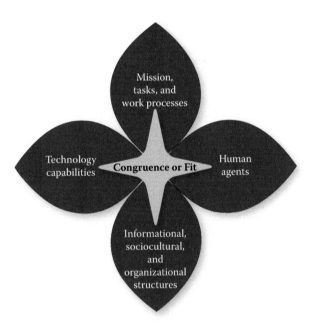

FIGURE 7.1 Four key dimensions of complex sociotechnical systems. The goal of the A2C2 program was to develop and test ways to optimize fit or congruence among (1) the human agents involved in a task; (2) the mission, tasks, and work processes they were conducting; (3) the technical capabilities they were utilizing to do that work; and (4) the informational, sociocultural, and organizational structures that the sociotechnical organization is embedded in.

7.2   TOGETHER FOR ADAPTIVE AND RESILIENT PERFORMANCE

In 1995, the U.S. Office of Naval Research (ONR) launched a program of research to investigate the fit among the four key aspects of sociotechnical systems shown in Figure 7.1. The Adaptive Architectures in Command and Control (A2C2) program used methods from experimental psychology, systems engineering, mathematical modeling and simulation, social network theory, and anthropology. A2C2 focused on the fit or congruence of organizational structures/processes with changing mission requirements. The history, results, and lessons learned from this program provide a concrete illustration of what research in CSE looks like in practice, and how CSE can be used for both understanding and design. No single review of this 16-year research program can do justice to all of the findings (and new methodologies) that were produced. Instead, this chapter identifies the major breakthroughs and innovations of the program and summarizes the results.

7.3   THE PROBLEM: UNDERSTANDING ORGANIZATIONAL FIT AND ADAPTATION

Advances in networked communication are rapidly changing the structure, processes, and capabilities of organizations, enabling new structures and making others obsolete. It has been argued that changes in communication will facilitate flatter organizational structures—ones having higher horizontal communication rates in contrast to more traditional hierarchical structures based primarily on vertical communication (Fukuyama 2000). Tapscott and Williams (2006) argued that technology now supports new forms of mass collaboration, and that this mass collaboration has major implications for organized human endeavors in a host of areas: "In an age where mass collaboration can reshape an industry overnight, the old hierarchical ways of organizing work and innovation do not afford the level of agility, creativity, and connectivity that companies require to remain competitive in today's environment" (p. 31).

Military thinkers have always viewed information as a resource. Military command organizations have a tradition of strict hierarchy and top-down control with well-defined roles at each level. However, military organizations have been experiencing the transformational effects of network connectivity. Alberts and Hayes (2003) argued that new military organizational forms are required for (and enabled by) the information age and that "work processes and organizational structures need to be adapted to allow greater innovation and flexibility at all levels" (p. 68).

It was with this understanding of technology change that the A2C2 research program was originally conceived. The basic premise of the program was that high-performing mission-critical organizations cope with stress (generated by both internal and external factors) by a series of adaption mechanisms. Those adaptation mechanisms can be conceptualized as nested feedback loops, as shown in Figure 7.2, with increasing complexity, as a function of the nature and intensity of the "stressors," ranging from individual work adaptation (e.g., decision-making strategies) to organizational process adaptation (e.g., teamwork), to structural adaptation (e.g., hierarchies-to-networks). The premise illustrated in Figure 7.2 is that high-performing organizations cope

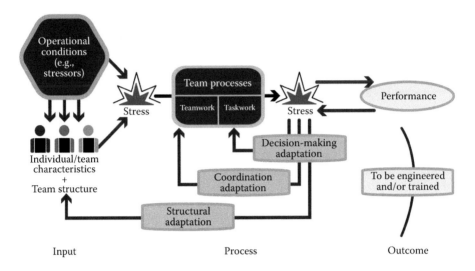

FIGURE 7.2   Nested feedback loops of adaptation in high-performing organizations.

with stress through internal mechanisms of decision strategy adaptation, coordination strategy adaptation, and structural reconfiguration in order to keep performance at the required level while maintaining stress below an acceptable threshold. Structural adaptation, the outer loop shown in Figure 7.2, was the main focus of the A2C2 program.

The goal of the program was to investigate how military organizations can best function, change, and adapt in a world of exploding network connectivity and access to information. The research plan was to proceed in two phases. In the first phase, organizational models would be developed and empirically validated for a series of Navy organizations. In the second phase, these models would be used to conduct computational experiments that would address the most critical design and performance questions confronting military command and control organizations. While the A2C2 program did not proceed in such a simple linear fashion (itself a lesson in CSE), it did result in considerable progress in both of these areas.

The A2C2 program not only produced a wealth of research findings but also expanded the frontier regarding how such research can be conducted—through a transdisciplinary effort involving multiple organizations, and using computational models and computational experimentation as a key research tool to augment organizations-in-the-loop empirical studies. The research efforts led to insights about how to craft effective networked organizations and how to investigate new organizational forms from a CSE perspective. The first question was where to begin.

7.4 MODES OF INVESTIGATION TO UNDERSTAND SOCIOTECHNICAL SYSTEMS

It is a daunting task to achieve a thorough empirical understanding and conceptual model of complex sociotechnical systems (such as air traffic control, military operations, and large scale disaster relief efforts) from all of the perspectives shown in Figure 7.1. A multitude of variables affect work system performance, and many of them are not well defined or well understood. Taking the unit of analysis as the organization requires that the methods of inquiry shift as well as the questions to be addressed (Rasmussen 1986; Woods and Hollnagel 2006).

7.4.1 EXPERIMENTAL METHODS

Although controlled factorial experimentation is useful for identifying and isolating cause–effect relations between dependent and independent variables, such research tends to be reductionist and thus a long series of experiments is required to achieve an understanding of the dynamics of change, especially where there are complex interactions and emergent phenomena.

7.4.2 ETHNOGRAPHIC METHODS

Ethnographic analysis is a useful method for understanding how sociotechnical systems are structured and how they perform (see Hoffman and Militello 2008). But this analysis runs the risk of the "*n* of 1" phenomenon in which the results from study of one work system may not generalize to others (Berry et al. 1997; Woods and Hollnagel 2006). Ethnographic analysis without abstraction can provide little insight into how a work system can and should adapt to changing conditions and capabilities (see Lave 1988).

7.4.3 COMPUTATIONAL METHODS

As computational modeling methods have become increasingly powerful, models can be used as tools for understanding large-scale sociotechnical systems (Woods 1990). Conceptual and formal models can capture complex dynamics but depend on many assumptions to make computation (or mathematical modeling) tractable. Ecologically valid modeling requires context (Lipshitz 2000) to determine structure or set parameters. Additionally, the assumptions made in the models often have no basis in the ecology of the work domain or questionable "psychological reality" as mental representations.

No single method for investigating cognitive systems on an organizational level is sufficient, but, arguably, all are necessary. One needs to establish relations between dependent and independent variables, one needs to achieve a rich and meaningful understanding of the work domain, and one must create conceptual models and push them toward formal models (Potter et al. 2000); multiple methods must be used in concert. Multiple empirical perspectives enable researchers to understand and improve sociotechnical work system performance, as no one method can provide all the answers (Woods and Hollnagel 2006).

The A2C2 program broke new ground by combining organizational modeling, controlled empirical data collection, and field observation in order to understand how military organizations can and should adapt. The output of the models, generally, was organizational constructs that were then abstracted and brought into human-in-the-loop laboratory experiments to validate and modify those results. This approach was called "Model-Based Experimentation."

7.5 MODEL-BASED EXPERIMENTATION

What might be the most effective organization of military forces in the age of network connectivity? At the start of the A2C2 program in 1995, there was no assured path to progress on this question. It seemed clear that the ability to adapt rapidly was critical to leverage information that would be arriving at an unprecedented rate. Confronting the tactics of a rapidly evolving adversary demands agility and variety (Ashby 1956; Woods 1995; Woods and Hollnagel 2006), but what organizational structure could facilitate this "agile adaptation?" What does it mean for an organization to be more adaptive? How could research be carried out on these issues? The questions to be addressed did not fit neatly into any academic discipline or research approach. While there were fundamental research issues to be addressed, there were also practical problems to be solved—how should the Navy think about organizing its forces for adaptability? Was reorganization the answer, or some ability to rapidly re-reorganize?

To solve the practical and applied challenges of the A2C2 program, the A2C2 team employed a mode of investigation shown in Figure 7.3. The approach started with the organizational structures in use by the Navy for command and control activities and the organizational structures being contemplated for the future.

Focused on naval engagement between U.S. Navy organizations (e.g., a Carrier Battle Group, Expeditionary Strike Group) and enemy assets, inquiry started with a theoretical and operational concept—a question from Navy leadership related to a problem they were facing in the organizational structures of major Navy components. What roles and responsibilities should be given to particular decision makers in an organization? How should the organization adapt to changing external factors (e.g., changes in adversary activities)?

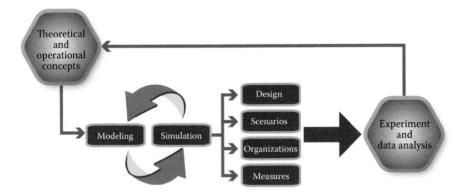

FIGURE 7.3 Model-based experimentation: Theoretical and operational concepts spur modeling and simulation. Insights gained from those computational efforts are then tested in human-in-the-loop laboratory experiments. Findings from the analysis of those data are used to update the operational doctrine.

From there, models were created to capture all of the elements in Figure 7.1. Models of the relevant Navy platforms were created, with a focus on their capabilities vis-à-vis enemy assets. Navy platforms (e.g., aircraft carriers, unmanned vehicles) had capabilities that approximated their real-world analogs, and the capabilities of adversarial assets were similarly developed. Models also represented the human capabilities required to control and operate Navy assets. The parameterization of the models was based on input from subject matter experts or ethnographic assessments by the A2C2 team. Some aspects of organizational structure were constrained, based on input from subject matter experts and observation, while others were allowed to vary and were optimized.

Simulation was then conducted using these models. In support of that simulation, scenarios were developed representing multi-day "battles" with different characteristics imagined to be better or worse for different organizational structures. Computational modeling approaches were used to determine "optimal" asset organizational control with respect to those adversary activities. For example, if there were four "commanders" in the scenario, each of whom might have control over a subset of assets, which assets should they control?

The results of these simulations provided answers to the following questions:

1. What organizational designs were feasible, given adversary behavior?
2. What changes in the composition of scenarios had the most effect on organizational performance?
3. Which organizational structures led to the highest outcome performance (i.e., success of friendly forces at defeating adversary forces)?
4. What measures of performance should be used to evaluate those organizations?

This motivated the careful design of a program for "organization-in-the-loop" experimentation. A subset of the organizational designs was recreated in a laboratory setting, with Navy personnel filling the roles of decision makers in various positions. The same scenarios used in the simulation activities were used in the experimentation—activities that reflected potential battlespace events involving U.S. and adversary naval forces. The structure of the organizations reflecting good performance in the simulations was abstracted for the experiments; different decision-makers had control of different asset types. Measures of performance (e.g., task completion time and accuracy) could be normalized and then compared with the simulation results.

7.6   FINDINGS ON ORGANIZATIONAL CONGRUENCE AND ADAPTATION

The focus of the A2C2 program was how organizations can and should adapt to changing missions and changing technologies. In order to address these questions, however, it was necessary to address a host of supporting questions, such as what events trigger adaptation and what factors make adaptation more or less difficult for an organization. We organize the major theoretical results of the program around two major themes: (1) how the nature of the mission to be performed drives the need for change, and (2) how the path of adaptation affects organizational performance.

7.7 HOW THE MISSION DRIVES THE NEED TO ADAPT

Organizations change (or need to change) when their mission changes. The first step in the A2C2 program was to characterize and represent military missions in order to understand the effect of organizational structure on mission success. The key insight was that missions have a structure that can be characterized mathematically and this characterization can be used to design an effective organizational structure for a given mission. Initial organizational models in the A2C2 program characterized military missions using a joint cognitive systems perspective (Woods and Hollnagel 2006) along three major categories: the primary tasks to be accomplished (including their timing and their interdependencies), the resources available to perform these tasks, and the human beings who control those resources to perform those tasks.

Using a multi-objective optimization approach, organizational designs were developed that mapped tasks to resources in order to create roles in a way that was predicted to increase mission effectiveness with respect to efficiency of engagement (Levchuk et al. 2004; Pattipati et al. 2002). Mathematical optimization techniques were developed to simultaneously maximize efficiency and minimize information loss, communication overhead, and delays, while staying within cost and capacity constraints. Figure 7.4 shows the stages of the modeling technique developed to

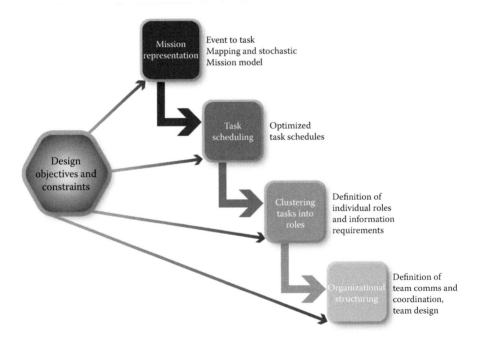

FIGURE 7.4 Stages of the computational technique developed to optimize organizational structure in the A2C2 program.

optimize organizational structure. Because these models considered the interdependencies among tasks (e.g., a ship must first destroy a missile site before launching airplanes), and the coordination requirements (e.g., two decision makers had to bring assets to bear simultaneously to prosecute a target) that were created by these interdependencies, models were able to make predictions about the process, not just the structure and outcomes, of the organization. For example, the computational models were able to predict how much coordination (e.g., volume of communication, use of multiple simulated assets) would be required to perform a mission effectively under a specific organizational structure and mission scenario, and these differences in coordination requirements proved to have a major role in generating differences in organizational performance (Levchuk et al. 2003).

The effective/efficient organizational structures were then tested in scenario-based simulation experiments to see if teams of military officers performing under the innovative model-based organizational designs would outperform teams using more traditional military organizational designs for roles and responsibilities. The model-based organizational structures were indeed found to outperform more traditional structures (Entin 1999, 2000).

One of the ways in which the model-based organizational designs led to higher mission performance levels was by reducing the *need for coordination* among team members—a factor in determining efficiency. To the extent possible, the model algorithms assigned tasks to team members in a way that allowed them to act independently, reducing their need to coordinate, which, in turn, reduced their need to communicate (MacMillan et al. 2002). Coordination activities and communication rates were lower under the model-based designs, as was subjective workload, suggesting that the model-based organizational design had reduced the "communication overhead" associated with effective organizational performance.

The finding that it was possible to achieve higher performance levels by changing an organizational structure to conform to the demands of the mission, as characterized by a task graph, led to the development and testing of the concept of the fit or *congruence* between an organization and its mission, with the mission being defined by the tasks that are required to complete it. Congruence is defined mathematically using a network representation of the tasks to be performed, the resources needed to perform them, and the control of those resources by the decision makers in the organization. Incongruence is defined by the extent to which the assignment of resources (controlled by various decision makers) to tasks increases the amount of coordination required among decision makers (e.g., several different decision makers control the resources needed for a task so the task cannot be completed without their coordinating) or creates workload imbalance among decision makers (e.g., one decision-maker controls all of the resources needed for multiple tasks).

Increases in incongruence cause increases in bottlenecks and delays in task processing. (Levchuk et al. 2003). Using this definition, congruence affects organizational performance through its effects on workload balance, communication requirements, and interdependence among personnel. The lack of congruence between an organization's roles and responsibilities and the mission it is attempting to perform was hypothesized to result in measurable "leading indicators" that would indicate to the organization that it needed to change its structure.

Computational experiments showed that it was possible to manipulate the degree of congruence between an organization and its mission by using a model of the organization to design mission scenarios that were congruent or incongruent with the organization's structure (Levchuk et al. 2003). In subsequent human-in-the-loop experimentation, this lack of congruence led to significantly worse organizational performance as measured by task performance in the scenario (an omnibus score based on task completion efficiency; Diedrich et al. 2003; Entin et al. 2003). These results are shown in Figure 7.5. Furthermore, the poorer performance in the incongruent organization-mission conditions was associated with observable and measurable differences in behavior—most notably, an increase in the amount of coordination-related communication was observed when the organization's structure did not match the mission it was attempting to accomplish. The increase in unnecessary communication associated with incongruence began early in the mission and persisted throughout (Entin et al. 2003).

The increase in coordination-related communication caused by incongruence between the organization structure and the mission echoes the finding from earlier experiments that organizational structures that are clearly nonoptimal are associated with greater communication requirements and efforts. The allocation of tasks and resources to individuals in the organization, when matched against the requirements of the tasks to be performed, creates the need for members of the organization to coordinate; this coordination requires communication, and this increased communication "overhead" can be associated with reduced performance. In comparison, in an organization that has been designed for "congruence," the responsibilities for tasks and the control of resources allow members to act more independently, coordinate effectively, and communicate only when necessary, with each member anticipating the need of the others, and providing information and resources before being asked to

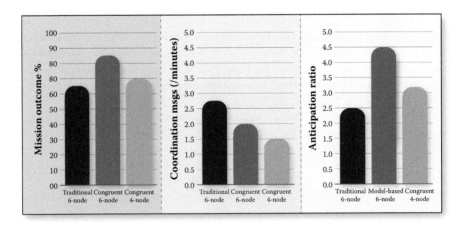

FIGURE 7.5 Results of human-in-the loop experimentation (Diedrich et al. 2003; Entin et al. 2003). Lack of organization/mission congruence led to decreased organizational performance as measured by task performance in the scenario (an overall score based on task completion efficiency).

do so. This behavior is captured by a measure of performance called the anticipation ratio as shown in Figure 7.5. It was shown that increases in anticipation ratios also enable smaller organizations (4 vs. 6 nodes in Figure 7.5) to make up for the loss of staffing through the use of more efficient communications strategies.

This was a consistent finding of the A2C2 program, from multiple perspectives and multiple experiments—if there is a way to structure the organization that reduces the need for coordination while maintaining the ability to perform the necessary tasks, then this structure is likely to perform better than an organizational structure that requires more coordination—at least when task efficiency is the objective function.

Some members of the A2C2 team argued that reduced communication rates might lead to unanticipated negative effects on performance. Organizations in which individuals have a greater need to negotiate with each other can be more adaptive over the long run (Carley and Ren 2001). Thus, there may be a trade-off between the efficiency of an organization (characterized by lower communication rates and higher performance) in the short term and the ability of the organization to adapt in the long term (Woods 2006). The possible trade-off between this type of immediate efficiency and the longer-term adaptability of the organization remains an open question.

7.8   HOW THE PATH OF ADAPTATION AFFECTS ORGANIZATIONAL PERFORMANCE

No single organizational structure will be best for all missions and all individuals/ roles in an organization. For example, a functional structure in which the subunits of the organization are specialized and must coordinate closely to accomplish their mission may perform best in predictable task environments, whereas divisional structures, which support independence with respect to certain responsibilities, may perform best in less predictable situations or contexts (Hollenbeck et al. 1999a,b). A2C2 experimentation showed that congruence between the mission type and the organizational structure leads to better organizational performance related to task completion efficiency (Diedrich et al. 2003; Entin et al. 2003). However, might it matter what path the organization takes to change its structure, and could certain types or directions of change be more or less difficult than others?

One of the areas investigated in the A2C2 program was the *path* by which an organization can and should shift from one structure to another. A path, in this case, refers to the incremental changes to an organization's roles and responsibilities from an initial state to a new state (new roles, responsibilities, echelons, etc.). One thread of the A2C2 research modeled organizations as discrete event dynamic systems, implemented as Colored Petri Nets, and used these models to analyze the paths by which organizations could best transition from one organizational structure to another (Handley and Levis 2001; Perdu and Levis 1999). Adaptation was studied as a *morphing* process in which an organization transitioned from one structure to another through incremental steps that sought to preserve its overall functionality while adaptation was taking place.

A series of laboratory experiments with undergraduate participants demonstrated that the *direction* of organizational change can affect the levels of performance that can be achieved after a change occurs (Hollenbeck et al. 1999a,b; Moon et al. 2000). The experiments used a simulated command and control task in which participants needed to enforce a demilitarized zone that was traversed by both enemy and friendly aircraft and ground vehicles. The participants' task was to identify the nature of the aircraft and vehicles and attack enemy targets without damaging friendly vehicles and aircraft. Resources under the control of the decision makers included jet aircraft, AWACS surveillance aircraft, helicopters, and tanks. Each type of resource had its strengths and weaknesses, for example, tanks moved slowly, had little surveillance capability, and were ineffective against high-speed aircraft, but were an effective means of attacking enemy ground vehicles. Organizational structure was created in the experiments by giving decision makers control over different types of assets.

It was hypothesized that some directions of change in organizational structure might be more difficult than others. Specifically, A2C2 experiments found that it is easier for an organization to move from a functional organizational structure in which individuals or units are specialists and control only one type of resources (e.g., one decision maker controls all jet aircraft) to a divisional structure in which individuals are not specialists and control multiple types of resources (e.g., one decision maker controls all of the jets, AWACS aircraft, helicopters, and tanks in a specific area), than it is for them to go in the opposite direction (e.g., starting with broad responsibilities and moving to a position that requires concentrated knowledge).

In the laboratory experiments, participants found that moving from a functional structure in which they controlled only one type of resource (e.g., jets), even though several types of resources were needed to successfully complete the mission, to a divisional structure in which they controlled all of the resources (jets, AWACS aircraft, helicopters, and tanks) needed for a specific task (identifying an enemy aircraft or ground vehicle and attacking it) meant that individuals needed to learn to coordinate *less*. Moving in the opposite direction meant that they had to learn to coordinate *more*, which turned out to be the more difficult change to make. Once individuals were accustomed to acting independently, they found it difficult to learn to coordinate their actions.

Experiments also showed an asymmetry in structural change in moving between a centralized and a decentralized organizational structure (Ellis et al. 2003; Woods 2006). Groups had better performance when moving from a centralized structure (in which team leaders direct and/or approve actions) to a decentralized structure (in which individuals could act more autonomously) than when moving in the opposite direction. Apparently, once one has experienced independence, it is hard to move to a structure where there is more centralized control.

7.9   EMERGING CHALLENGES AND NEW HORIZONS

At the conclusion of the A2C2 program, it is appropriate to take a step back, gauge its successes, evaluate its failures, and look at the path ahead from a CSE perspective. Today, the original challenge of A2C2 as shown in Figure 7.6—how to design complex organizations more agile and more adaptive—remains of vital importance for both military and civilian organizations (Alberts 2011).

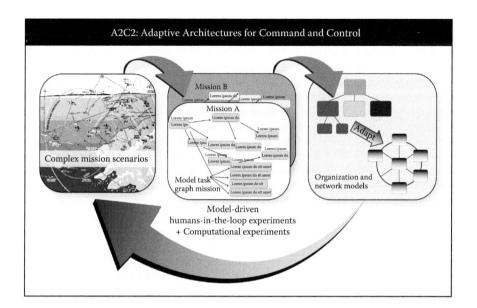

FIGURE 7.6 High-level view of the A2C2 program. Complex military campaigns are decomposed and represented in graph format. Through both computational and humans-in-the-loop experimentation, concepts for organizational structure and process alignment can be developed, tested, and fed back to military decision makers.

7.10   THE NEED FOR STRONGER THEORIES OF ORGANIZATIONAL ADAPTATION

As organizations introduce technological capabilities, and automated technologies play an increasingly large role, there is an increasing concern that organizations will lose rather than gain in adaptability (Alberts 2011). Human beings are adaptable— they learn, grow, and change as their environment changes. Automated technology has yet to demonstrate the ability to learn that we take for granted in human beings. To prevent automation from making organizations less rather than more adaptive, further research is needed on the triggers and causes of adaptation (Woods 2006). Theories are needed to guide the design of organizations so that increased connectivity, access to information, and advanced automation will make them more flexible and more resilient in a rapidly changing environment.

7.11   EMERGING METHODS OF INVESTIGATION

The A2C2 program came up against the limits of controlled empirical testing of organizational structures with human participants. It proved logistically infeasible to run the multi-day experiments required to test complex concepts with teams of more than seven to eight participants. This has limited the size of the organizations that could be studied in experiments, and many questions of interest arise only in larger organizations. In the last years of the project, the A2C2 team explored the idea of "hybrid" experiments in which a few of the participants in the experiment are humans and others are software agents. Conducting such experiments requires adapting traditional modes of investigation. Questions to be addressed include the following: Which roles should the human play? What levels of fidelity and performance are needed (and what is even possible!) in the software agents for the experiment to be useful? How can meaningful hypotheses be formulated? How can scenarios and agent behavior best be formed to create the situations in which those hypotheses can be tested? How are the concepts of variability and control defined in this environment? How can organizational performance data that result from a combination of human and agent behavior be interpreted?

It was an ongoing challenge for A2C2 experimentation that large amounts of detailed data from experiments would be reduced to performance averages in order to make comparisons between teams. For example, an experiment with two teams of 12 people produces an "n" of 2, not 24. Methods are needed to gain power without increasing the number of experiment participants, so that what is lost in replication is gained in the complexity of the analysis that can be performed. Clearly, methods are needed to evaluate "practical significance" divorced from the traditional focus on statistical significance in parametric testing based on means and variances (Hoffman et al. 2010).

7.12 THE ONGOING CHALLENGE OF CROSS-DISCIPLINARY RESEARCH

The A2C2 effort to develop mathematical and computational models that capture the key principles underlying successful networked organizations and the effort to test these models empirically have, from one perspective, been a success. As outlined in this chapter, models were used to develop organizational structures that were demonstrated in the laboratory experiments to outperform the task allocations of traditional military echelons, and to do so in the way that the models predicted (e.g., Carley and Ren 2001; Handley and Levis 2001; Levchuk et al. 2004). These models proved useful during the program for addressing a variety of organizational issues of relevance to the Navy as well as the other services.

Problems resulting from the lack of a common cross-disciplinary (experimental psychology, electrical engineering, operations research) vocabulary and a common conceptual structure plagued the program from its inception, however, and did not abate with time. Cross-disciplinary communication requires time and patience that are sometimes in short supply. And there is no practical way to bring someone outside a discipline up the level of knowledge that has been gained from graduate training and perhaps decades of research experience. Constant pressure to meet a common goal is required to prevent within-discipline subgroups from drifting back to the intellectual territory with which they are most comfortable and familiar, creating a growing rift between the disciplines.

The mapping of meaning from words to equations is an ongoing challenge, especially for team members not fluent in the languages of equations and algorithms. Perhaps as a result of this challenge, the arrow in Figure 7.3 that flows back from experiment results to improved theories and models was not as robust in A2C2 as was originally hoped. Concepts that are so clearly defined in equations may not survive the sometimes imprecise process of developing scenarios that are meaningful to experiment participants. The practical constraints of data collection in experiments with humans may produce data that are not completely relevant to the theoretical constructs captured in the model. Model developers can translate their equations into words for the benefit of experimenters, but the experimenters may interpret those words in a way that is far afield from the original intent.

ACKNOWLEDGMENTS

We would like to thank Chad Weiss and Anita Costello for contributions to this chapter. We would also like to express our deep appreciation for the support and review of the late Willard Vaughan and Gerald Malecki at the ONR. And of course, we are indebted to David Woods for his pioneering work in CSE, which helped us examine this work from a new vantage point.

REFERENCES

Alberts, D.S. (2011). *The Agility Advantage: A Survival Guide for Complex Enterprises and Endeavors*. Washington, DC: Command and Control Research Program.

Alberts, D.S. and Hayes, R.E. (2003). *Power to the Edge: Command and Control in the Information Age*. Washington, DC: Command and Control Research Program.

Ashby, W.R. (1956). *An Introduction to Cybernetics*. London: Methuen and Co.

Berry, J.W., Poortinga, Y.H., and Pandey, J. (1997). *Handbook of Cross-Cultural Psychology, Vol. 1: Theory and Method* (2nd ed.). Boston: Allyn & Bacon.

Carley, K.M. and Ren, Y. (2001). Tradeoffs between performance and adaptability for C3I architectures. In *Proceedings of the 2000 International Symposium on Command and Control Research and Technology*. Washington, DC: International Command and Control Institute.

Diedrich, F.J., Entin, E.E., Hutchins, S.G., Hocevar, S.P., Rubineau, B., and MacMillan, J. (2003). When do organizations need to change (Part I)? Coping with incongruence. *Proceedings of the 2003 Command and Control Research and Technology Symposium*. Washington, DC: International Command and Control Institute.

Ellis, A.P.J., Hollenbeck, J.R., Ilgen, D.R., and Humphrey, S.E. (2003). The asymmetric nature of structural changes in command and control teams: The impact of centralizing and decentralizing on group outcomes. *Proceedings of the 2003 Command and Control Research and Technology Symposium*. Washington, DC: International Command and Control Institute.

Entin, E.E. (1999). Optimized command and control architectures for improved process and performance. *Proceedings of the 1999 Command and Control Research and Technology Symposium*. Washington, DC: International Command and Control Institute.

Entin, E.E. (2000). Performance and process measure relationships in transitioning from a low to high fidelity simulation environment. *Proceedings of the 2000 IEEE Systems, Man, and Cybernetics Conference*, San Diego, CA.

Entin, E.E., Diedrich, F.J., Kleinman, D.L., Kemple, W.G., Hocevar, S.G., Rubineau, B., and Serfaty, D. (2003). When do organizations need to change (Part II)? Incongruence in action. *Proceedings of the 2003 Command and Control Research and Technology Symposium*. Washington, DC: International Command and Control Institute.

Fukuyama, F. (2000). Social and organizational consequences of the information revolution. In R.O. Hundley, R.H. Anderson, T.K. Bikson, J.A. Dewar, J. Green, M. Libicki, and C.R. Neu (Eds.), *The Global Course of the Information Revolution: Political, Economic, and Social Consequences. CF-154-NIC*. Washington, DC: RAND and the National Defense Research Institute.

Handley, H.A. and Levis, A.H. (2001). A model to evaluate the effect of organizational adaptation, *Computational and Mathematical Organization Theory*, 7, 5–44.

Hoffman, R.R., Marx, M., Amin, R., and McDermott, P. (2010). Measurement for evaluating the learnability and resilience of methods of cognitive work. *Theoretical Issues in Ergonomic Science*, 11(6), 561–575. http://www.tandfonline.com/doi/abs/10.1080/14639220903386757.

Hoffman, R.R. and Militello, L.G. (2008). *Perspectives on Cognitive Task Analysis: Historical Origins and Modern Communities of Practice*. Boca Raton, FL: CRC Press/Taylor & Francis.

Hollenbeck, J.R., Ilgen, D.R., Moon, H., Shepard, L., Ellis, A., West, B., and Porter, C. (1999a). Structural contingency theory and individual differences: Examination of external and internal person-team fit. *Proceedings of the 31st SIOP Convention*. Bowling Green, OH: Society for Industrial and Organizational Psychology.

Hollenbeck, J.R., Ilgen, D.R., Sheppard, L., Ellis, A., Monn, H., and West, B. (1999b). Person-team fit: A structural approach. *Proceedings of the 1999 Command and Control Research and Technology Symposium* (11–32). Washington, DC: International Command and Control Institute.

Hollnagel, E., Woods, D.D., and Leveson, L. (Eds.) (2006). *Resilience Engineering: Concepts and Precepts*. Burlington, VT: Ashgate.

Lave, J. (1988). *Cognition in Practice: Mind, Mathematics, and Culture in Everyday Life*. Cambridge, UK: Cambridge University Press.

Levchuk, G.M., Kleinman, D.L., Ruan, S., and Pattipati, K.R. (2003). Congruence of human organizations and mission: Theory versus data. *Proceedings of the 2003 International Command and Control Research and Technology Symposium*. Washington, DC: International Command and Control Institute.

Levchuk, G.M., Yu, F., Levchuk, Y., and Pattipati, K.R. (2004). Networks of decision-making and communicating agents: A new methodology for design and evaluation of organizational strategies and heterarchical structures. *Proceedings of the 9th International Command and Control Research and Technology Symposium*. Washington, DC: International Command and Control Institute.

Lipshitz, R. (2000). There is more to seeing than meets the eyeball: The art and science of observation. *Proceedings of the 5th International Conference on Naturalistic Decision Making*. Stockholm, Sweden.

MacMillan, J., Paley, M.J., Levchuk, Y.N., Entin, E.E., Serfaty, D., and Freeman, J.T. (2002). Designing the best team for the task: Optimal organizational structures for military missions. In M. McNeese, E. Salas, and M. Endsley (Eds.), *New Trends in Cooperative Activities: System Dynamics in Complex Settings*. Santa Monica, CA: Human Factors and Ergonomics Society Press.

Moon, H., Hollenbeck, J., Ilgen, D., West, B., Ellis, A., Humphrey, S., and Porter, A. (2000). Asymmetry in structure movement: Challenges on the road to adaptive organization structures. *Proceedings of the 2000 Command and Control Research and Technology Symposium* (11–32). Washington, DC: International Command and Control Institute.

Pattipati, K.R., Meirina, C., Pete, A., Levchuk, G., and Kleinman, D.L. (2002). Decision networks and command organizations. In A.P. Sage (Ed.), *Systems Engineering and Management for Sustainable Development, Encyclopedia of Life Support Systems*. Oxford, UK: Eolss Publishers.

Perdu, D.M. and Levis, A.H. (1999). Adaptation as a morphing process: A methodology for the design and evaluation of adaptive organizational structures. *Computational and Mathematical Organization Theory*, 4, 5–41.

Potter, S.S., Roth, E.M., Woods, D.D., and Elm, W. (2000). Bootstrapping multiple converging cognitive task analysis techniques for system design. In J.M.C. Schraagen, S.F. Chipman, and V.L. Shalin (Eds.), *Cognitive Task Analysis*. Hillsdale, NJ: Lawrence Erlbaum.

Rasmussen, J. (1986). *Information Processing and Human–Machine Interaction: An Approach to Cognitive Engineering*. New York: North-Holland.

Tapscott, D. and Williams, A.D. (2006). *Wikinomics: How Mass Collaboration Changes Everything*. New York: Portfolio Publications.

Woods, D.D. (1990). Modeling and predicting human error. In J. Elkind, S. Card, J. Hochberg, and B. Huey (Eds.), *Human Performance Models for Computer-Aided Engineering*. New York: Academic Press.

Woods, D.D. (1995). Towards a theoretical base for representation design in the computer medium: Ecological perception and aiding human cognition. In J. Flach, P. Hancock, J. Caird, and K. Vicente (Eds.), *An Ecological Approach to Human Machine Systems I: A Global Perspective*. Hillsdale, NJ: Lawrence Erlbaum.

Woods, D.D. (2006). Essential characteristics of resilience for organizations. In E. Hollnagel, D.D. Woods, and N. Leveson (Eds.), *Resilience Engineering: Concepts and Precepts*. Aldershot, UK: Ashgate.

Woods, D.D. and Hollnagel, E. (2006). *Joint Cognitive Systems: Patterns in Cognitive Systems Engineering*. Boca Raton, FL: CRC Press.

8 A Taxonomy of Emergent Trusting in the Human–Machine Relationship

Robert R. Hoffman

CONTENTS

8.1 INTRODUCTION

The ideas presented here emerged as a consequence of my work with experts. I was first invited to consider the issue of trust in computers some 30-odd years ago, at a time when the then new cognitive systems engineering was focused on such issues as depth versus breadth of pull-down menus, and was using such methods as keystroke analysis (see Hoffman 1989). As my own work became more "applied"—in that I was looking at domains of expertise—I started asking experts whether or not they trusted their computers. Nearly always, the response to this question was a chuckle. That told the tale. Of course, they did not trust their computers. Why would I ask such a stilly thing?

Now, run the clock forward. Trust in automation, trust in the Internet, and trust in cyber systems, and so on, are of great concern in computer science and cognitive systems engineering (e.g., Chancey et al. 2015; Hoff and Bashir 2015; Hoffman et al. 2009; Huynh et al. 2006; Naone 2009; Merritt and Ilgen 2008; Merritt et al. 2013, 2015a; Pop et al. 2015; Shadbolt 2002; Wickens et al. 2015; Woods and Hollnagel 2006). Trust is of particular concern as more so-called autonomous systems are being developed and tested (i.e., self-driving automobiles, intelligent robots) (Schaefer et al. 2016).

Most salient to me is the fact that the vast majority of the human factors research on trust in automation involves college students as participants, task situations that are new to the participants (e.g., process control or baggage screening tasks), tasks that are simplified and sometimes superficial (e.g., game-like emulations), one or perhaps two measures of performance (e.g., accuracy, latency), procedures that can be conducted in the academic laboratory, lasting about the duration of a college class (e.g., 5 minutes of practice followed by about 50 minutes of trials), and, finally, few attempts to reveal participants' reasoning. This is by no means surprising given the nature of the academic research "paradigm of convenience," and we do need research of this kind. But in light of what we know about the effects of extended practice and expertise, and the characteristic complexities of "real world" human–machine work systems, it is astounding that the published reports almost always reach conclusions about "people." It is taken for granted that the findings are widely generalizable and the results carry over to the design of technologies intended for use in the "real world."

In support of such generalizations, it has been claimed that automation bias is characteristic of experts. Among others, Wickens et al. (2015) made this claim, citing a study by Skitka et al. (2000). These latter researchers reported that a process for making individuals socially accountable for their errors served to reduce automation bias. But wait… the participants were college students. Any finding concerning the reduction of automation bias, to any degree, does not justify any claim that experts, generally, are biased or that they are just as biased as college freshmen, or that the findings from the academic laboratory carry over to the "real world" without any qualification.

This chapter reflects on this trust *Zeitgeist*, but this is not a literature review. As O'Hara (2004) noted, any comprehensive account of the concept of trust would have to "plunder many sources; the philosophy of Socrates and Aristotle, Hobbes and Kant; the sociology of Durkheim, Weber, and Putnam; literature; economics; scientific methodology; the most ancient of history and the most current of current affairs" (p. 5). No book chapter could possibly digest all that. Instead, I ask two questions that are constitutional of the scientific method: What is the subject matter for theories and models of trust in technology? What do we need theories of models of trust in machines to do for us? My purpose is to call out the assumptions underlying the

Zeitgeist in the study of human–machine trust. I assert that those assumptions have led to a reductive understanding of the complexities that are involved, although individual researchers are indubitably aware of the complexities.

In the tradition of naturalistic taxonomics, this chapter surveys the various factors, dimensions, and varieties of trust that have been proposed; exposes the metaphors and assumptions of the models; and attempts to make sense of the complications and complexities, both philosophical and pragmatic. Each of the first eight short sections of this chapter presents a challenge to the *Zeitgeist*, posed as rhetorical questions. These culminate in a view of human trust in machines that regards trusting as a process, that emphasizes dynamics over states, and that distinguishes a considerable variety of trusting relationships. Fundamentally, it sees those relationships as potentially simultaneous. The chapter concludes with considerations of applications and measurement.

8.2 IS TRUST IN MACHINES TO BE UNDERSTOOD IN TERMS OF INTERPERSONAL TRUST?

The vast literature on trust primarily concerns interpersonal trust. There are clear differences between interpersonal trust and trust in automation in terms of the influencing factors and the ways in which trusting develops (see Lee and See 2004). For example, Kucala's (2013) discussion of trust from the perspective of business management lists the qualities of leaders that foster trust in employees: honesty, respect, integrity, humility, justice, honor, and courage. None of these is much considered in the literature on human–machine trusting (but see Johnson et al. 2014), because it is hard to see how any of them could be qualities describing the human–machine relation. Other factors that Kucala lists, such as timeliness of decisions, understandability of communications, and interdependence, do seem to be applicable to trust in automation.

Some research in cognitive systems engineering has been premised on the idea that concepts or dimensions of interpersonal trust carry over for the analysis of human–computer trusting (e.g., Barber 1983; Muir 1987, 1994), or human–robot teaming (Atkinson et al. 2014). Consistent with this premise, research has demonstrated that certain interpersonal trust factors can apply to trust in automation. This includes reliability, understandability, and predictability (Lee and Moray 1992; Merritt and Ilgen, 2008; Muir and Moray 1996; Parasuraman and Riley 1997; Pritchett and Bisantz 2002; Seong and Bisantz 2002, 2008; Sheridan 1980). Individual personality traits and affective state can affect trust and delegation of authority to autonomous agents (Cramer et al. 2008, 2010; Stokes et al. 2010). The tendency of people to anthropomorphize computers certainly affects human–computer interaction (e.g., Lewandowsky et al. 2000; Nass and Moon 2000). Many people readily attribute mental states to computers (Parlangeli et al. 2012). Android technologies and social robotics also introduce a bidirectionality to trusting, where the machine might be viewed as partner rather than as a tool.

Unlike interpersonal trust, which is recoverable after a failure, when a mistake is made by a machine, people can lose confidence in its predictability and reliability (Beck et al. 2002). While apology and forgiveness are common enough in human interactions, what would be the role of apology and forgiveness in human–machine relationships? To date, this has only been explored tentatively, in the context of online transaction systems (Vasalou et al. 2008). Even more extreme is the possibility of reciprocation—of how, whether, and why a machine might "trust" a human. This remains largely the province of science fiction (but see Crispen and Hoffman 2016; Hancock et al. 2011; Johnson et al. 2011). Also outside of consideration because it lies in the realm of science fiction is a form of negative trust that might be called Malevolence: The human is highly confident that the machine will do bad things.

It is fairly clear that trusting in the human–automation interaction differs significantly from that in human–robot interaction (Schaefer et al. 2016), related to the fact that many robots are anthropomorphic (see Wagner 2008). Future research and meta-analyses need to separate studies in which the technology is a "box" from studies in which the technology is an emulated human or an anthropomorphic robot. Despite the convenience of anthropomorphism, trust in computers involves factors that relate strongly to the limitations and foibles of computational technology and the design intent that underlies the algorithms (Corritore et al. 2003; Desai et al. 2009; Oleson et al. 2011). Thus, I regard trusting as a unique and directional relation between humans and machines.

8.3   TRUST IN AUTOMATION IS COMPLEX, SO WHY ARE OUR CONCEPTUAL MODELS SO SIMPLE?

One class of models of trust in automation is those that are essentially listings of variables or factors that have or are believed to have a direct causal influence on trust (e.g., Muir 1987): cultural differences, operator predispositions, operator personality, knowledge about the automation, and so on. The goal of such models is to capture all of the variables, or at least the most important variables, that might have causal influence in how humans come to trust in and thereby rely upon computational technologies (e.g., Oleson et al. 2011; Rempel et al. 1985). Literature reviews have revealed a number of context factors that influence trust, such as the type of technology, the complexity of the task, perceived risks, and so on (Schaefer et al. 2016). Thus, for example, under high-risk conditions, some people may reduce their reliance on complex technology but increase their reliance on simple technology (Hoff and Bashir 2015). There are individual differences in beliefs about automation reliability and trustworthiness (Merritt et al. 2015b; Pop et al. 2015). Some individuals have an "all-or-none" belief, that automation either performs perfectly, or that it always makes errors (Wickens et al. 2015). In their model of trust in automation, Hoff and Bashir list 29 variables and illustrate the many interactions that have been reported in the literature. In their model of trust in automation (and robots), Schaefer et al. (2016) list 31 factors.

The listed factors that these researchers adduce are said to influence the development of trust. Furthermore, all the variables are said to interact, which is indubitably the only possible correct conclusion for such theoretics and meta-analytics. To give just three examples, all just from the one 2016 meeting of the Human Factors and Ergonomics Society: a valid recommendation from a computer is less appreciated if the operator is capable of performing the task on their own (Yang et al. 2016); operator fatigue interacts with the reliability of the technology in influencing actual reliance (Wohleber et al. 2016); and cultural differences on such factors as individualism and power relations manifest as differing tendencies in the trust of automation (Chien et al. 2016).

An attempt to exhaustively list the variables that have a causal influence on trust is important, especially with regard to the difference between the analysis of legacy contexts (one operator–one machine) and the analysis of emerging contexts in which human–machine teaming, collaboration, and interdependence are crucial (see Choiu and Lee 2015).

A second class of models of trust are process models. This includes mathematical instantiations (using such approaches as linear modeling) that are designed to predict or estimate values or levels of trust, or point-like values of automation-dependent judgments (cf. Seong and Bisantz 2002).This class also includes conceptual models that depict a process by which trust results from the causal influence of mediating variables, and in turn leads to action (i.e., reliance).

Exemplifying this second class of models is the seminal and highly influential conceptual model of Lee and See (2004). It has the characteristics of both a causal (mediating variables) model and a process model; that is, it combines a causal diagram with a list. The model has this primary causal chain:

1. Information →
2. Operator's belief →
3. Operator's trust →
4. Operator's intention →
5. Operator's action

and a secondary causal chain loops back on the primary one:

1. Operator's action →
2. Automation's action →
3. Display of resultants →
4. Information

But in addition, leading into some of the nodes (1 through 8) is a list of factors that are held to have a causal influence on trust. This renders the Lee and See model as somewhat oppositional: It reflects both the tendency of theorists to reduce complex cognitive processes to simple linear chains, on the one hand, and the inclination to adduce long lists of causal variables and their interactions, on the other hand.

8.4   CAN TRUST BE "DESIGNED-IN?"

The mathematical models referenced above are designed to generate "trust metrics" that might indeed inform the design of intelligent decision aids. They might also be used as performance thresholds in analyses of the efficiency or quality of technology-supported work. They might even roughly scale the understandability of algorithms.

The concept of trust as a single state and the rostering of variables that affect trust have culminated in ideas about how to ensure trustworthiness and hence lead to appropriate reliance (Riley 1996). For example, based on their literature review, Hoff and Bashir (2015) present a number of general recommendations for the creation of trustworthy automation. Examples are "provide accurate, ongoing feedback," use "simplified interfaces," "increase the anthropomorphism of the automation," "consider the chin shape of embodied computer agents," and "increase the transparency of the automation." These and other recommendations reduce two basic things: feedback and observability/understandability. While these are good targets for design, the underlying analyses are incomplete. The stance is that trustworthiness can be built in, and if it is built in, appropriate "calibration" will be the result. In other words, trust is a state to be instilled.

8.5 IS TRUST A STATE?

Researchers of trust in automation generally acknowledge that trust is dynamic; that it develops. But this is usually meant in the sense that trust builds or increases over time. Dynamics are certainly inherent in the emergence of interpersonal trust (e.g., Lewicki et al. 1998), and the adoption of interpersonal trust as a reference frame (see Section 8.2 above) contributes to a particular assumption about the dynamics of trust. Threshold effects and contingent information availability illustrate the dynamics of trust. The level of trust as specified in some momentary judgment is reflected in reliance and the tendency to maintain a certain level of reliance even as the level of trust changes. This can result in a dichotomous pattern of reliance. In a closed loop, the reliance, in turn, can affect the information an operator has regarding the performance of the automation, because the performance of the automation is only perceivable when the person is relying on the automation (Gao and Lee 2006).

Despite the acknowledgement of dynamics, models of trust in automation mostly regard trust as a state. Dynamics are involved only in the achievement of, or progress to, that state. The purpose of models is to predict the values of trust, and from that anticipate reliance, disuse, and so on.

I assert that neither trusting (as a relation), nor trustworthiness (as an attribution), nor reliance (as a decision or an activity) is a state. Trust judgments certainly are episodic, but trust is typically discussed as if it were a stable, semi-persistent, or even a final state, almost always on the assumption of some single fixed task or goal (e.g., Khasawneh et al. 2003; Muir 1987, 1994).

8.6 DOES TRUST GET "CALIBRATED?"

A factor contributing to the apparent "single state fixation" in the research literature is the utilization of the concept of trust calibration. This term became prominent in the literature when Lee and See (2004) used the word to denote that state, as they said, in which the human has an accurate knowledge of the capabilities of the technology. There is an almost painful irony here, that of turning this semantic concept (i.e., the user develops a mental model) into a metrical concept—this metaphor treats the human as a form of automation, specifically a mechanical measuring device rather like an altimeter. The following are misleading assumptions of the metaphor:

1. Trust is measured on a single scale.
2. There is a point on that scale that serves as a metric; that is, trust is sufficient or falls at the "right" level.
3. When trust reaches that metric point, we want it to stay there. We want the trust to not change once it calibrates.

We know that continued practice with automation that provides the operator with dynamic information concerning the system status can lead to appropriate reliance. In a sense, but only in a sense, this is calibration. McGuirl and Sarter (2006) had pilot instructors fly in a simulator that ran scenarios involving icing. A decision support system provided the participants with dynamic information concerning changes in the aircraft performance and stability that could be attributed to icing.

> [T]his trend information would enable pilots to determine more accurately when they could trust the system and apply the ice monitoring and diagnosis task to the automation versus when they should become more involved and get ready to perform the task on their own … [T]his was hypothesized to result in fewer stalls and improved stall recovery. (p. 657)

The important aspect of this work is not the use of the word "calibration." One could effectively swap the phrase "trusting relation" for the phrase "trust calibration" with absolutely no loss of content. The important aspect of this work is the view that trust is a dynamic, since the world that is being operated upon by the worker is itself dynamic. Reliance does not come to converge on some stable state or level. *The work converges on an appropriate procedure for actively managing reliance.*

> Pilots who received information about changes in the system confidence reported that they adopted one of two strategies for deciding how to respond to the decision aid's recommendations. In the first case, once the pilot noticed that the system confidence became unacceptably low, he or she would invariably take the action opposite or contrary to that which the decision support system was recommending… The second strategy could be referred to as "hedging": In these cases, at the first indication of icing, the pilot would place the power setting at the boundary for wing and tail recover and not commit… until further evidence was available. (p. 664)

This dual-strategy phenomenon that McGuirl and Sarter (2006) observed leads to a next challenge question.

8.7   IS TRUST A SINGLE STATE?

The dual-strategy phenomenon shows that workers can hold multiple distinct trusting relations to their machines. The pilot's strategies involved trusting that was contingent or context-dependent and trusting that was skeptical. In domains such as information technology security and weather forecasting, workers can depend on a dozens of software suites, each offering multiple functionalities, some suites running on the same computer, some running on different computers or servers, some using the same operating system, and some using different operating systems. In my own relation to my word processing software, I am positive that it will perform well in the crafting of simple documents, but I am simultaneously confident it will crash when the document gets long, or when it has multiple high-resolution images. And every time that there is a software upgrade, the trusting of many of the functions becomes tentative and skeptical.

Trust is not a single state. This leads to what might appear to be an even more radical suggestion.

8.8 DOES TRUST DEVELOP?

Studies that nod to the dynamics of trust generally refer to the "development" of trust (e.g., Schaefer et al. 2016). While it is indubitably true that trusting often *changes* over time, it should not be assumed trusting *always* develops in the sense of matu-rational convergence on some single state, level, or stable point. In some cases, it can appear as if trust is developing, but this is perhaps the exception and should not be elevated to the prototype.

For example, trust and mistrust often develop swiftly (Meyerson et al. 1966). Some people show a bias or disposition to believe that automation is more capable and reliable than it actually is, and such high expectations result in swift mistrust when the automation makes an error (Merritt et al. 2015b; Pop et al. 2015; Wickens et al. 2015). Swift trust is when a trustor immediately trusts a trustee on the basis of authority, confession, profession, or even exigency. Naive belief in the infallibility of computers is also an instance of swift (and perhaps unjustified) trust. Swift trust can be prominent early in a relationship, with contingent trust emerging over time as people experience automation in different circumstances. In other words, trust does not "develop," it morphs.

8.9 IS "TRUST" JUST A CONVENIENT WORD?

A number of what are believed to be different kinds of trust have been noted in the pertinent literature. Meyerson et al. (1966) popularized the notion of "swift trust," originally referencing the initial trust relation that is assumed when teams are formed within organizations.

Bobko et al. (2014) discussed the state of "suspicion," in which there emerges a glimmering perception or feeling of mistrust (or malicious intent), followed by rea-soning to apprehend what is going on, and the attempt to explain perceived dis-crepancies. The construct of suspicion adds another flavor to the dynamics of trusting. It links to the notion of the understandability of automation (Muir 1987; Woods et al. 1990) and also the notion that workers are engaged in the attempt to make sense of their technology at the same time that they are using the technology to make sense of the world that they are attempting to control or observe. A key point that Bobko et al.

make is that suspicion is *simultaneously* more than one cognitive state: a feeling of uncertainty and a worry about malicious intent.

Merritt et al. (2013) demonstrated a way to study what might be called "default trust," in an experiment in which college students were engaged in a simple baggage screening task. In addition to an explicit self-rating measure of propensity to trust automation, a test of implicit associations measured reaction time in a "good–bad" judgment for concept terms related to automation (among other control categories). The researchers found that implicit attitude toward automation influenced the propensity to trust in the automation, and when the automation made errors (i.e., in the categorization of images showing weapons), implicit trust and propensity to trust were additive. Yet, implicit attitude and propensity as measured *a priori* were not correlated. If they encounter a reason to distrust the automation, or have less trust in the automation, they are more likely to continue trusting it if they have both implicit trust and an explicit propensity to trust it.

In other words, some people enter into the task just trusting the automation—default trust. It is safe to assert that much of the time, people do not pause to deliberately think about whether they trust their technology. Without compelling evidence to the contrary, and as a consequence of inertia of many kinds, the tendency is to just continue business as usual. People have to get on with their work. For example, you might not think about whether or not an external drive will automatically back up your machine overnight. It has always done that. On occasion, you think about it and worry that you might lose all your stuff. But generally, you just move on or shrug it off.

Default trust is different, albeit subtly so, from automation bias. Automation bias is when an individual enters into a dependency on a machine with the belief that the machine can (or cannot) be trusted precisely because it is a computer. Default trust is when an individual enters into a dependency on a machine with the expectation that the machine will do what it is intended to do, and no explicit ruminations on whether the trustworthiness of the machine hinges on the fact that the machine is a computer.

8.10 INTERMEDIATE CONCLUSIONS

The seven rhetorical questions posed above serve to capture the *Zeitgeist*:

1. The view is held that trust in machines is like interpersonal trust, despite our knowledge that there are crucial differences.
2. The view is held that our conceptual models need to be simple, despite our knowledge that there are many causal variables, all of which interact.
3. The view is held that trust can be designed-in to the machines, despite our knowledge that people enter the human–machine relation with attitudes, and despite our knowledge that appropriate trusting and relying take time to emerge.
4. The view is held that trust is a state, despite our knowledge that trusting and relying on machines is essentially dynamic.

5. The view is held that trust is a variable that gets calibrated to some desirable metric, despite our knowledge that the human is not a mechanical measuring instrument.

6. The view is held that trust is a single state, despite its manifestation as multiple, simultaneous, and different forms.

7. The view is held that trust develops in that it builds and converges or matures, despite our knowledge that this is not universally the case, and indeed, trust can come and go.

8. The view is held that trust is some single human–machine relation, despite our knowledge that there are distinguishable varieties of trust.

Trust is like a knitted sweater: Once you start pulling on a thread, it continues to unravel. The most widely cited definition of trust is that of Mayer et al. (1995, p. 712). If we were to adapt their definition of interpersonal trust to the topic of trust in automation, we would say that trust is the willingness of a human to be vulnerable to the actions of a machine based on the expectation that the machine will perform a particular action that is important to the human. But this assumes that trust is a single thing, and that it is a state. Trust is a convenient word provided to us by our language, and used as such hides a number of distinguishable, simultaneous, and dynamic processes of the human–machine relationship. I now present the taxonomics that escape this "convenient word" trap, by calling out the complexities, unafraid of them.

8.11 TAXONOMICS

Trust in automation has been described as an attitude (positive or negative valuation of the machine by the human), as an attribution (that the machine possesses a quality called trustworthiness), as an expectation (about the machine's future behavior), as a belief (faith in the machine, its benevolence, and directability), as an intention (of the human to act in a certain way with respect to the machine), as a trait (some people are trusting, perhaps too trusting in machines), and as an emotional state (related to affective factors such as liking or familiarity). But these are not exclusive—a trusting relation can be, and usually is, some mixture of these, all at once.

Given the manifest dynamics and complexities of trust, I refer to "trusting" as a gerund rather than "trust" as a noun. I conceive of trusting not as a state, as a single mediating variable, or even as a goal, but as a continual process of active exploration and evaluation of trustworthiness and reliability, within the envelope of the ever-changing work and the ever-changing work system (Fitzhugh et al. 2011).

Simultaneously, I regard trusting as a number of distinguishable categories of agent–agent relationships and interdependencies, which can be seen as an admixture of both types and dimensions, in that some of the types can be distinguished on some of the dimensions. An example dimension is depth; that is, trust may be shallow (swiftly established for the immediate purpose of accomplishing a task at hand) or deep (based on substantial evidence acquired over time with experience). Rempel et al. (1985) identified three dimensions of trust that influence people's acceptance of information provided by an external source: Predictability, Dependability, and Faith. My approach to getting the taxonomics of trust off the ground is to develop categories, understanding that these can simultaneously be thought of as partaking in the nature of dimensions. For instance, trust that is colored by a feeling of skepticism or suspicion is but a fuzzy boundary away from mistrust (see Bobko et al. 2014).

I begin with the major distinctions of the taxonomy.

8.11.1   REFERENCE

Trusting can be *about* a number of things, which might at least for convenience be placed into broad categories of Information (sensor data, inferences and information, knowledge claims, etc.), Resources (capabilities, valuables, etc.), and Actions (intentions, activities, processes, decisions, directives, etc.).

8.11.2   EVIDENCE, BELIEF, AND REASON

The "why" of trusting is that it can derive its justification or provenance from a variety of sources, such as experience, evidence, belief, or authority.

8.11.3   CONTEXT

Trusting can be context or situation independent or dependent. Trusting can be assured for only certain circumstances, or it can be assured for most circumstances. Trusting can be relatively stable across time or it can be contingent as circumstances dictate.

Tables 8.1 and 8.2 provide some specification of these possible contextual dependencies and the categories they entail.

Some of the distinctions in Tables 8.1 and 8.2 may be regarded as mere convenience—distinctions that complexity and thereby just muddy the water. While some of the distinctions are subtle, subtlety does not negate their empirical validity. Take for example the subtle difference between justified trusting and faith-based trusting. On the former, the human might be open or even sensitive to empirical evidence that contradicts their reasons for trusting the machine, whereas in the latter, the human might not be open, and if any such evidence were manifest, it might be dismissed. This could be seen as confirmation bias, that is, the tendency to not seek evidence that contradicts one's beliefs or hypotheses. Skeptical Trusting seems very similar to Tentative Trusting. But skepticism strongly entails that the human will be particularly sensitive to or even actively seek information that undermines the trusting relation. Over-trusting, or trusting the machine precisely because it is a machine (focusing on the *Why*), could be considered a form of Absolute Trusting, which is focused on the *When*.

But calling out these subtle differences shows value because the distinctions can be used in combinations of the *Whys* and *Whens*. For example, trusting that is Absolute, Justified, and Stable would of course be the ideal: *The human can almost always be certain that what the machine does in this particular context is good because the human has empirical evidence for this, and because what the machine has done is*

TABLE 8.1
Trusting Contextual Dependencies: Trusting Is *When*

Trusting Is *When*: Absolute Trusting

Information | The human takes the machine's assertions (data, claims) as valid and true in all circumstances.

Resources | The human can allot resources to the machine with directives as to their disposition, and can be certain that the resources will be utilized as directed in all circumstances.

Actions | The human can give the machine directives (plans or goals) and be certain that they will be carried out in all circumstances.

Trusting Is *When*: Contingent Trusting

Information | The human can take some of the machine's presentations or assertions as valid and true under certain circumstances.

Resources | The human can allot resources to the machine with directives as to their disposition under certain anticipated circumstances, and be certain that the machine will attempt to utilize the resources as directed.

Actions | The human can give the machine directives under certain anticipated circumstances, and be certain that the machine will attempt to carry them out.

Trusting Is *When*: Progressive Trusting

Information | Over time and across experiences, the human takes more of the machine's presentations or assertions as valid and true.

Resources | Over time and across experiences, the human allots more resources to the machine with directives as to their disposition and be certain that the machine will attempt to utilize the resources as directed.

Actions | Over time and across experiences, the human gives the machine more directives and is certain that the machine will attempt to carry them out.

Trusting Is *When*: Stable Trusting

Information | The human can take some of the machine's presentations or assertions as true most of the time.

Resources | The human can give the machine resources with directives as to their disposition and be certain most of the time that the resources will be utilized as directed.

Actions | The human can give the machine directives and be certain most of the time that they will be carried out.

Trusting Is *When*: Tentative Trusting

Information | For the time being, the human can take some of the machine's presentations or assertions as valid and true.

Resources | The human can give the machine resources with directives as to their disposition and for the time being be certain that the machine will utilize the resources as directed.

Actions | The human can give the machine directives and for the time being be certain the machine will carry them out.

Trusting Is *When*: Skeptical Trusting

Information | The human takes some of the machine's presentations or assertions as valid and true but is particularly sensitive to evidence that they might not be.

(Continued)

TABLE 8.1 (CONTINUED)
Trusting Contextual Dependencies: Trusting Is *When*

Resources	The human can give the machine resources with directives as to their disposition but is particularly sensitive to evidence that they might not be utilized as directed.
Actions	The human can give the machine directives but is particularly sensitive to evidence that they might not be carried out.

Trusting Is *When*: Digressive Trusting

Information	Over time and across experiences, the human takes fewer of the machine's presentations or assertions as valid and true.
Resources	Over time and across experiences, the human allots fewer resources to the machine with directives as to their disposition and become less certain that the machine will attempt to utilize the resources as directed.
Actions	Over time and across experiences, the human gives the machine fewer directives and becomes less certain that the machine will attempt to carry them out.

Trusting Is *When*: Swift Trusting

Information	Evidence or assertions of the authority or trustworthiness of the machine are taken by the human immediately as reason to take the machine's presentations or assertions as valid and true.
Resources	Evidence or assertions of the authority or trustworthiness of the machine are taken immediately by the human as reason to allot resources to the machine with directives as to their disposition, and be certain that the machine will attempt to utilize the resources as directed.
Actions	Evidence or assertions of the authority or trustworthiness of the machine are immediately taken by the human as reason to give the machine directives, and be certain that the machine will attempt to carry them out.

Trusting Is *When*: Default Trusting

Information	The human does not deliberate on whether the machine's assertions (data, claims) as valid and true, but acts as if they are.
Resources	The human allots resources to the machine with directives as to their disposition, and does not deliberate on whether the resources might not be utilized as directed.
Actions	The human gives the machine directives and does not deliberate on whether they might not be carried out.

nearly always good in this context. However it is justified, such trust is nevertheless an abductive inference. Since it is based on the understandability and perceived predictability of the machine, it is a defeasible inference (i.e., it is potentially fallible).

As another example, Stable Justified Trusting is when the human can take the machine's presentations or assertions as true most of the time, or over some time span, and the human can give the machine directives and be certain most of the time that they will be carried out. This can be taken as a refined definition of what is meant by the notion of trust calibration (McGuirl and Sarter 2006; Parasuraman and Riley 1997) but avoids an implication that the human is a measuring instrument (a machine).

As a third example, when uncertainty about the machine becomes salient (as in an unfamiliar context), then trusting becomes Contingent, Faith-based, and Tentative— a situation requiring close attention, but perhaps is actually the norm: *The human will*

TABLE 8.2

Trusting Contextual Dependencies: Trusting Is *Why*

Trusting Is *Why*: Justified Trusting

Information	The human has empirically derived reasons to believe that the machine's presentations or assertions are valid and true.
Resources	The human has empirically derived reasons to believe that the machine will utilize resources as directed.
Actions	The human has empirically derived reasons to believe that the machine will carry out its directives.

Trusting Is *Why*: Faith-Based Trusting

Information	The human has the belief that the machine's presentations or assertions are valid and true.
Resources	The human has the belief that the machine will utilize resources as directed.
Actions	The human has the belief that the machine will carry out its directives.

Trusting Is *Why*: Authority-Based Trusting

Information	The human has the belief, based on assertions from some authority, that the machine's presentations or assertions are valid and true.
Resources	The human has the belief, based on assertions from some authority, that the machine will utilize resources as directed.
Actions	The human has the belief, based on assertions from some authority, that the machine will carry out its directives.

Trusting Is *Why*: Over-Trusting (Automation Bias)

Information	The human takes the machine's assertions (data, claims) as valid and true because they come from a machine, or from overgeneralization based on limited past experience or assumptions about the machine's creators.
Resources	The human allots resources to the machine with directives as to their disposition, and feels certain that the resources will be utilized as directed because the utilization will be by a machine.
Actions	The human gives the machine directives and is certain that they will be carried out because it is a machine.

trust only some of the machine's presentations or assertions, and just for now, unless circumstances change or the human acquires additional pertinent evidence.

All I have considered so far is what we might call "positive trusting." There are also varieties of negative trusting (Hoffman et al. 2013; Merritt and Ilgen 2008). These are just as important as positive trust in human–machine work systems. And again, these are adding to the complexity, which (again) we should not shy away from.

8.11.4 Varieties of Negative Trusting

Were we to impose a continuum, weak positive trusting—trusting that is Contingent, Skeptical, and Tentative—is still a form of trusting. It is not a form of mistrusting. Eventually, however, skeptical trust (*I'll trust you, but I'm not so sure, and I'm on the lookout*) can and does give way to mistrust (*Sorry, but I just don't trust you anymore*). Varieties of Negative Trusting are defined in Table 8.3.

TABLE 8.3

Some Varieties of Negative Trusting

Mistrusting	The belief that the machine *might* do things that are *not* in the human's interest.
Distrusting	The belief that the machine *may or may not* do things that *are* in the human's interest.
Anti-trusting	The belief that the machine *will* do things that are *not* in the human's interest.
Counter-trusting	The human believes that the machine must not be relied upon because the machine is presenting information that suggests it should be trusted.

While mistrust might be thought of as falling at one end of a continuum, negative varieties of trusting seem to be associated with phenomena that are qualitatively different from those associated with positive varieties of trusting, and are not just the converse of phenomena associated with positive trust. This breakdown in parity, captured in the italicizations in the Table 8.3 definitions, motivates the consideration of the positive and negative varieties as independent rather than as poles on dimensions.

Counter-trusting can occur under deception and is a theme to science fiction lore in human–robot interaction (see Hancock et al. 2011). Counter-trusting is the judgment that one should consider doing something contrary to, perhaps even precisely the opposite of, what the machine suggests one should do. The machine must *not* be relied upon especially because it is presenting information that suggests it *should* be trusted. Counter-trusting would, presumably, accompany Reliable Absolute Mistrusting and Reliable Absolute Distrusting.

Getting more down-to-earth, studies of how people deal with the user-hostile aspects of software (Koopman and Hoffman 2003) reveal a variety of reasons why people have to create work-arounds and kludges, and why people are frustrated by their computers, even to the point of committing automation abuse (Hoffman et al. 2008). A human might feel positive trust toward a machine with respect to certain tasks and goals and simultaneously feel mistrusting or distrusting when other tasks and goals are engaged. Indeed, in complex sociotechnical systems, this is undoubtedly the norm. The machine might do some things that are not in my interest but it may do some things that are. This would be Contingent, Tentative Mistrusting. In fact, this is likely the commonest case. For example, software routinely triggers automation surprises (Hoffman et al. 2014; Sarter et al. 1997), which leave the human not knowing whether, what, when, or why to mistrust.

I began this chapter with a reference to how domain experts feel about their computers. In a study of expert weather forecasters, they were asked whether they trusted what their technology was showing them (Hoffman et al. 2017). Their responses showed that they are, uniformly, cautious and sometimes very skeptical. The expert forecasters would routinely seek converging evidence for their prognoses, including a deliberate search for potentially disconfirming evidence. And unlike apprentices, the experts would be skeptical of what the computer models showed, relying on their understanding of how the models worked and their knowledge of how and when the computer models are biased. Here is an example excerpt from an interview.

Question: *Can you remember a situation where you did not feel you could trust or believe certain data, such as a computer model or some other product—a situation where the guidance gave a different answer than the one you came up with?*

Response: *I'm always nervous in my first 3–5 years at a location. You have to build up your seasonal base. In the United Kingdom, you have to apprentice for five years in each area. In the U.S. we do it more quickly. As for the data in the computers, if it initializes wrong you know it will be off in all of its forecasts. You develop your own favorites anyhow. [Different models] have strong and weak points. Some handle different levels [of the atmosphere] better. In the short range, they may all be about the same, but after 48 to 72 hours or so you see differences.*

In the taxonomy presented here, this is Stable Contingent Mistrusting, a judgment of confidence that the machines can be trusted to present bad or misleading information under certain known and recurring circumstances.

The Contextual Dependencies outlined in Tables 8.1 through 8.3 apply to the varieties of Negative Trusting. Two examples are as follows:

Absolute, Justified, Stable Mistrusting: *The human can always be certain that what the machine presents or asserts is wrong because the human has evidence and because what the machine says is always wrong.*

Contingent, Unjustified, Tentative Mistrusting: *The human will inappropriately mistrust some of the presentations of the machine, for now, unless circumstances change or the human acquires additional evidence.*

Of course, there can be and often there is Swift Mistrusting and Swift Distrusting. Trust in automation can rapidly break down under conditions of time pressure, or when there are conspicuous system faults or errors, or when there is a high false alarm rate (Dzindolet et al. 2003; Madhavan and Wiegmann 2007). As noted earlier, the trusting of machines can be hard to reestablish once lost. There may be an asymmetry: Swift interpersonal trusting can be generated as a result of a confession, which is an assertion (1) that is immediately believable (perhaps but not necessarily on the basis of authority); (2) that leaves the trustee vulnerable by admission of some flaw, fault, weakness, or limitation; and (3) that conveys a shared intent with regard to the referential variables. The question of whether a machine could instill Swift Trusting to mitigate Mistrust or Swift Mistrust—and the general question of whether and how a machine might be benevolent (or malevolent?)—is a topic of ongoing research (Johnson et al. 2011, 2014).

8.12 INTERMEDIATE CONCLUSIONS

- The human's stance toward the machine is *always* some mixture of justified and unjustified trusting and justified and unjustified mistrusting.

- In macrocognitive work, *multiple trusting relations exist simultaneously*, with the human trusting the machines to some extent, for some tasks but not others, in some circumstances but not others, and for various kinds of reasons.

These two assertions are the inevitable answer to the eight rhetorical questions that were used to compose the first sections of this chapter. Indeed, they might be regarded as principles of macrocognition (see Hoffman and Woods 2011).

8.13   RELIANCE

Each of the line entries in Tables 8.1 through 8.3 entails the kinds of actions that the human might or might not take in terms of reliance on the machine in the conduct of the work. In addition, each of the line entries entails the kinds of activities in which the human might engage to further evaluate and continually reevaluate the machine's performance and trustworthiness. A clear example of this is Skeptical Trusting. The varieties of positive and negative trusting can be associated with different reliance stances. For instance, Progressive Trusting about actions could be understood as entailing reliance. Additional examples are presented in Table 8.4. This does not take into account the fact that there are contingencies between reliance and the information received subsequent to the reliance that guides the further morphing of trust (Gao and Lee 2006). The purpose here is merely to show how the taxonomics can be applied to the evaluation of the nature of the reliance.

8.14   MEASUREMENT

In sociotechnical work, there are always multiple goals, but to approach the question of trust measurement, we can begin by considering the work with respect to some "principal task goal." All goals are necessarily tied to circumstance or context, so we can ask about trusting relationships with respect to a principal task's primary goal given some context. The more common situations are likely to be as follows:

1. The human has a Contingent, Justified, and Stable Trusting relation with the machine; that is, the human trusts the machine with respect to certain tasks or goals in certain contexts or problem situations.
2. The human has a Contingent, Justified, and Tentative Mistrusting relation with the machine; that is, the human mistrusts the machine with respect to certain tasks or goals in certain contexts or problem situations, but the mistrust might be repaired or mitigated by additional evidence.

In an ideal situation, the machine should be relied upon, the machine is relied upon, and the reliance leads to good outcomes. This would be Justified Positive Trusting. Table 8.5 presents the remainder of this straight forward combinatoric.

In research on trust in automation, the researchers or external observers typically know or can determine the correctness and value of the machine's presentations (e.g., information displays) and the correctness or goodness of the machine's actions.

TABLE 8.4

Reliance States Associated with Some of the Varieties of Positive and Negative Trusting

Trust Variety	Belief #1: What the Machine Does	Belief #2: What the Machine Does Not Do	Reliance Stance
Positive trusting (in general)	The machine does things that are in the human's interest.	The machine does not do things that are not in the human's interest.	Confidence that the machine will do good things.
Over-trusting	The machine always does things that are in the human's interest.	The machine will never do things that are not in the human's interest.	Confident reliance.
Benevolence	The machine may do things that are not in the machine's interest.	The machine will not do things that are not in the human's interest.	Confidence that the machine will do good things even if that means self-sacrifice.
Distrusting	The machine does things that might not be in the human's interest.	The machine might not do things that are in the human's interest.	Concern and vigilance; lack of confidence that the machine will do good things; suspicion the machine might do bad things.
Mistrusting	The machine will do things that are not in the human's interest.	The machine might not do things that are in the human's interest.	Confidence that the machine will do bad things; fear, skepticism, vigilance.
Anti-trusting	The machine does things that are not in the human's interest.	The machine will not do things that are in the human's interest.	Fear, skepticism, avoidance.

TABLE 8.5

Considering Trusting and Mistrusting with Respect to the Justification of Reliance

	Trusting	Mistrusting
Justified	The machine should be relied upon, the machine is relied upon, and reliance leads to good outcomes.	The machine should not be relied upon, the machine is not relied upon, and nonreliance leads to good outcomes.
Unjustified	The machine should not be relied upon, the machine is relied upon, and reliance leads to bad outcomes.	The machine should be relied upon, and nonreliance may lead to bad outcomes.

Whenever we have information that allows us to scale the trusting relationship, it would be possible to index trusting by the ratio:

$$\frac{\textbf{Justified Trust + Justified Mistrust}}{\textbf{Trust (Justified and Unjustified) + Mistrust (Justified and Unjustified)}}$$

This can be thought of as a way of scaling the notion that is ordinarily referred to as trust calibration. Such an evaluation would be relative to the task or goal, and to the problem type or context, but it would be possible to aggregate over tasks and contexts to derive a general Coefficient of Trusting, for the assessment of human–machine work systems. Figure 8.1 shows what is perhaps the desirable trajectory for the morphing of trust.

Multiple measures taken over time could be integrated for overall evaluations of human–machine performance, but episodic measures would be valuable in tracking such things as: How do humans maintain trust and cooperation in teams when they are working through a technological intermediary? What is the trend for desirable movement—from the quadrants of Unjustified Trusting + Unjustified Mistrusting to the quadrants of Justified Trusting + Justified Mistrusting? How can we derive metrics for Swift Trusting and Swift Mistrusting, perhaps identifying early indicators or mistrust signatures?

In general, the calculation of values given individual task-context duals, along with the dynamic conditions of human–machine interdependence required for successful task outcomes, could provide a means for displaying the competence envelope of the human–machine work system and allow the human to see progress toward goals within that envelope across time and circumstance.

This would work for known and fixed tasks, but the capability might be extensible. Running assessments might support rapid recovery in circumstances in which actions have been taken or decisions made on the basis of information or machine actions that

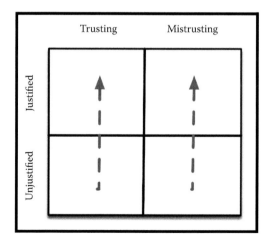

FIGURE 8.1   Dashed arrows show desirable trajectories in the dynamics of trusting.

are subsequently found to be untrustworthy. Running assessments would support a trustworthiness trace-back capability for hindsight analyses of factors or events that contributed to increases or decreases in the measurements. This explanatory process could support the identification of signatures that signal imminent loss of trust or over-trust.

The above measurement ideas are, admittedly, a promissory note. I assert, however, that something along these lines *must* be undertaken to move modeling to a new level.

The present analysis suggests that the creation of models to predict final trust states is perhaps not the most important thing that should be measured and evaluated. Not only is the goal of predicting point-like state values ill-conceived, additional factors at the work systems level must be taken into consideration, especially the fact that trusting involves multiple simultaneous relations and is continually morphing.

8.15 TRUST AS AN EMERGENT IN MACROCOGNITIVE WORK SYSTEMS

I began this chapter with a reference to expertise, highlighting the fact that the majority of the research on trust in automation involves college student-participants and tasks that are significantly reduced relative to the "real world" of human–machine work systems. Referencing professional domains, there is far more to trusting than just the relation of the individual human to the individual machine or software tool. Trusting of a machine can involve trusting developers of the machine. If the worker knows from experience that the developer does a good job with some functions or technologies and therefore concludes that the developer probably did a good job with the other functions or technologies, that might be regarded as "faith-based."

Trusting of a machine involves far more than trusting some single machine or software system, because the technology is a moving target (Ballas 2007; Bradshaw et al. 2005). "At the same time that we are struggling to understand trust and automation, automation itself is very far from a stationary or unitary entity" (Schaefer et al. 2016, p. 393).

Trusting in the work system brings in the matter of teams and organizations, so interpersonal trust cannot be divorced from trust in automation. Trusting at the team level can be thought of as the expectation of reciprocity from others (Ostrom 2003). The parties involved in joint activity enter into a Basic Compact, that is, an agreement (often tacit) to facilitate coordination, work toward shared goals, and prevent coordination breakdowns (Klein et al. 2004). In an interdependence relationship, some degree of benevolence is involved. One human or machine relies on another human or machine to reciprocate in the future by taking an action that might give up some benefit and yet make both better off than they were at the starting point (Hoffman et al. 2013; Johnson et al. 2011). Coordination and synchronization in distributed systems requires reciprocity; otherwise, distributed systems are brittle and exhibit one or more maladaptive behavior patterns.

Trusting in work systems emerges from knowledge about the resilience (and brittleness) of the work system—how work systems composed of multiple humans and machines adapt when events challenge boundary conditions (Hollnagel et al.

2006; Woods 2011). Trust in work systems can be thought of as confidence in this ability of different units at different echelons to act resiliently despite uncertainty. To achieve resilience, the technology and work methods must be created so as to support directability, responsiveness, reciprocity, and responsibility (Klein et al. 2004).

To achieve robustness, adaptivity, and resilience in a macrocognitive work system, humans must develop Contingent, Justified, Stable Trusting *and* Contingent, Justified, Stable Mistrusting in all components of the work system. Only in this circumstance can the work system respond adaptively to disrupting events that alter plans and activities (Watts-Perotti and Woods 2009). When one agent risks counting on others but anticipates little reciprocity or responsiveness, the result will be unstable, and the responsible party will tend to shift to more conservative, independent strategies that threaten cooperation and collaboration. Unjustified Trusting and Unjustified Mistrusting and Distrusting among human and/or machine agents in a work system would not support adaptation or resilience activities (see Hoffman and Hancock 2017).

Given these considerations, both empirical and theoretical—and especially the view that trusting is a process—it can be further asserted that trusting of machines is always exploratory (Woods 2011). What workers need to know is not the values of some momentary all-encompassing trust state. Humans need to be able to actively probe their work systems. Active exploration of trusting–relying relationships cannot and should not be aimed at achieving single stable states or maintaining some decontextualized metrical value, but must be aimed at maintaining an appropriate and context-dependent expectation. Active exploration, on the part of the human, of the trustworthiness of the machine within the competence envelope of the total work system will involve (1) verification of reasons to take the machine's presentations or assertions as true, (2) verification of reasons why directives given by the human will be carried out, and (3) an assessment of situational uncertainty that might affect the livelihood of favorable outcomes.

Active exploration of human-machine interdependence would be a new focus and justification for trust modeling. For this, there must be a usable, useful, and understandable method built into the cognitive work that permits the systematic evaluation of and experimentation on the human–machine relationship. The goals would include the following:

- Enabling the worker to identify and mitigate Unjustified Trusting and Unjustified Mistrusting situations.
- Enabling the worker to discover indicators to mitigate the impacts and risks of unwarranted reliance, or unwarranted rejection of recommendations, especially in time-pressured or information-challenged (too much, too little, or uncertain) situations.
- Enabling the worker to adjust their reliance to the task and situation.
- Enabling the worker to develop Justified Swift Trust and Justified Swift Mistrust in the machine. The worker needs guidance to know when to trust, or when not to trust, early and "blindly" (Roth 2009).
- Enabling the worker to understand and anticipate circumstances in which the machine's recommendations will not be trustworthy, and the machine's

recommendations should not be followed even though they appear trust-worthy.

- Enabling the worker to understand and anticipate circumstances (i.e., unforeseen variations on contextual parameters) in which the software should not be trusted even if it is working as it should, and perhaps especially if it is working as it should (Woods 2011).
- Enabling the worker to develop Justified Trusting over long time spans of experience with a machine in a variety of challenging situations.

The notions of directability and observability mean that we must escape the traditional distinction between the operational context and the experimentation context, especially given the ever-changing nature of the challenges that confront macrocognitive work systems. To accomplish these desirable capabilities, the competence envelope of the technology must be made reasonably explicit in descriptions of what the technology cannot do or cannot do well in different circumstances (Mueller and Klein 2011). Given that the state space intrinsic to computerized work systems has indeterminate boundaries, this represents a significant challenge. Using an active exploration system as envisioned here, the worker could actively probe the technology (probing the world *through* the technology) in order to test hypotheses about trusting and mistrusting, and then use the results to adjust subsequent human–machine activities (i.e., reliance).

8.16   FINAL CONCLUSIONS

- Human–machine interdependent macrocognitive work does not hinge upon or converge upon a single trust state.
- Trust is dynamic, though it can be temporarily stable.
- Trust is always context-dependent, though it might seem invariant.
- Trust can appear to be insulated, though it is actually contingent.
- Trust has to be maintained, and even managed.
- Likewise, mistrust is dynamic; it too can be maintained (which is unfortunate) and it too should be actively managed.

While models of trust calculate values of trust states, on the assumption of predictive potential, the taxonomics of trusting suggests a far richer and fertile ground for empirical investigation and provides a terminology to support such investigation. The dynamics of trusting imply that *transitions among the varieties of trust* are very important. The dynamic nature of trusting manifests itself in these transitions. This has to be combined with various methods of delegation/reliance and interdependency. Having the elements organized in this way would go far toward enabling a computational model of trust with broader functionality and utility.

The taxonomy presented here might, above all else, help the research community to ask expansive questions when they are evaluating trust issues. How do we integrate judgments of trustworthiness across humans representing different core values, cultural norms, perspectives, specialties, and backgrounds? How do we model and measure the resilience of a macrocognitive work system in context? How do we synthesize trust judgments from multiple human–machine teams to improve decision-making and support the establishment and maintenance of common ground? How do we evaluate trusting when the cognitive work is forced to the edge of its competence envelope? How can we evaluate trusting as workers try to coordinate across echelons in order to adapt and achieve common goals? The taxonomy presented here can help us categorize and thereby analyze over such broader circumstances.

Intelligent technology for macrocognitive work must be understandable as well as usable and useful (Hollnagel et al. 2006; Pritchett and Bisantz 2002; Roth 2009; Seong and Bisantz 2003, 2008). Understandability emerges only if the operator can experience the variation of the technology's behavior when it is working at the edges of its competence envelope. As Woods (2009) phrased it, "Trust emerges from the dynamics of reciprocity, responsibility and resilience in networked systems." In fact, "trust" is merely an entry point for the analysis of the usability, usefulness, understandability, and observability of the technology in macrocognitive work systems (Hoffman et al. 2009, 2013). It is these features of work systems that have to be measured, and such measurement must embrace the complexities that I have outlined.

REFERENCES

Atkinson, D.J., Clancey, W.J., and Clark, M.H. (2014). Shared awareness, autonomy and trust in human–robot teamwork. In *Artificial Intelligence and Human–Robot Interaction: Papers from the 2014 AAAI Fall Symposium* (pp. 36–38).

Ballas, J.A. (2007). Human centered computing for tactical weather forecasting: An example of the "Moving Target Rule." In R.R. Hoffman (Ed.), *Expertise out of Context: Proceedings of the Sixth International Conference on Naturalistic Decision Making* (pp. 317–326). Mahwah, NJ: Lawrence Erlbaum.

Barber, B. (1983). *The Logic and Limits of Trust.* New Brunswick, NJ: Rutgers University Press.

Beck, H.P., Dzindolet, M.T., and Pierce, L.G. (2002). Operators' automation usage decisions and the sources of misuse and disuse. *Advances in Human Performance and Cognitive Engineering Research, 2,* 37–78.

Bisantz, A.M., and Seong, Y. (2001). Assessment of operator trust in and utilization of automated decision-aids under different framing conditions. *International Journal of Industrial Ergonomics, 28,* 85–97.

Bobko, P., Barelka, A.J., and Hirshfield, L.M. (2014). The construct of state-level suspicion: A model and research agenda for automated information technology (IT) contexts. *Human Factors, 56,* 498–508.

Bradshaw, J.M., Jung, H., Kulkarni, S., Johnson, M., Feltovich, P., Allen, J., Bunch, L., Chambers, N., Galescu, L., Jeffers, R., Suri, N., Taysom, W., and Uszok, A. (2005). Toward trustworthy adjustable autonomy in KAoS. In R. Falcone (Ed.), *Trusting Agents for Trustworthy Electronic Societies* (pp. 18–42). Lecture Notes in Artificial Intelligence. Berlin: Springer.

Chancey, E.T., Bliss, J.P., Proaps, A.B., and Madhavan, P. (2015). The role of trust as a mediator between system characteristics and response behaviors. *Human Factors, 57,* 947–958.

Chien, S.-Y., Sycara, K., Liu, J.-S., and Kumru, A. (2016). Relation between trust attitudes toward automation, Hofstede's Cultural dimensions, and Big Five personality traits.

In *Proceedings of the Human Factors and Ergonomics Society 2016 Annual Meeting* (pp. 840–845). Santa Monica, CA: Human Factors and Ergonomics Society.

Choiu, E.R., and Lee, J.D. (2015). Beyond reliance and compliance: Human-automation coordination and cooperation. In *Proceedings of the Human Factors and Ergonomics Society 59th Annual Meeting* (pp. 195–200). Santa Monica, CA: Human Factors and Ergonomics Society.

Corritore, C.L., Kracher, B., and Wiedenbeck, S. (2003). On-line trust: Concepts, evolving themes, a model. *International Journal of Human–Computer Studies, 58,* 737–758.

Cramer, H., Evers, V., Kemper, N., and Wielinga, B. (2008). Effects of autonomy, traffic conditions and driver personality traits on attitudes and trust towards in-vehicle agents. In *Proceedings of the IEEE/WIC/ACM Int. Conference on Web Intelligence and Intelligent Agent Technology, 3,* 477–482.

Cramer, H., Goddijn, J., Wielinga, B., and Evers, V. (2010). Effects of (in)accurate empathy and situational valence on attitudes towards robots. In *Proceedings of the 5th ACM/IEEE International Conference on Human–Robot Interaction* (pp. 141–142). New York: Association for Computing Machinery.

Crispen, P., and Hoffman, R.R. (2016, November/December). How many experts? *IEEE Intelligent Systems,* 57–62.

Desai, M., Stubbs, K., Steinfeld, A., and Yanco, H. (2009). Creating trustworthy robots: Lessons and inspirations form automated systems. In *Proceedings of the Artificial Intelligence and Simulation of Behavior Convention: New Frontiers in Human–Robot Interaction.* Edinburgh, Scotland: University of Edinburgh.

Dzindolet, M.T., Peterson, S.A., Pomranky, R.A., Pierce, L.G., and Beck, H.P. (2003). The role of trust in automation reliance. *International Journal of Human–Computer Studies, 58,* 697–718.

Fitzhugh, E.W., Hoffman, R.R., and Miller, J.E. (2011). Active trust management. In N. Stanton (Ed.), *Trust in Military Teams* (pp. 197–218). London: Ashgate.

Gao, J., and Lee, J.D. (2006). Extending the decision field theory to model operators' reliance on automation in supervisory control situations. *IEEE Systems, Man, and Cybernetics, 36,* 943–959.

Hancock, P.A., Billings, D.R., and Schaeffer, K.E. (2011, July). Can you trust your robot? *Ergonomics in Design,* 24–29.

Hoff, K.A., and Bashir, M. (2015). Trust in automation: Integrating empirical evidence on factors that influence trust. *Human Factors, 57,* 407–434.

Hoffman, R.R. (1989). Whom (or what) do you trust: Historical reflections on the psychology and sociology of information technology. In *Proceedings of the Fourth Annual Symposium on Human Interaction with Complex Systems* (pp. 28–36). New York: IEEE Computer Society.

Hoffman, R.R., Coffey, J.W., Ford, K.M., and Carnot, M.J. (2001, October) STORM-LK: A human-centered knowledge model for weather forecasting. In J.M. Flach (Ed.), *Proceedings of the 45th Annual Meeting of the Human Factors and Ergonomics Society* (p. 752). Santa Monica, CA: Human Factors and Ergonomics Society.

Hoffman, R.R., and Hancock, P.A. (2017). Measuring resilience. *Human Factors, 59,* 564–581.

Hoffman, R.R., Hawley, J.K., and Bradshaw, J.M. (2014, March/April). Myths of automation. Part 2: Some very human consequences. *IEEE Intelligent Systems,* 82–85.

Hoffman, R.R., Johnson, M., Bradshaw, J.M., and Underbrink, A. (2013, January/February). Trust in automation. *IEEE Intelligent Systems,* 84–88.

Hoffman, R.R., LaDue, D., Mogil, H.M., Roebber, P., and Trafton, J.G. (2017). *Minding the Weather: How Expert Forecasters Think.* Cambridge, MA: MIT Press.

Hoffman, R.R., Lee, J.D., Woods, D.D., Shadbolt, N., Miller, J., and Bradshaw, J.M. (2009, November/December). The dynamics of trust in cyberdomains. *IEEE Intelligent Systems*, 5–11.

Hoffman, R.R., Marx, M., and Hancock, P.A. (2008/March-April). Metrics, metrics, metrics: Negative hedonicity. *IEEE Intelligent Systems*, 69–73.

Hoffman, R.R., and Woods, D.D. (2011, November/December). Beyond Simon's slice: Five fundamental tradeoffs that bound the performance of macrocognitive work systems. *IEEE Intelligent Systems*, 67–71.

Hollnagel, E., Woods, D.D., and Leveson, N. (Eds.) (2006). *Resilience Engineering: Concepts and Precepts*. Aldershot, UK: Ashgate.

Huynh, T.D., Jennings, N.R., and Shadbolt, N.R. (2006). An integrated trust and reputation model for open multi-agent systems. *Autonomous Agents and Multi-Agent Systems, 13*, 119–154.

Johnson, M., Bradshaw, J.M., Hoffman, R.R., Feltovich, P.J., and Woods, D.D. (November/December 2014). Seven cardinal virtues of human–machine teamwork. *IEEE Intelligent Systems*, 74–79.

Johnson, M., Bradshaw, J.M., Feltovich, P.J., Hoffman, R.R., Jonker, C., van Riemsdijk, B., and Sierhuis, M. (2011, May/June). Beyond cooperative robotics: The central role of interdependence in coactive design. *IEEE Intelligent Systems*, 81–88.

Khasawneh, M.T., Bowling, S.R., Jiang, X., Gramopadhye, A.K., and Melloy, B.J. (2003). A model for predicting human trust in automated system. In *Proceedings of the 8th Annual International Conference on Industrial Engineering—Theory, Applications and Practice* (pp. 216–222). Sponsored by the International Journal of Industrial Engineering [http://ijietap.org].

Klein, G., Woods, D.D., Bradshaw, J.D., Hoffman, R.R., and Feltovich, P.J. (November/December 2004). Ten challenges for making automation a "team player" in joint human-agent activity. *IEEE: Intelligent Systems*, 91–95.

Koopman, P., and Hoffman, R.R. (November/December 2003). Work-arounds, make-work, and kludges. *IEEE: Intelligent Systems*, 70–75.

Kucala, D. (2013, June). The truthiness of trustworthiness. *Chief Learning Officer*, 57–59.

Lee, J.D. and Moray, N. (1992). Trust, control strategies, and allocation of functions in human–machine systems. *Ergonomics*, 35, 1243–1270.

Lee, J.D., and See, K.A. (2004). Trust in automation: Designing for appropriate reliance. *Human Factors*, 46, 50–80.

Lewandowsky, S., Mundy, M., and Tan, G. (2000). The dynamics of trust: Comparing humans to automation. *Journal of Experimental Psychology-Applied*, 6, 104–123.

Lewicki, R.J., McAlister, D.J., and Bias, R.J. (1998). Trust and distrust: New relationships and realities. *Academy of Management Review*, 23, 438–445.

Madhavan, P., and Wiegmann, D.A. (2007). Effects of information source, pedigree, and reliability on operator interaction with decision support systems. *Human Factors,* 49(5), 773–785.

Mayer, R.C., Davis, J.H., and Schoorman, F.D. (1995). An integrative model of organizational trust. *Academy of Management Review, 20*, 709–734.

McGuirl, J.M., and Sarter, N. (2006). Supporting trust calibration and the effective use of decision aids by presenting dynamic system confidence information. *Human Factors,* 48, 656–665.

Merritt, S.M., Heimbaugh, H., LaChapell, J., and Lee, D. (2013). I trust it, but don't know why: Effects of implicit attitudes toward automation in trust in an automated system. *Human Factors,* 55, 520–534.

Merritt, S.M., and Ilgen, D.R. (2008). Not all trust is created equal: Dispositional and history-based trust in human–automation interactions. *Human Factors,* 50, 194–201.

Merritt, S.M., Lee, D., Unnerstall, J.L., and Huber, K. (2015a). Are well-calibrated users effective users? Associations between calibration of trust and performance on an automation-aided task. *Human Factors, 57*, 34–47.

Merritt, S.M., Unnerstall, J.L., Lee, D., and Huber, K. (2015b). Measuring individual differences in the perfect automation schema. *Human Factors, 57*, 740–753.

Meyerson, D., Weick, K., and Kramer, R. (1966). Swift trust and temporary groups. In T.R. Tyler and R. Kramer (Eds.), *Trust in Organizations: Frontiers of Theory and Research* (pp. 166–195). Thousand Oaks, CA: Sage.

Mueller, S.T., and Klein, G. (2011, March/April). Improving users' mental models of intelligent software tools. *IEEE Intelligent Systems*, 77–83.

Muir, B.M. (1987). Trust between humans and machines, and the design of decision aids. *International Journal of Man–Machine Studies, 27*, 527–539.

Muir, B.M. (1994). Trust in automation Part 1: Theoretical issues in the study of trust and human intervention in automated systems. *Ergonomics, 37*, 1905–1922.

Muir, B.M., and Moray, N. (1996). Trust in automation. *Part II Experimental studies of trust and human intervention in a process control simulation. Ergonomics, 39*, 429–460.

Naone, E. (4 September 2009). Adding trust to Wikipedia, and beyond. *Technology Review* [http://www.technologyreview.com/web/23355/?a=f].

Nass, C., and Moon, Y. (2000). Machines and mindlessness: Social responses to computers. *Journal of Social Issues, 56*, 81–103.

O'Hara, K. (2004) *Trust: From Socrates to Spin*. Cambridge, UK: Icon Books.

Oleson, K.E., Billings, D.R., Chen, J.Y.C, and Hancock, P.A. (2011). Antecedents of trust in human–robot collaborations. In *Proceedings of the IEEE International Multi-Disciplinary Conference on Cognitive Methods in Situation Awareness and Decision Support* (pp. 175–178). New York: Institute of Electrical and Electronics Engineers.

Parasuraman, R., and Riley, V. (1997). Humans and automation: Use, misuse, disuse, abuse. *Human Factors, 39*, 230–253.

Parlangeli, O., Chiantini, T., and Guidi, S. (2012). A mind in a disk: The attribution of mental states to technological systems. *Work, 41*, 1118–1123.

Pop, V.L., Shrewsbury, A., and Durso, F.T. (2015). Individual differences in the calibration of trust in automation. *Human Factors, 57*, 545–556.

Pritchett, A.R., and Bisantz, A.M. (2002). Measuring judgment interaction with displays and automation. In *Proceedings of the Human Factors and Ergonomics Society 46th Annual Meeting* (pp. 512–516). Santa Monica, CA: Human Factors and Ergonomics Society.

Rempel, J.K., Holmes, J.G., and Zanna, M.P. (1985). Trust in close relationships. *Journal of Personality and Social Psychology, 49*, 95–112.

Riley, V. (1996). Operator reliance on automation: Theory and data. In R. Parasuraman and M. Mouloua (Eds.), *Automation Theory and Applications* (pp. 19–35). Mahwah, NJ: Erlbaum.

Roth, E.M. (2009). Facilitating 'calibrated' trust in technology of dynamically changing 'trustworthiness'. Presentation at the Working Meeting on Trust in Cyberdomains. Institute for Human and Machine Cognition, Pensacola, FL. Supported by the Human Effectiveness Directorate, Air Force Research Laboratory, Wright–Patterson AFB, OH.

Sarter, N., Woods, D.D., and Billings, C.E. (1997). Automation surprises. In G. Salvendy (Ed.), *Handbook of Human Factors/Ergonomics*, 2nd ed. (pp. 1926–1943). New York, NY: Wiley.

Schaefer, K.E., Chen, J.Y.C., Szalma, J.L., and Hancock, P.A. (2016). A meta-analysis of factors influencing the development of trust in automation: Implications for understanding autonomy in future systems. *Human Factors, 58*, 377–400.

Seong, Y., and Bisantz, A. (2002). Judgment and trust in conjuction with automated aids. In *Proceedings of the Human Factors and Ergonomics Society 46th Annual Meeting* (pp. 423–428). Santa Monica, CA: Human Factors and Ergonomics Society.

Seong, Y., and Bisantz, A.M. (2008). The impact of cognitive feedback on judgment performance and trust with decision aids. *International Journal of Industrial Ergonomics,* 38, 608–625.

Shadbolt, N. (January/February, 2002). A matter of trust. *IEEE Intelligent Systems,* 2–3.

Sheridan, T. (1980). Computer control and human alienation. *Technology Review,* 83, 61–73.

Skitka, L.J., Mosier, K.L., and Burdick, M. (2000). Accountability and automation bias. *International Journal of Human–Computer Studies,* 52, 701–717.

Stokes, C., Lyons, J., Littlejohn, K., Natarian, J., Case, E., and Speranza, N. (2010). Accounting for the human in cyberspace: Effects of mood on trust in automation. In *Proceedings of the 2010 International Symposium on Collaborative Technologies and Systems* (pp. 180–187). New York: Institute for Electrical and Electronics Engineers.

Vasalou, A., Hopfensitz, A., and Pitt, J. (2008). In praise of forgiveness: Ways for repairing trust breakdowns in one-off online interactions. *International Journal of Human–Computer Studies,* 66, 466–480.

Wagner, A. 2009. *The Role of Trust and Relationships in Human–Robot Social Interaction.* Doctoral Dissertation, Georgia Institute of Technology, Atlanta, GA.

Watts-Perotti, J., and Woods, D.D. (2009). Cooperative advocacy: A strategy for integrating diverse perspectives in anomaly response. *Journal of Collaborative Computing,* 18, 175–198.

Wickens, C.D., Clegg, B.A., Vieane, A.Z., and Sebok, A.L. (2015). Complacency and automation bias in the use of imperfect automation. *Human Factors,* 57, 728–739.

Wohleber, R.W., Calhoun, G.L., Funke, G.J., Ruff, H., Chiu, C.-Y.P., Lin, C., and Matthews, G. (2016). The impact of automation reliability and operator fatigue on performance and reliance. In *Proceedings of the Human Factors and Ergonomics Society 2016 Annual Meeting* (pp. 211–216). Santa Monica, CA: Human Factors and Ergonomics Society.

Woods, D.D. (2009). Trust Emerges from the Dynamics of Reciprocity, Responsibility and Resilience in Networked Systems. Presentation at the *Working Meeting on Trust in Cyberdomains. Institute for Human and Machine Cognition*, Pensacola, FL. Supported by the Human Effectiveness Directorate, Air Force Research Laboratory, Wright–Patterson AFB, OH.

Woods, D.D. (2011, September). Reflections on 30 years of picking up the pieces after explosions of technology. Presentation at the AFRL Autonomy Workshop, Air Force Research Laboratory, Wright–Patterson Air Force Base, OH.

Woods, D.D., and Hollnagel, E. (2006). *Joint Cognitive Systems: Patterns in Cognitive Systems Engineering.* Boca Raton, FL: CRC Press.

Woods, D.D., Roth, E.M., and Bennett, K. (1990). Explorations in joint human–machine cognitive systems. In W. Zachary and S. Robinson (Eds.), *Cognition, Computing, and Cooperation* (pp. 123–158). Norwood, NJ: Ablex.

Yang, X.J., Wickens, C.D., and Hölttä-Otto, K. (2016). How users adjust trust on automation: Contrast effect and hindsight bias. In *Proceedings of the Human Factors and Ergonomics Society 2016 Annual Meeting* (pp. 196–200). Santa Monica, CA: Human Factors and Ergonomics Society.

9 Improving Sensemaking through the Design of Representations

John M. Flach and Kevin B. Bennett

CONTENTS

9.1 REDEFINING THE INTERFACE PROBLEM

… performance is a joint function of the nature of the cognitive demands posed by the domain, the nature of the representation of that domain available to the problem solver and the characteristics of the problem solving agent…

Woods
1991, p. 175

Woods (1984, 1991) was one of the first to begin to formulate the problem of interface design from the perspective of the impact of representations on problem solving or sensemaking. Up until that time, the interface design problem was typically framed in terms of human information processing limitations (e.g., perceptual thresholds, memory limitations, speed–accuracy trade-offs). However, Woods made a compelling case that problems such as "getting lost" and "keyhole" effects were not inevitable consequences of human information processing limitations. Rather, such problems represented a failure to recognize that "perception is an *active, inherently selective* process of data gathering, that is, *part of cognitive processing* rather than a process of passive reception and transmission prior to cognitive processing" (1991, p. 233).

Woods's insights reflected the beginnings of a paradigm shift in the way we think about humans, technology, and work (Hollnagel and Woods 1983; Norman 1986;

Norman and Draper 1986; Rasmussen 1986). This paradigm shift was motivated in part by safety concerns associated with the development of nuclear power and advanced avionics. In particular, the accident at Three Mile Island made it very obvious to many people that a "knobs and dials" approach to interface design was not sufficient. Safety in these work systems was not simply a matter of accessing data and following predetermined procedures or rules.

In the aviation domain, automation has increased safety, but has also increased operational complexity. It became clearly evident that automation alone would not be the ultimate solution for safer systems and, in fact, people realized that increased automation created new categories of errors (Billings 1997; Sarter et al. 1997). Ultimately, it became clear that safety would depend on cooperation between humans and automated systems. Thus, the goal for cognitive systems engineering (CSE) is to enhance the ability of humans and technologies to collaborate to make sense of novel situations that emerge from dynamic interactions among the components of complex sociotechnical systems (Christoffersen and Woods 2002).

For those framing the CSE paradigm, it was evident that the "unanticipated variability" that inevitably emerges in complex sociotechnical systems (e.g., nuclear power, aviation, modern combat operations) would demand that operators be able to creatively muddle through to discover satisfactory solutions (e.g., Bennett et al. 2015; Flach 2015; Lindblom 1959, 1979; Roth et al. 1987; Woods et al. 1990). Thus, designers needed to begin exploring how information and display technologies could support effective muddling. In other words, the goal was to design representations that supported creative problem solving (or productive thinking) to help workers anticipate, diagnose, and mitigate the problems that inevitably result from the dynamic complexity in such dynamical systems. For CSE, humans' ability to diagnose complex problems that could not be anticipated in the design of either procedures or automatic control systems, and to invent creative solutions, was seen as a valuable resource.

At about the same time that Woods and others were exploring the interface demands posed by complex sociotechnical systems, Shneiderman (1982, 1983, 1992) and others (Hutchins et al. 1986; Lave 1988) were beginning to explore the opportunities that graphical interface technologies offered for supporting richer and more direct couplings between perception and action. Shneiderman used the term "direct manipulation" to describe interactive graphical systems that allowed representations to be manipulated through physical actions that provided continuous feedback about consequences, rather than through complex syntaxes. Flach and Vicente (1989) noted that "direct manipulation" went hand in hand with "direct perception" in order to bridge the gulfs of execution and evaluation that Norman (1986) identified as critical targets for the emerging field of cognitive engineering.

Thus, as illustrated in Figure 9.1, a central problem for CSE is to design interface representations that can be effective bridges between human experience and complex sociotechnical processes, where "effective" is not simply about access to data, but where "effective" refers to facilitating active sensemaking processes that involve anticipating, discovering, and diagnosing problems in order to create satisfying solutions. In bridging this gulf, it is typically not sufficient to match existing mental models, because typically the problems will be novel—requiring experts to apply

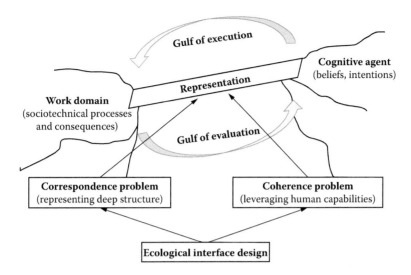

FIGURE 9.1 In order to bridge the gulfs of execution and evaluation, it becomes necessary to design representations that reflect the deep structure of work domains (the correspondence problem) in ways that are comprehensible to the humans who are ultimately responsible for success (the coherence problem).

their knowledge in a creative fashion, as opposed to a rote application of predetermined solutions. Thus, it often becomes necessary that interactions with the representation help to shape or tune the mental models of agents to keep pace with dynamic changes in the "deep structure" of the sociotechnical problems to be solved.

In the new paradigm, understanding the kinds of psychophysical and manual control constraints reflected in information processing models of human performance remains an important concern. However, understanding these constraints is not sufficient. In addition, it becomes necessary to consider the human's ability to think productively with respect to the particular problem domain. Thus, the research literature evaluating human problem solving (e.g., Duncker 1945; Wertheimer 1959) and expertise (e.g., Ericsson et al. 2006) becomes particularly relevant for interface or representation design. For example, Ericsson and Charness (1994) concluded that "acquired skill can allow experts to circumvent basic capacity limits of short-term memory and of the speed of basic reactions, making potential limits irrelevant" (p. 731).

Thus, while cognitive systems engineers recognize human limitations such as the limited capacity of working memory, they evidence a much deeper appreciation for the power of humans to adapt to this constraint through recoding and integration of information into larger "chunks" that allow experts to efficiently manage the demands of complex problems (e.g., to quickly focus attention on the most promising options in a chess match). CSE recognizes that this capability reflects familiarity with the deep structure of the problem domain and the ability to leverage that structure toward productive ways of thinking (e.g., Chase and Simon 1973; de Groot 1965; Miller 1956; Reynolds 1982; Vicente and Wang 1998; Wertheimer 1959).

In sum, the term "cognitive systems engineering" reflects a shift in emphasis with respect to the interface design problem compared with the "traditional" information processing approach that inspired conventional human factors approaches. Rather than emphasizing the information processing limitations of humans, attention has shifted to the cognitive capabilities of humans. Rather than viewing the human operators as limiting constraints on performance, human operators are increasingly seen as a resource for dealing with unanticipated variability. Rather than framing the design goal as constraining humans to follow predetermined rules or procedures, the design goal is framed as supporting exploration and discovery with the ultimate goal to facilitate efficient control under normal conditions and creative problem solving in unusual conditions.

9.2   ECOLOGICAL INTERFACE DESIGN

The growth of computer applications has radically changed the nature of the man–machine interface. First, through increased automation, the nature of the human's task has shifted from an emphasis on perceptual-motor skills to an emphasis on cognitive activities, e.g., problem solving and decision making.... Second, through the increasing sophistication of computer applications, the man–machine interface is gradually becoming the interaction of two cognitive systems.

Hollnagel and Woods
1983, p. 384

As reflected in this quotation from Hollnagel and Woods (1983), the CSE community recognized that both the challenges associated with the need to support creative problem solving and the opportunities provided by advanced computing and interactive graphical display technologies had clear implications for interface design. Rasmussen and Vicente (1989; Vicente and Rasmussen 1992) framed the problem into what they called *Ecological Interface Design (EID)*. The term "ecological" was chosen to emphasize a component of the representation design problem that had been neglected in classical human factors approaches to interface design—the correspondence problem. In essence, the *correspondence problem* referred to the need to identify the "deep structure" of the problem domain (or ecology), so that this structure could guide the design of representations.

The classical human factors approaches had emphasized the need to match the interface to the information processing limitations of the human (one aspect of the *coherence problem*). But the problem of identifying the functional demands of the work ecology tended to get much less consideration, if any. There seemed to be an implicit assumption that the correspondence problem had already been solved by domain experts or domain-related sciences (e.g., physical sciences in relation to vehicle and industrial process control). Thus, while classical human factors approaches to interface design were well aware that the prime function of an interface was to communicate with the human, there was a tendency to emphasize the psychophysical (e.g., signal detectability) and syntactical aspects of elemental displays (e.g., motion compatibility). It was typically assumed that the problem of data integration and decision making involved higher-level cognitive processes (e.g., internal models) that would be addressed through training (e.g., making people aware of potential deviations from logical norms) (e.g., Kantowitz and Sorkin 1983).

Through association with Gibson's (1979) ecological approach to perception, the term "ecological" also emphasized the active coupling between perception and action and the potential for functional constraints (i.e., boundaries to field of safe travel) to be specified in terms of visual patterns (i.e., invariants in optical flow fields). In some sense, the representation design problem was seen as building visual geometries that mimicked the nested structure of optical flow fields in order to specify functional relations within the events that are either observed or controlled by sociotechnical work systems. In this context, well-designed configural visual graphics might allow operators to directly "see" important functional relations (e.g., energy constraints) that were not directly specified in either natural flow fields or in traditional interfaces.

In the next two sections of this chapter, we will consider each of the two components of an EID approach (i.e., the correspondence and coherence problems) and a third section will consider how these problems interact to ultimately determine how effectively a representation will bridge the gap between the demands of the work domain and the cognitive capabilities of expert humans.

9.3   CORRESPONDENCE PROBLEM

There is no question that workers in complex sociotechnical systems … have extraordinarily difficult jobs. It is also undeniable that to support people in these challenging roles we must first understand the nature of their work.

Naikar
2013, p. xiii

As noted above, the correspondence problem involves linking the representations used in interfaces to aspects of the target problem or work domain. Thus, as noted in the above quote from Neelam Naikar (2013), it becomes important to understand the nature of the problem or work domain. There are two distinctions that are important to fully appreciate the implications of a CSE approach to the correspondence problem: (1) the distinction between descriptions of a problem based on objective analyses as might generally be provided by physical or engineering sciences versus functional descriptions of a problem, and (2) the distinction between *task analysis* and *work analysis*.

9.3.1   Functional versus Objective Physical Descriptions

The key distinction between functional descriptions of a problem and more classical physical analyses is the emphasis of functional descriptions on means–ends relations, where *means* refers to the possibilities for action and *ends* refers to the goals or purposes (or more generally values) that determine whether the consequences of an action are satisfactory. Thus, functional descriptions require that the classical "objective" perspective (e.g., the physics of a power plant or the aerodynamics of an aircraft) be reframed to emphasize the fit with the functional goals and values, on the one hand (i.e., the ends), and the alternative action capabilities, on the other hand (i.e., the means).

This does not necessarily imply that an operator in a nuclear control room or a pilot has to know the physics of their systems in the same way as a physicist might. However, they do need to know how the physical constraints governing the system bound and shape the possibilities and consequences of action. Often, interfaces can be improved by organizing the display to directly reflect the underlying physical laws that constrain the relations among the various data and controls available in an interface. For example, Vicente's (1999; Vicente et al. 1995) DURESS interface organized information about a feedwater control system to make the mass and energy balances more explicit. Beltracchi (1987) designed an interface for feedwater control that made the entropy constraints explicit to the operator. In the aviation domain, Amelink et al. (2005) developed a representation that made the energy constraints associated with controlling the approach to landing explicit. This interface provides new insights into the relative contributions of the throttle and stick (elevators) for controlling speed and altitude when landing.

In some domains such as medicine and military command and control, the physical constraints are not as well documented or understood as in others. In these domains, expert judgment and heuristics may provide inspiration for ways to organize information so that the means–ends relations are better represented. For example, McEwen et al. (2012, 2014) used correlational models linking data from medical records (e.g., cholesterol levels, blood pressure, etc.) to models of health risk (Framingham Model) and to published standards of care from expert panels (Adult Treatment Panel III Guidelines or ATP III) as inspiration for representing the data to support evidence-based clinical decision making with respect to cardiovascular health. Also, in designing an interface to support tactical command and control, Bennett et al. (2008; Hall et al., 2012) organized information to make the force ratio (an indicator of relative strength between opposing military forces) explicit, based on input from military experts.

In addition to physical constraints, it is important that other constrains on work system performance, such as safety regulations and laws, also be considered in developing an effective representation. For example, EID displays developed at TU Delft integrated regulations on the separation between aircraft (Ellerbroek et al. 2013a,b) or on ground clearance (Borst et al. 2010; Comans et al. 2014) into designs for in-cockpit air traffic control interfaces. In the case of ground clearance, Borst et al. (2010) found that when the physical constraints were made explicit utilizing EID principles, pilots would sometimes cut corners that violated legal limits, because they felt that they could do so with little risk. This tendency for increased risk with the EID interface was reduced when the regulatory boundaries were incorporated into the representation (Comans et al. 2014).

The main point is that in exploring the deep structure with respect to the correspondence problem, it is critical to consider the physical and regulatory constraints with respect to the means of achieving the functional goals for the work domain. Inspiration often comes from exploring the physical laws and expert heuristics associated with the control processes. However, some laws and relations are more important than others. The key is to emphasize those constraints and relations that are most relevant to linking action (means) with the functional goals (ends).

The purpose of the representation is not to prescribe a single best way to reach the goal, but rather it should reflect the field of possibilities, so that humans can make smart selections. The point is for the display to enhance situation awareness so that the human operators will choose ways that are satisfactory for alternative situations. This leads naturally to the distinction between task and work analysis.

9.3.2 TASK VERSUS WORK ANALYSIS

The key distinction between task analysis and work analysis is that task analysis is typically used to describe *activities* (either observable, inferred, or required) and work analysis is used to describe *constraints.* Vicente (1999) used Simon's metaphor of the ant on the beach to clarify this distinction. Task analysis focuses on the activity of the ant—specific trajectories over the beach. Work analysis, on the other hand, focuses on the beach, describing the objects on the beach with respect to implications for the ant (e.g., whether objects are attractive, desirable vs. repelling or undesirable, or whether objects impede or facilitate locomotion) (e.g., see Bisantz and Burns 1988; Naikar 2013).

Note that terms like *desirable* and *undesirable* reflect a value system, as noted above, that would typically not be part of an objective physical description of the beach. Rather, these dimensions reflect specific preferences of the ant (i.e., values in a particular work domain). Also, note that the implications for locomotion require that the objects be evaluated or scaled relative to the ant's action capabilities (i.e., controls available in a particular work domain). Gibson's (1979) term *affordance* is now routinely used by designers and cognitive engineers to refer to the functional properties of work domains—the action possibilities and consequences.

In order to understand the deep structure of a problem, however, it is not sufficient to list the many constraints or affordances that will be present in a sociotechnical system. The challenge is to discover potential structural relations among the constraints that allow complex domain problems to be parsed in a way that the pieces will add up to insights about the whole. A leading hypothesis about how to do this is Rasmussen's (1986) *Abstraction-Decomposition Method* (also see Flach et al. 2008).

The Abstraction Hierarchy suggests five nested levels for describing work domains. Each level is designed to emphasize different sets of constraints that interact to shape performance. The top level in this nested structure focuses on constraints related to purposes, intentions, and values reflecting the ultimate reason *why* the work is being done. In control theoretic terms, a central question at this level is to consider what criteria will be used to score success of any particular solution (e.g., the cost function in optimal control or the payoff matrix in signal detection theory).

The next level in the Abstraction Hierarchy considers the space of possibilities in terms of physical and regulatory constraints on the possible paths for achieving the goals specified in the level above. This level concerns questions about what is physically or legally possible. In control theoretic terms, this level of analysis focuses on identifying the underlying *state dimensions of the problem.* That is, what are the critical variables for specifying the field of possibilities? For example, the physics of motion (e.g., inertia) will determine what variables (e.g., position, velocity, etc.) will be necessary for vehicular control.

The next level considers the general functional *organization constraints of the work system.* This level describes the work system in terms of general physical or information processing functions and their interconnections (e.g., the feedback loops). In control theoretical analyses, block diagrams showing the flow of product or information between processes are commonly used to represent means–ends constraints at this level. Note that the state variables identified at the previous level will have clear implications for the feedback requirements at this level. For example, the aerodynamic constraints (at the abstract function level) suggest what state variables must be fed back (at the general function level) in order for a pilot to anticipate changes in the trajectory of the aircraft relative to the desired or safe paths (at the functional purpose level).

At the next level of analysis, the general functions identified at the previous level are associated with specific physical processes, in order to consider the constraints that a particular process contributes. For example, an important question at this level might be whether the loop around certain variables is closed through a human (e.g., a pilot) or through a computer (e.g., autopilot). Different solutions introduce different design constraints (e.g., humans require oxygen; computers require electrical power source).

Finally, the "lowest" level within the Abstraction Hierarchy examines the physical spatiotemporal relations among the components. For example, this might be in terms of the schematics of the layout of components. This level of analysis considers interactions due to spatial or temporal proximity that might not be evident in other forms of representations (e.g., human being burned due to location of hot power source or electrical source being too far to reach the computer).

We chose not to give names to the five levels, because we believe the spirit of decomposition suggested by the Abstraction Hierarchy is more important than the details. In fact, we suggest that in searching for the deep structure of a work domain, it is best NOT to overly constrain the search (e.g., prescribe fixed levels). Rather, the key is to become immersed in the domain and to discover those levels of constraint that make sense for the specific domain. We suggest Rasmussen's Abstraction Hierarchy as a useful heuristic, but it should not become a procrustean bed that every analysis should fit.

The main point is that in searching for the deep structure of a work domain, one must look beyond specific microscale activities to consider the constraints that shape those activities. This involves not only considering what people do or how they do it, but it requires consideration of the question of what people could do, or should do, and why one or another way might be preferred or valued in a specific situation. It is typically necessary to consider multiple nested sets of constraints in which constraints at one level set boundaries on constraints at other levels. For example, values might set boundary conditions that partition the physical state space into desirable (e.g., goals or attractors) and undesirable (e.g., risks or dangers) regions, and physical laws may place constraints on the input/output, feedback, and stability properties among component subprocesses. One way to visualize the interactions among levels is as a nesting of control loops with outer loops setting conditions (e.g., target set points, degrees of freedom) on inner loops (Powers 1973).

As far as designing effective representations, the key to the correspondence problem is to ensure that the representation is grounded in the pragmatics of the work domain. The goal is to make the relations among the data that are most important for satisfying the functional goals for the system salient in the representation.

9.4 COHERENCE PROBLEM

There are no a priori neutral representations.... The central question is what are the relative effects of different forms of representation on the cognitive activities involved in solving domain problems. HCI [Human Computer Interaction] research then needs to investigate representational form as opposed to merely visual form, to investigate the referential functions that are performed by HCI tokens within a symbol system, and to investigate the interface as a coherent representational system rather than as a collection of independent parts, e.g., display pages.

Woods
1991, p. 175

Cognitive science has long recognized that humans are not passive receptors of information. Rather, they actively filter and organize information based on biases and expectations shaped by prior experiences. Piaget (1973) referred to this process in which new experiences are actively shaped by *schemas* resulting from prior experiences as *assimilation*. In more **modern** terminology, we might say that humans enter the situation with a collection of mental models that guide attention and that actively shape and filter the information that is processed. The coherence problem has to do with the fit between the representation and the expectations and biases that the humans bring to the system. This aspect of the interface design problem has long been recognized by human factors and is the basis for the conventional design mantra to "match the user's mental model."

However, in complex work domains, especially those where failures bring potentially catastrophic consequences, CSE has recognized that matching existing mental models or schemas is NOT adequate. In fact, a major challenge in these work systems is for the humans to intervene in those situations where the process violates prior assumptions and expectations. In these situations, it is critical that humans are able to discover and correct the weaknesses in prior mental models, so that the expectations become better aligned with the functional realities of the processes that they are attempting to manage. Thus, effective representations should facilitate the learning and discovery process involved in correcting and improving mental models. Piaget (1973) referred to the process of revising mental models to better adapt to new challenges as *accommodation*.

To effectively address the coherence problem, it is generally good to start where people are—that is, to begin with an understanding of the skills and knowledge that people bring to the work. This then becomes a foundation from which to build new and better ways to think about the work domain—where "productive" means ways that are better attuned to the constraints of the work domain.

Two means that interface designers use to do this is through analogs and metaphors. In both cases, the strategy is to map knowledge and intuitions from a familiar context (i.e., the base domain) into a new context (e.g., the target work domain) in such a way that the knowledge and intuitions from the familiar context help to elucidate relations in the new context (Bennett and Flach 2011).

Analog representations take advantage of the expectations that arise from people's experience interacting with physical objects in space and time. The base domain is human experience moving through the world and manipulating physical objects. In a word, analog representation taps into people's skills of pattern recognition that result from "natural" physical experiences. In analog representations, the geometry and motion of objects in a graphical interface represent variables and relations among variables associated with the work processes. Thus, spatiotemporal patterns in the representation reflect changes and interactions among variables in the work domain being represented.

A classical example of an analog representation is the octagon geometry used by Woods et al. (1981) in the design of a safety parameter display for a nuclear control room. In this display, the position of the vertices of an octagon represented the levels of eight parameters of the process. The mapping of the parameters to positions was also scaled so that the overall symmetry of the octagon represented relations among the parameters. Different deformations of the symmetry would specify different types of failures. Thus, the geometric form of the display was an analog to the underlying processes relations.

Another familiar example of the use of analog representations is maps. Hutchins (1995) discussed how different maps are designed to support different types of cognitive problem solving. For example, the Mercator projection is designed to support rhumb-line sailing (i.e., maintaining constant compass heading). Thus, lines on a Mercator projection represent routes with constant compass headings (i.e., rhumb-lines). Another common analog that has been used is a line and fulcrum to represent the balance between two variables, such as mass and energy (Vicente 1999), or between the power of opposing military forces (Bennett et al. 2008).

In the aviation domain, analogs to natural optical flow fields have been an important source of inspiration. For example, Amelink et al. (2005) used a "highway in the sky" to show the position and velocity of the aircraft with respect to targets for a landing approach. In addition, they used the projective geometry of splay angle to show position relative to a target total energy path. This provided specific information about the role of the throttle (correct errors with respect to the energy path) and stick (achieve the appropriate balance between kinetic [speed] and potential [altitude] energy) for managing the approach to landing. Also, Mulder et al. used projective geometry to represent collision zones as shaded regions on an in-cockpit display for managing air traffic. This allowed pilots to resolve collisions by placing their flight vector into the open spaces on the display. Thus, shaded regions represented potential collisions and openings represented safe fields of travel.

Metaphors tap into forms of knowledge (i.e., base domains) other than pattern recognition skills. The most familiar use of metaphor is the desktop metaphor that played a significant role in the development of personal computing. In this case, familiarity with how to manage files in an office together with analogical relations associated with movements using the mouse control were employed as the referents to facilitate intuitions about how to manage information in a computer. Thus, similar data files can be organized into "folders" and files can be deleted by placing them in the "trash can," and they are only permanently deleted when the "trash is emptied."

The key to both analogs and metaphors is to allow existing schemas to facilitate the assimilation of new information in a way that facilitates accommodation to new contexts. In other words, the goal is to facilitate smart generalizations from familiar experiences to new situations. Or in still other terms, the goal is to support adaptation and learning. This is the essence of the problem of designing interfaces to complex work domains that are coherent.

9.5   INTERACTIONS

… the difficult and critical [display] design problem … is what is a useful partial isomorphism between the form and the content of a representation. To accomplish this the designer first must determine what constraints between the symbol and what is symbolized should be established, and what techniques for analyzing the domain semantics can specify these constraints… A second design problem is how to map the chosen aspects of domain structure and behavior into characteristics of the representation so that the domain seman-tics are directly visible to the problem solver.

Woods and Roth
1988, p. 29

It is critical to realize that the correspondence and coherence problems go hand in hand. Focusing on either, without due consideration to the other, is not sufficient. Although the law of requisite variety must be respected (i.e., the problem must be fully described), exclusive focus on solving the correspondence problem will typi-cally result in interfaces that completely overload and baffle the humans with a booming, buzzing confusion of data with little meaning. Although the capacity limitations of humans must be respected, exclusive focus on solving the coherence problem will typically lead to easy-to-use interfaces that trivialize complex problems, leading to a false sense of security (Feltovich et al. 2004). In both cases, the eventual result will be catastrophic failures of the system. To achieve the goal of supporting productive thinking in order to solve complex problems, it is necessary that both the correspondence and coherence problems be addressed.

The success of the safety parameter display developed by Woods et al. (1981) inspired many people to consider using octagons or other geometric forms in their interface designs. But too often, the attention was exclusively on improving coherence. The motivation was that "objects" supported parallel processing of multiple variables. However, often designers failed to give the same care that Woods et al. (1981) gave to the correspondence problem—that is, in choosing the parameters to map into the geometric form. It is critical to appreciate that the eight polar axes incorporated into the safety parameter display represented the integration of hundreds of variables. The parameters were chosen specifically because they represented key states of the process (e.g., mass balances) that were directly relevant to judging the stability of the process. The parameters were then positioned and scaled so that distinctly different deforma-tions were associated with categorically different types of failures to facilitate fault diagnosis. Thus, the utility of the octagon shape as a representation of the process depended both on the capacity of people to "see" the patterns (i.e., coherence) and on the fact that the patterns on the interface were grounded in the deep structure of the work processes (i.e., correspondence).

It is also important to note that the octagon display was a component of a larger interface that included other representations. For complex processes, it is impossible, and not merely difficult, to create a single representation or display that fully meets both the goals for correspondence (i.e., comprehensively covers the deep structure of the process) and the goals for coherence (i.e., can be easily apprehended). Thus, the interfaces in sociotechnical work systems typically require a collection of representations (each providing different insights into the process). In these cases, it becomes necessary for designers to consider the mappings from one representation to another.

Woods (1984) used an analogy to editing a film so that transitions from one scene to another allow viewers to easily follow the narrative. In the same way, interface designers need to consider transitions between multiple representations and the impact on human problem-solving processes. Woods (1984) suggested multiple ways that designers can help operators to structure their search through multiple representations in order to maintain a coherent view of the complex processes (e.g., including landmarks or long shots—global views or maps). Woods (1984) used the term "visual momentum" to reference the degree that transitions between representations were coherent—high visual momentum means high coherence across multiple representations (Bennett and Flach 2012).

The key point is that the correspondence and coherence problems are intimately related, such that each simultaneously constrains the other. Thus, a question that is posed about one is ill-posed without reference to the other. For example, it is meaningless to consider whether any particular type of display (e.g., various polygons) is better than any other type (e.g., bar graphs) without reference to specific constraints within a problem domain. With regard to the design of representations, semantics (i.e., meaning) and syntax (i.e., form) are intimately linked. In fact, that is exactly the challenge, to build meaningful forms. The goal of EID is that the structure or form of the representation directly conveys the meaning (i.e., deep structure) of the domain being represented. Interacting with the representation should facilitate learning about the processes being represented.

9.6 CONCLUSION

The characteristics of man as a cognitive system, primarily his adaptability, should not be used as a buffer for bad designs, but rather as a beacon for good designs.

Hollnagel and Woods
1983, p. 597

CSE and EID reflect paradigmatic changes in how we view the role of humans in complex work systems and how we approach the challenge of interface design. CSE views humans as experts capable of discovering creative solutions to complex problems. EID views the challenge of interface design as the development of representations that facilitate both skilled control and creative problem solving.

To support skilled control and creative problem solving, it is necessary to address both the correspondence problem and the coherence problem. In addressing the correspondence problem, it is necessary to analyze the work domain to identify the functional constraints associated with means and ends. In addressing the coherence problem, it is necessary to explore ways to leverage human pattern recognition skill and existing knowledge to facilitate smart generalizations to novel situations. A well-designed interface needs to simultaneously leverage existing skills and knowledge (support assimilation) and shape those skills and knowledge to meet the demands of novel situations (support accommodation). Thus, interactions with the representation should provide direct insight into the potentially changing dynamics of the process.

It may be surprising to some to come to the end of a chapter on interface design without seeing any pictures of interfaces. However, static images rarely capture the power of EID displays, where many of the patterns that are most important to supporting productive thinking are in motions associated with process changes and actions of the operator. Additionally, without extensive explanation of the underlying deeper processes, it will be difficult for readers to appreciate the value of a particular visual form. Thus, pictures typically inspire naïve generalizations based on the form or syntax of the picture, rather than on the semantic mapping of that form to a problem. Those readers who need to see pictures can find lots of pictures in the papers cited as examples throughout this chapter. These papers also provide details about how components in the representations map to the particular work domains. In many of the references cited, the mapping to the work domain was explicitly linked to levels in a means–ends Abstraction Hierarchy. In an earlier publication, where space was less constrained we have provided many pictures of EID interfaces along with discussions of the motivations behind the representations (Bennett and Flach 2011).

Thus, the goal of this chapter is not to suggest any particular form of representation. Rather, the goal is to reframe the interface design problem from one of information processing to one of meaning processing (Flach and Voorhorst 2016; Flach et al. 2011, 2015). That is, the goal is to inspire designers to consider ways to support productive thinking through the design of representations that make the deep structure of work domains accessible.

REFERENCES

Amelink, H.J.M., Mulder, M., van Paassen, M.M., and Flach, J.M. 2005. Theoretical foundations for total energy-based perspective flight-path displays for aircraft guidance. *International Journal of Aviation Psychology* 15:205–231.

Beltracchi, L. 1987. A direct manipulation interface for heat engines based upon the Rankine cycle. *IEEE Transactions on Systems, Man, and Cybernetics* SMC-17:478–487.

Bennett, K.B., and Flach, J.M. 2011. *Display and Interface Design: Subtle Science, Exact Art*. London: Taylor & Francis.

Bennett, K.B., and Flach, J.M. 2012. Visual momentum redux. *International Journal of Human–Computer Studies* 70:399–414.

Bennett, K.B., Flach, J.M., McEwen, T., and Fox, O. 2015. Enhancing creative problem solving through visual display design. In *The Handbook of Human-Systems Integration (HSI)*, eds. D. Boehm-Davis, F. Durso, and J.D. Lee, 419–433. Washington, DC: American Psychological Association. ISBN-13: 978-1433818288.

Bennett, K.B., Posey, S.M., and Shattuck, L.G. 2008. Ecological interface design for military command and control. *Journal of Cognitive Engineering and Decision Making* 2 (4):349–385.

Billings, C.E. 1997. *Aviation Automation: The Search for a Human-Centered Approach.* Mahwah, NJ: Erlbaum.

Bisantz, A.M., and Burns, C.M. eds. 2008. *Applications of Cognitive Work Analysis.* London: Taylor & Francis.

Borst, C., Mulder, M., and Paassen, M.M. van. 2010. Design and simulator evaluation of an ecological synthetic vision display. *Journal of Guidance, Control, and Dynamics* 33 (5):1577–1591.

Chase, W.G., and Simon, H.A. 1973. The mind's eye in chess. In *Visual Information Processing*, ed. W.G. Chase, 215–281. New York: Academic Press.

Comans, J. Borst, C., van Paassen, M.M., and Mulder, M. (2014). Risk perception in ecological information systems. In *Advances in Aviation Psychology*, eds. M.A. Vidulich, P.S. Tsang, and J.M. Flach, 121–138. Aldershot, UK: Ashgate.

Christoffersen, K., and Woods, D.D. 2002. How to make automated systems team players. In *Advances in Human Performance and Cognitive Engineering Research*, ed. E. Salas. 2:1–12. St. Louis, MO: Elsevier Science.

de Groot, A. 1965. *Thought and Choice in Chess.* The Hague, Netherlands: Mouton Press.

Duncker, K. 1945. On problem-solving. *Psychological Monographs* 58(5):1–113.

Ellerbroek, J., Brantegem, K.C.R., van Paassen, M.M., de Gelder, N., and Mulder, M. 2013a. Experimental evaluation of a co-planar airborne separation display. *IEEE Transactions on Human–Machine Systems* 43(3):290–301.

Ellerbroek, J., Brantegem, K.C.R., van Paassen, M.M., and Mulder, M. 2013b. Design of a co-planar airborne separation display. *IEEE Transactions on Human–Machine Systems* 43(3):277–289.

Ericsson, K.A., and Charness, N. 1994. Expert performance: Its structure and acquisition. *American Psychologist* 49(8):725–747.

Ericsson, K.A., Charness, N. Feltovich, P.J., and Hoffman, R.R. eds. 2006. *Cambridge Handbook of Expertise and Expert Performance.* New York: Cambridge University Press.

Feltovich, P.J., Hoffman, R.R., and Woods, D. 2004. Keeping it too simple: How the reductive tendency affects cognitive engineering. *IEEE Intelligent Systems* May/June:90–95.

Flach, J.M. 2015. Supporting productive thinking: The semiotic context for cognitive systems engineering (CSE). *Applied Ergonomics*, available on line: http://dx.doi.org/10.1016/j.apergo.2015.09.001

Flach, J.M., Bennett, K.B., Stappers, P.J., and Saakes, D.P. 2011. An ecological approach to meaning processing: The dynamics of abductive systems. In *Human Factors of Web Design*, 2nd Edition, eds. R.W. Proctor and K.-P.L. Vu, 509–526. Mahwah, NJ: Erlbaum.

Flach, J.M., Bennett, K.B., Woods, D.D., and Jagacinski, R.J. 2015. Interface design: A control theoretic context for a triadic meaning processing approach. In *Handbook of Applied Perceptual Research*, eds. R.R. Hoffman, P.A. Hancock, R. Parasuraman, J.L. Szalma, and M. Scerbo, 647–668. New York: Cambridge University Press.

Flach, J.M., Schwartz, D., Bennett, A., Russell, S., and Hughes, T. 2008. Integrated constraint evaluation: A framework for continuous work analysis. In *Applications of Cognitive Work Analysis*, eds. A.M. Bisantz and C.M. Burns, 273–297. London: Taylor & Francis.

Flach, J.M., and Vicente, K.J. 1989. *Complexity, Difficulty, Direct Manipulation, and Direct Perception.* Tech Report: EPRL-89-03. Engineering Psychology Research Laboratory, University of Illinois.

Flach, J.M., and Voorhorst, F.A. 2016. *What Matters?* Dayton, OH: Wright State University Library. http://corescholar.libraries.wright.edu/books/127/.

Gibson, J.J. 1979. *The Ecological Approach to Visual Perception.* Boston: Houghton Mifflin.

Hall, D.S., Shattuck, L.G., and Bennett, K.B. 2012. Evaluation of an ecological interface design for military command and control. *Journal of Cognitive Engineering and Decision Making* 6(2):165–193.

Hollnagel, E., and Woods, D.D. 1983. Cognitive systems engineering: New wine in new bottles. *International Journal of Man–Machine Studies* 18:583–600. [originally Riso Report M2330, February 1982] (Reprinted *International Journal of Human–Computer Studies*, 51(2):339–56, 1999 as part of special 30th anniversary issue).

Hutchins, E.L. 1995. *Cognition in the Wild*. Cambridge, MA: MIT Press.

Hutchins, E.L, Holland, J.D., and Norman, D.A. 1986. Direct manipulation interfaces. In *User Centered System Design*, 87–124. Hillsdale, NJ: Erlbaum.

Kantowitz, B.H., and Sorkin, R.D. 1983. *Human Factors: Understanding People–System Relationships*. New York: Wiley.

Lave, J. 1988. *Cognition in Practice: Mind, Mathematics, and Culture in Everyday Life*. Cambridge: Cambridge University Press.

Lindblom, C.E. 1959. The science of "muddling through." *Public Administration Review* 19 (2):79–88.

Lindblom, C.E. 1979. Still muddling, not yet through. *Public Administration Review* 39 (6):517–526.

McEwen, T.R., Flach, J.M., and Elder, N.C. 2012. Ecological interface for assessing cardiac disease. *Proceedings of the ASME 2012 11th Biennial Conference on Engineering Systems Design and Analysis*, ESDA2012, July 2–4:881–888. Nantes, France. ASME ESDA2012-82974.

McEwen, T., Flach, J.M., and Elder, N. 2014. Interfaces to medical information systems: Supporting evidence-based practice. *IEEE: Systems, Man, and Cybernetics Annual Meeting* Oct 5–8:341–346. San Diego, CA.

Miller, G.A. 1956. The magic number seven, plus or minus two: Some limits on our capacity for processing information. *Psychological Review* 63:81–97.

Naikar, N. 2013. *Work Domain Analysis*. Boca Raton, FL: CRC Press.

Norman, D.A. 1986. Cognitive engineering. In *User Centered System Design*, eds. D.A. Norman and S.W. Draper, 31–61. Hillsdale, NJ: Erlbaum.

Norman, D.A., and Draper, S.W. eds. 1986. *User Centered System Design*. Hillsdale, NJ: Lawrence Erlbaum.

Piaget, J. 1973. *The Child and Reality*. Translated by R. Arnold. New York: Grossman. Original edition: *Problemes de Psychologie Genetique*.

Powers, W.T. 1973. *Behavior: The Control of Perception*. Chicago: Aldine de Gruyter.

Rasmussen, J. 1986. *Information Processing and Human–Machine Interaction: An Approach to Cognitive Engineering*. New York: Elsevier.

Rasmussen, J., and Vicente, K.J. 1989. Coping with human errors through system design: Implications for ecological interface design. *International Journal of Man–Machine Studies* 31:517–534.

Reynolds, R.I. 1982. Search heuristics of chess players of different calibers. *American Journal of Psychology* 95:373–392.

Roth, E.M., Bennett, K.B., and Woods, D.D. 1987. Human interaction with an 'intelligent' machine. *International Journal of Man–Machine Studies* 27:479–526.

Sarter, N., Woods, D.D., and Billings, C. 1997. Automation surprises. In *Handbook of Human Factors/Ergonomics*, 2nd edition, ed. G. Salvendy, 1926–1943. New York: Wiley.

Shneiderman, B. 1982. The future of interactive systems and the emergence of direct manipulation. *Behavior and Information Technology* 1:237–256.

Shneiderman, B. 1983. Direct manipulation: A step beyond programming languages. *IEEE Computer* 16(8):57–69.

Shneiderman, B. 1992. *Designing the User Interface: Strategies for Effective Human Computer Interaction*. Reading, MA: Addison-Wesley.

Vicente, K.J. 1999. *Cognitive Work Analysis*. Mahwah, NJ: Erlbaum.

Vicente, K.J., Christoffersen, K., and Pereklita, A. 1995. Supporting operator problem solving through ecological interface design. *IEEE Transactions on Systems, Man, and Cybernetics*, SMC-25:589–606.

Vicente, K.J., and Rasmussen, J. 1992. Ecological interface design: Theoretical foundations. *IEEE Transactions on Systems, Man, and Cybernetics*, SMC-22:589–606.

Vicente, K.J., and Wang, J.H. 1998. An ecological theory of expertise effects in memory recall. *Psychological Review* 105:33–57.

Wertheimer, M. 1959. *Productive Thinking*. New York: Harper & Row.

Woods, D.D. 1984. Visual momentum: A concept to improve the cognitive coupling of person and computer. *International Journal of Man–Machine Studies* 21:229–244.

Woods, D.D. 1991. The cognitive engineering of problem representations. In *Human–Computer Interaction and Complex Systems*, eds. G.R.S. Weir and J.L. Alty, 169–188. London, UK: Academic Press.

Woods, D.D., and Roth, E.M. 1988. Cognitive systems engineering. In *Handbook of Human–Computer Interaction*, ed. M.G. Helander, 1–43. Amsterdam: North-Holland.

Woods, D.D., Roth, E.M., and Bennett, K.B. 1990. Explorations in joint human–machine cognitive systems. In *Cognition, Computation, and Cooperation*, eds. S.P. Robertson, W. Zachary, and J.B. Black, 123–158. Norwood, NJ: Ablex.

Woods, D.D., Wise, J.A., and Hanes, L.F. 1981. An evaluation of nuclear power plant safety parameter display systems. *Proceedings of the Human Factors Society Annual Meeting* 25:110–114. Santa Monica, CA: Human Factors and Ergonomics Society.

10 Making Brittle Technologies Useful

Philip J. Smith

CONTENTS

One of the continuing challenges for cognitive systems engineering is the need to develop a deeper understanding of how cognitive tools influence operator performance (Hollnagel and Woods 2005; Smith et al. 2012; Woods and Hollnagel 2006), consistent with one of the fundamental precepts of CSE that "automation does not simply supplant human activity but rather changes it, often in ways unintended and unanticipated by the designers of automation" (Parasuraman and Manzey 2010, p. 381). Equally important is the need for CSE as a field to influence the design decisions of system developers (Roth et al. 1987; Woods and Roth 1988).

One of the barriers to achieving these goals has been the temptation to apply labels that easily resonate with audiences, but that fail to provide the insights necessary to guide the design and functioning of operational systems. As an example from the annals of cognitive psychology, we have a misleading label such as "confirmation bias" (Mynatt et al. 1978; Wason 1960). Within the CSE literature itself, I find similar issues with labels such as "complacency."

In this chapter, I first describe the history of the label "confirmation bias" to illustrate how such a misleading label can focus attention on motivational influences on behavior without adequately considering the adaptive cognitive processes that, under very narrow and unusual circumstances, can *incorrectly* make it appear *as if* people are *motivated* to avoid those data that could reject a hypothesis that they have generated. The goal of this initial discussion is to highlight the need for designers to more deeply understand the cognitive processes that drive and support human performance and to emphasize that such cognitive processes need to be understood in context.

This initial discussion then serves as a lead-in to a more detailed consideration of labels such as "compliance" and "overreliance" as they have been applied to the impact of brittle technologies on human performance. This broader focus emphasizes the need to understand joint human–machine systems in terms of the interplay of the task and broader environmental context with the design of a new technology (cognitive tool) and its impacts on the cognitive processes of the human operator(s). In this discussion, models based on attention, perception, memory, and problem-solving processes are outlined, along with consideration of design solutions that support these cognitive processes more effectively.

10.1 RESONATING WITH "CONFIRMATION BIAS" AS A LABEL (NOT!)

Clever experiments by psychologists such as Wason (1960) and Mynatt et al. (1978) have demonstrated that we can create situations where, when trying to complete an inductive reasoning task, there is a tendency for some people to form a hypothesis based on the initially available data and to then collect data that are consistent with the question "What are typical findings that I would expect to see if this hypothesis is correct?" rather than asking "What data can I collect that would be most informative for further supporting my hypothesis or for disconfirming my hypothesis relative to competing hypotheses?"

More specifically, Wason's (1960) experiment studied the performances of college students, examining whether they seek:

> (i) confirming evidence alone (enumerative induction) or (ii) confirming and disconfirming evidence (eliminative induction), in order to draw conclusions in a simple conceptual task. The experiment is designed so that use of confirming evidence alone will almost certainly lead to erroneous conclusions because (i) the correct concept is entailed by many more obvious ones, and (ii) the universe of possible instances (numbers) is infinite. (p. 129)

In this study, students were presented with three numbers 2, 4, 6, and told that these numbers "conformed to a simple relational rule and that their task was to discover it by making up successive sets of three numbers, using information given after each set to the effect that the numbers conformed, or did not conform, to the rule" (p. 130). The set of three numbers was actually generated based on the rule that the three numbers were in increasing order of magnitude.

The results concluded that 9 out of the 29 students "were unable, or unwilling to test their hypotheses" (p. 130) because they generated rules that were special cases of the correct rule (e.g., a hypothesized rule such as "The set of 3 numbers is being generated by picking some initial number and adding 2 to each successive number") and then repeatedly tried to test this hypothesized rule by asking about exemplars consistent with this hypothesized rule (such as "Is 1, 3, 5 consistent with the rule?").

This behavior has come to be labeled a confirmation bias, noting that "inferences from confirming evidence (Bacon's "induction by simple enumeration") can obviously lead to wrong conclusions because different hypotheses may be compatible with the same data. In their crudest form such inferences are apparent in the selection of facts to justify prejudices" (Wason 1960, p. 130).

This latter statement tends to fixate on motivation as the underlying causal factor, even though there is no clear reason why participants in studies such as the one reviewed above should have "prejudices." There also is more broadly an assumption by many that these results with college students represent humanity writ large.

In contrast with this characterization, Klayman and Hah (1987) provide a more useful framing of such findings for system designers, noting that:

> many phenomena labeled confirmation bias are better understood in terms of a general positive test strategy. With this strategy, there is a tendency to test cases that are expected (or known) to have the property of interest rather than those expected (or known) to lack that property. This strategy is not equivalent to confirmation bias in the first sense; we show that the positive test strategy can be a very good heuristic for determining the truth or falsity of a hypothesis under realistic conditions. It can, however, lead to systematic errors or inefficiencies. The appropriateness of human hypotheses-testing strategies and prescriptions about optimal strategies must be understood in terms of the interaction between the strategy and the task at hand. (p. 211)

This framing suggests that designers need to understand the following:

- One strategy that is a natural phenomenon of human decision making involves a "generate and test strategy" in which the person quickly generates a hypothesized answer that is consistent with initially observed data (Elstein et al. 1978; Smith et al. 1986) and then proceeds to test that hypothesis without simultaneously considering alternatives. (Note that this "generate and test" model of decision making has strong analogies to recognition primed decision-making [Klein 1993] and case-based reasoning [Kolodner 1993], as well as the Miller et al. 1960 TOTE model, except that these other models of decision-making tend to focus on more complex task environments.)
- An expert's schema associated with this hypothesis will include a "pre-compiled" indication of the test (or tests) to run to effectively evaluate that hypothesis ("the task at hand") as a result of training or past experience (Smith et al. 1986). This test may or may not be consistent with Klayman and Ha's positive test strategy depending on the "task at hand." The expert doesn't think: I've got three competing hypotheses that I need to evaluate. What is the most diagnostic test that I can run to discriminate among these hypotheses? The expert often simply thinks: "What test should I run to test this (one) hypothesis? However, if the activated schema adequately represents the expertise necessary to respond to the "task at hand" (i.e., if the person is truly an expert for this task), the knowledge embedded in the associated schema will often indicate a test that implicitly considers discriminating among the likely alternatives based on past experience.
- The less-than-expert performer, on the other hand, may not have "pre-compiled" knowledge regarding what test would be most effective and therefore may be forced to rely on a weak method (Newell and Simon 1972) such as a more deliberative attempt to apply a positive test strategy in order to evaluate the hypothesis that has been generated, where the "positive test" selected has not been vetted by past experience or training. Furthermore, even though this is a more deliberative memory-driven process, it may be completed without awareness that a "positive test strategy" is being applied and therefore completed without reflection regarding potential weaknesses in applying this strategy.
- Expertise is context dependent. A person may be an expert when confronted with certain situations but less-than-expert when confronted with an anomalous scenario that is outside his or her range of training or experience. Thus, even experts within some domain may be confronted with an anomalous scenario that makes it necessary for them to rely on weak methods.

More concisely, the point of this first example is that

- Simplistic labels such as confirmation bias seem to resonate with many listeners (who seem to take overgeneralizations at face value) and can reinforce a shallow understanding of human performance by system designers, emphasizing constructs such as inappropriate motivation that are not sufficiently informative in making design decisions.

- The alternative suggested by CSE is to understand the cognitive processes of both experts and less-than-experts for the "task at hand" and to use this knowledge to guide the development of cognitive tools. The designers of such cooperative problem-solving systems (Jones and Jacob 2000; Jones and Mitchell 1995; Smith et al. 1997) need to consider the underlying cognitive processes in order to support and amplify the performances of operators with different levels of expertise for the range of anticipated and unanticipated tasks/scenarios that could be encountered, and over which operators may exhibit varying levels of expertise (Feltovich et al. 1997).

10.2 RESONATING WITH "COMPLACENCY" AS A LABEL (NOT!)

There is a substantial literature focusing on the impact of brittle technologies on worker performance. This includes technologies designed to present alerts, decision aids that provide recommendations for actions that must be initiated or approved by the human operator, and automated systems where specific tasks are delegated to the technology, with the results monitored by a human supervisor (Sheridan 2002) in case intervention is required.

Much of this literature uses the label "automation complacency." One concern with this terminology is that, like the term "confirmation bias," in common parlance, "complacency" as a label seems to focus on motivation as the cause. Indeed, my experience has been that often even advanced graduate students have difficulty getting past the shallow perspective encouraged by this label, fixating on motivational issues rather than considering the influences that system design can have on the operator's or decision-maker's cognitive processes.

The cognitive engineering literature within which this label appears does, however, provide a considerably deeper exploration of how brittle technologies can influence workers, among other things making a distinction among designs that can induce "errors of omission" versus "errors of commission." Wickens et al. (2015), for example, describe complacency as a phenomenon "associated simply with the failure to be vigilant in supervising automation prior to the automation failure," associated with an "assumption that 'all is well' when in fact a dangerous condition exists" (p. 960). They further describe overreliance as performance associated with "an increased likelihood of the human's failing to identify and respond to a true system issue" (an error of omission) (p. 959) and describe overcompliance in terms of false alarms where "the operator can incorrectly believe that the automation has correctly identified a fault," typically leading to "incorrect, unnecessary actions" (p. 959) (an error of commission). The literature also explores the impacts of brittle technologies in different task contexts such as detection, planning, diagnosis, and process control.

Within this framework, the literature attempts to characterize the cognitive processes that result in outcomes that are categorized by these definitions, focusing on both errors of omission and errors of commission. Models that have been proposed can be categorized into two classes:

- Those that focus on attentional processes and associated decision-making processes to deal with the demands of multitasking (Mosier and Skitka 1996; Mosier et al. 2001; Parasuraman and Manzey 2010; Skitka et al. 1999).
- Those that produce outcomes that look like hypothesis fixation (Fraser et al. 1992; Woods and Hollnagel 2006) resulting from perceptual processes (Layton et al. 1994; Smith et al. 1992, 1997), from constructive memory processes that can result in memory distortions (Mosier and Skitka 1996), and from influences of the brittle technology on problem-solving processes (Smith et al. 1997). Note that these are all explanations based on cognitive processes, not on inappropriate motivation. Below, I discuss how such cognitive processes are influenced by the introduction of technologies into different task contexts.

10.3 RELIANCE IN A MULTITASKING ENVIRONMENT: THE ROLE OF ATTENTIONAL PROCESSES

Parasuraman and Manzey (2010) concluded that "automation complacency occurs under conditions of multiple task load, when manual tasks compete with the automated task for the operator's attention" (p. 381). In laboratory studies of college students, an example of such a task has been the use of Comstock and Arnegard's (1992) Multiple Attribute Task Battery, a multitask flight simulation system that presents "a two-dimensional compensatory tracking and an engine fuel management task, both of which had to be carried out manually, and a third task involving engine monitoring that required participants to detect abnormal readings on one of four gauges" (Parasuraman and Manzey 2010, p. 383). The latter task was supported by technology that was not perfectly reliable and that sometimes failed to alert the operator of an abnormal reading. In the field, an analogous task would be monitoring aircraft subsystems (Wiener 1981) or working as an anesthesiologist in the operating room while performing other tasks (Aaron et al. 2015).

Studies of multitasking (Muthard and Wickens 2003; Parasuraman and Manzey 2010) make it very clear that we can design technologies (for real-world tasks or for laboratory studies) that demonstrate *reliance* on technology, showing an increase in the failure of system operators to detect abnormal readings when an alerting system fails to provide an indication of an abnormal reading (as contrasted with manual performance of the same tasks without any such technological support). Whether this performance should be labeled reliance versus overreliance in such multitasking environments, however, depends on some determination of the "optimal" allocation of attention across tasks (Moray and Inagaki 2000). Such *reliance* is also consistent with Mosier and Skitka's (1996) findings regarding automation compliance or automation bias: "The tendency to use automated cues as a heuristic replacement for vigilant information seeking and processing" (p. 205).

This difference in performance with or without support from a software alerting function can be at least partially accounted for using a model focusing on attentional processes. The fundamental assumption underlying such an attention model regarding the cause of reliance is that the human operator has limited attentional resources and therefore needs to distribute them across multiple competing tasks (Bailey and Scerbo 2007; Parasuraman and Manzey 2010). (Designers must keep in mind, however, that in extrapolating this to expert performance, it is important to keep in mind that with over-practice on a given task, attentional demands can be reduced.)

Figure 10.1, from Parasuraman and Manzey (2010), presents a qualitative model showing attention as a mediating construct affecting the degree of reliance on decision aids resulting from "(a) an actual, overt redirection of visual attention in terms of reduced proactive sampling of relevant information needed to verify an automated aid" or "(b) a more subtle effect reflected in less attentive processing of this information, perhaps because covert attention is allocated elsewhere" (p. 402). This figure also identifies factors that have been hypothesized to influence the allocation of attentional resources when interacting with technology. The result of such attention allocation could be overreliance, appropriate reliance, or underreliance on the technology.

Consistent with the "redirection of visual attention," Metzger and Parasuraman (2005) reported the results of a study that demonstrated that the introduction of alerting software can reduce the extent to which the operator (in this study, students) visually scans raw data sources in order to provide manual confirmation that the alerting software has not missed some abnormal state. Other studies, however, indicate that a construct such as inattentional blindness (Thomas and Wickens 2006) is necessary because reliance can be demonstrated (relative to completely manual performance) even when the visual scanning of key data is the same for manual and technology-supported operations (Duley et al. 1997). Bahner et al. (2008) indicate similar results supporting the phenomenon of inattentional blindness as associated with the duration/depth of processing of relevant information in a study of automation bias looking at information access and processing.

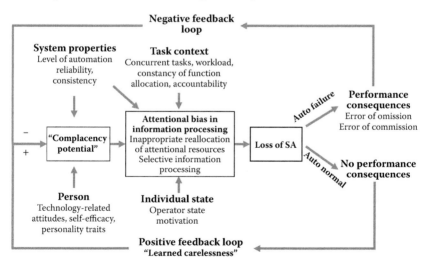

FIGURE 10.1 Factors influencing attentional bias. (From Parasuraman, R. and Manzey, D. (2010). Complacency and bias in human use of automation: An attentional integration. *Human Factors,* 52(3), 381–410. Copyright 2010 by Sage Publications, reprinted with permission of Sage Publications, Inc.)

Thus, consistent with the concept of bounded rationality (Simon 1957), this attentional model suggests that the distribution of limited attentional resources across multiple tasks can rationally result in *reliance* on technology based on the perceived risks associated with the trade-off of distributing more or less attention to an automation supported task versus other competing tasks that lack automation support. However, assuming this attentional model, *overreliance* also can result if, relative to some optimal distribution of attention across these tasks, attention has been suboptimally allocated (Moray and Inagaki 2000).

Most of the studies of automation complacency compare performance when an automated alerting system is present with performance when such technology support is not provided (manual performance) using testbeds such as the Multiple Attribute Task Battery (a multitask flight simulation system) described earlier. For example, Galster and Parasuraman (2001) found "clear evidence of a complacency effect ... with pilots detecting fewer engine malfunctions when using the EICAS [Engine Indicator and Crew Alerting System] than when performing this task manually" (Parasuraman and Manzey 2010, p. 387).

Furthermore, the instructions regarding priorities in such studies generally have been neutral: "Subjects were requested to give equal attention to all three tasks" (Parasuraman et al. 1993, p. 8). Thus, it is not really possible to infer the assumptions made by participants regarding how they should allocate attention when an automated alert system was provided. As a result, while the evidence for *reliance* on support from an alerting system is overwhelmingly strong, it is often unreasonable to assign the label *overreliance*. Without a clear determination in such studies regarding how participants should have been calibrated in order to optimally timeshare among the tasks, it is impossible to assert that overreliance has been observed. Note, however, that, regardless of whether appropriate reliance or overreliance could arise, design strategies that reduce the need for reliance, as well as design strategies that reduce the impact of reliance, should be considered.

Thus, while the evidence for operator reliance on brittle software in multitask environments is unequivocal (both in research studies and in practice), based on this conceptual framework, two questions arise:

- Under what circumstances is attention "inappropriately" allocated, resulting in over- or underreliance? The designer should, however, keep in mind that, as Dekker and Hollnagel (2004) point out, since the optimal distribution of attention is generally unknown in real-world task environments, it is incorrect to conclude that a system operator was over- or underreliant just because a bad outcome occurred—as even an optimal process for allocating attention can sometimes produce a bad outcome.
- What are the factors that influence this attention allocation strategy in appropriate or inappropriate ways?

Regarding this second point, note that the model characterized in Figure 10.1 indicates a variety of *potential influencing factors* associated with the person, individual state, system properties, and task context. The literature is less conclusive, however, regarding whether, when and how such factors will actually influence this attentional bias or degree of reliance. And in most cases, because the "optimal" level of reliance is unknown, the results should really be framed as indications of how different factors influence the degree of *reliance* rather than using the labels *complacency* or *overreliance*.

One such influencing factor suggested in this model is the *Task Context*. Here, the most significant factor highlighted is whether or not the operator must multitask. Parasuraman et al. (1993) found, for example, that "when participants had simply to 'back up' the automation routine without other duties, monitoring was efficient and near perfect in accuracy" and that "detection of automation failures was significantly poorer in the multitask condition than in the single task condition" (p. 384). Parasuraman et al. (1993) further indicated that vigilance was not a sufficient construct to account for their findings as there was "no evidence of within-session, time-on-task effects on detection rate of automation failures" on their tasks involving complex, dynamic displays (p. 16), although caution is warranted in overgeneralizing this finding, as the durations were short and event probabilities were high relative to the tasks involved in many vigilance studies.

A second influencing factor indicated by this model focuses on *System Properties*. Within this attentional model, increased system reliability should increase level of reliance (May et al. 1993). The results regarding this, however, are mixed. In a study with college students, Parasuraman et al. (1993), for example, found that detection for automation failures was significantly improved when automation detection performance changed over time. When automation detection probability was rotated between each of twelve 10-minute blocks in four 30-minute sessions, mean detection performance for automation failures by participants was 0.82. However, when detection probability was constant across all 12 blocks, mean detection probability was 0.33. These results (supported by other similar studies) are consistent with the attentional model where reliance should increase with the perceived reliability of the alerting software.

In contrast, though, this same set of studies found that "detection rate did not differ between conditions where the automation exhibited low-reliability (57.25%) vs. high-reliability (87.5%) for two different subgroups in the constant reliability condition" (p. 16), a finding that does not seem consistent with at least a simple understanding of the determinants of trade-offs in attention (keeping in mind the caveat that a wider range of system reliability might show a relationship since the "low" level of reliability in these studies was not really very low).

As a third example of the influencing factors indicated in the model shown in Figure 10.1, consider *Personal Factors*. As an illustration of this factor, one hypothesis has been that individuals will exhibit a "first-time failure effect," predicting that without prior experience, operators may have an "initial attitude toward high-reliability automation based on high trust" (Parasuraman and Manzey 2010, p. 386). This prediction suggests that an operator will perform more poorly the first time the software fails but will show improved performance on subsequent failures (assuming feedback after each such miss).

In terms of the model shown in Figure 10.1, the assumption is that such an initial belief leads operators to behave as if they believe a priori that the likelihood that the software will fail is low, resulting in the allocation of fewer attentional resources, and thus resulting in a decrease in the detection of automation failures. Such an a priori assumption is presumed to be based on a belief that the developers of the technology were smart and well-intentioned, supported by preliminary exploration/experience with the performance envelope. (See Hoffman, this volume, for a discussion of default trusting.)

Empirical results from tests of this hypothesis have been mixed. Some have found consistent evidence while other have not (Rovira et al. 2007; Wickens et al. 2000). Findings have been equally inconsistent in studies attempting to study the relationship between level of reliance and measures of trust (Bagheri and Jamieson 2004). These results collectively caution against oversimplification, as trusting is a not a simple state or thing that gets calibrated, but is an active process of exploration. (See Hoffman, this volume.)

10.3.1 INFLUENCES ON ATTENTIONAL PROCESSES—SUMMARY

It is clear that if the designer introduces software support to provide alerts within a multitasking environment where the task demands exceed the attentional capacity of the operator, he/she can influence the operator to show *reliance* on the technology. This *reliance* can lead to brittle performance at the system level, where the technology and the human operator as a joint cognitive system fail to respond appropriately to some situation. If the problem space that has been explored consists primarily of routine cases that fall well within the competence envelope, and if the cases that push the work system to the boundaries of its competence envelope are, by definition, infrequent or rare, then this result is to be expected. It is more difficult to assess whether and when *overreliance* will result, especially for unanticipated scenarios, unless the operator/decision maker can see the movement of the work system toward the boundaries of its competence envelope. If that trajectory is not visible, work failures can be expected, which, in hindsight, are often called overreliance. Thus, this possibility needs to be carefully considered in making design decisions. In addition, the designer needs to consider the potential influencing factors noted in Figure 10.1 with the understanding that findings to date are less definitive regarding whether and when one of these influencing factors will have a significant impact.

As described above, much of the research on the impacts of brittle technologies on performance of the work system (not just the human) focuses on attentional models. Such attentional models are not sufficient, however, to account for all of the findings regarding the influences of decision support technologies on worker performance. Below, I discuss how other perceptual, memory, and problem-solving processes can be influenced by brittle technologies, producing *outcomes* that can be characterized as hypothesis or plan fixation.

10.4   HYPOTHESES FIXATION ATTRIBUTED TO PERCEPTUAL, MEMORY, AND PROBLEM-SOLVING PROCESSES

Layton et al. (1994) and Smith et al. (1997) demonstrated that decision support systems can influence the performances of experienced practitioners in profound ways that lead to poor decisions even when no multitasking is involved. In this research, experienced commercial pilots and airline dispatchers were used as participants to evaluate the effects of different types of decision support tools for flight planning. The 27 dispatchers had an average of 9.2 years of dispatching experience with six commercial airlines. The 30 pilots had an average of 9300 flight hours flying for eight commercial airlines.

10.4.1 AUTOMATION-INDUCED PERCEPTUAL FRAMING AS A CAUSE OF HYPOTHESIS FIXATION

Smith et al. (1997) looked at the differences between the following:

- The performance of "sketching-only" participants (experienced commercial airline dispatchers and pilots who were given a graphical interface to manually sketch and evaluate alternative routes around convective weather). To sketch alternative routes, the graphical interface allowed the users to draw routes from one navigational fix or waypoint to the next on a map display that showed navigational fixes and convective weather as illustrated in Figure 10.2.
- The performance of participants using a tool that first generated and then displayed a recommended route around the weather. This is illustrated in Figure 10.3. After they first viewed the software's recommended reroute, these participants also had all of the same capabilities for manually sketching and evaluating alternative routes as the sketching-only group.

To evaluate alternative routes, in addition to viewing the routes overlaid on a weather map display, the dispatchers and pilots could display information about

FIGURE 10.2 Routing options available for sketching. (From Layton, C., Smith, P.J. and McCoy, C.E. Design of a cooperative problem-solving system for en-route flight planning: An empirical evaluation, *Human Factors*, 36, Figure 2, p. 81. Copyright 1994 by Sage Publications, reprinted with permission of Sage Publications, Inc.)

FIGURE 10.3 Computer-recommended orange reroute resulting in an automation-induced perceptual framing effect. (Modified from Layton, C., Smith, P.J. and McCoy, C.E. Design of a cooperative problem-solving system for en-route flight planning: An empirical evaluation, *Human Factors*, 36, Figure 5, p. 91. Copyright 1994 by Sage Publications, reprinted with permission of Sage Publications, Inc.)

winds and turbulence along each alternative route and view the associated air time and fuel burn.

For one of the scenarios studied by Smith et al. (1997), the participants were told:

> It's summer and the aircraft is eight minutes into a flight from Oakland to Joliet. You got off the ground at 1600 Zulu. You notice that there is a solid line of convective thunderstorms directly in your path. Decide what you think the aircraft should do. (p. 365)

The weather information was presented using standard weather displays that these dispatchers and pilots reviewed on a regular basis as part of their flight planning job responsibilities. The results for this particular scenario were striking: 35% of the participants who used the "sketching only" version sketched and selected a route to the south of the weather, while 0% of the participants who first saw the computer's northern reroute selected this southern route.

The concurrent verbal reports were very informative regarding the influence of simply displaying the computer's recommended solution before the user considered the situation. Immediately upon seeing the map display with the current route and weather:

- Participants using the "sketching-only" version made statements like "the cold front in this case is small and is not likely to develop problems to the south, so probably in this case I would elect to route the flight around the south end."

- Participants who immediately looked at the computer's recommended reroute to the north along with the current route and weather (see Figure 10.3) made statements such as "given the season of the year, I don't think it's going to build to the north. I think it's going to build in the south more."

Thus, the results indicate the influence of a powerful, automatic perceptual process that resulted in very different mental representations of the weather situation for the participants in the two groups, which, in turn, affected their selection of reroutes. The *outcome* of this mental process when seeing the computer's recommended solution illustrates an *automation-induced perceptual framing effect*. For without this influence, we would have expected the two groups to have selected the same distribution of reroutes to the north and south instead of a 35% difference. (It should be further noted that effects of this automation-induced perceptual framing effect persisted even though more than half of the participants who immediately saw the computer-recommended solution and selected the northern route proceeded to generate and evaluate routes to both the north and south of the storm. Thus, it would be incorrect to characterize those participants as overreliant.)

More broadly, the *outcome* of this phenomenon could be categorized as a form of *hypothesis fixation* (Fraser et al. 1992; Woods and Hollnagel 2006). Furthermore, since hypothesis fixation is an outcome rather than a cognitive process, the results suggest one way in which the early presentation of a recommended diagnosis or plan by a decision support system can result in hypothesis fixation—through an automatic perceptual process that influences the user to construct a mental representation of the situation (Johnson et al. 1981) that is consistent with the recommendation of the software. This interpretation is in tune with the caution given during the training of physicians and veterinarians that, when viewing a referral, the specialist should first review the case and evaluate the patient without knowledge of any initial diagnosis proposed by the referring doctor, suggesting that such an influence is not restricted to a recommendation generated by software.

In short, even without the effects of the attentional demands associated with multitasking, the presentation of a recommended solution by the computer has a direct influence on the perceptual and cognitive processes of the user, affecting the triggered mental processes in powerful ways and changing the ultimate decision that is made. Below, I discuss still other ways in which the worker's cognitive processes can be influenced by such technologies, resulting in an outcome that looks like hypothesis fixation.

10.4.2 MEMORY AND PERCEPTUAL DISTORTIONS AS A CAUSE OF HYPOTHESIS FIXATION

In the initial discussion of the misleading label "confirmation bias," I highlighted the central role that hypothesis generation plays in human expertise (Elstein et al. 1978). Mosier et al. (1998) cited another cognitive process that can produce an outcome that looks like hypothesis fixation due to the constructive nature of human memory

(Loftus 1975). Mosier et al. (1998) found evidence that pilots distorted their memories while trying to determine what was happening when a false alarm for an engine fire was presented without support from any of the data available from other instruments. Their findings indicated that two-thirds of the pilots incorrectly believed that they had viewed at least one other data source consistent with an engine fire.

Smith et al. (1986) described a similar phenomenon in a study of fault diagnosis by pilots:

> One pilot provided a very interesting example of how people ... distort their recall of other available data ... to make it consistent with the activated frame [hypothesis]. Immediately after hearing a scenario, he said: "Let me get this straight now: *Increasing airspeed*, decreasing altitude and you mean pitch as far as being above or below the horizon based on the artificial horizon?" He then hypothesized that the plane was in a nose-level descent and repeated his recall of the symptoms he had heard: "What's happening to my power? Very definitely we have a situation where we seem to be losing power. The fact that we're decreasing in altitude and our *airspeed is constant* indicates that we are basically in a situation where we are losing altitude. It would stay fairly constant if we're coming down." (italics added) (p. 711)

The interpretation by Smith et al. (1986):

> Within a time span of less than 30 seconds the pilot has distorted his recall. Originally he stated that there was an indication of increasing airspeed. After activating the frame for a nose-level descent, he stated that the airspeed was constant. This indicates a rather self-defeating process. The nose-level descent frame, which the pilot is trying to test by mentally reviewing the symptoms he has learned of, is being used to help "recall" or reconstruct the set of symptoms. The role of the activated frame is so powerful in this recall/reconstruction process that the pilot "remembers" symptoms consistent with that frame rather than the symptoms actually presented. (p. 711)

Both of these studies indicate the powerful effect that the (re)constructive nature of human memory can have on performance. If the system operator focuses on an initial hypothesis (such as a recommendation by a decision support system), the person may subsequently distort his or her memory regarding the content of other data displays to be consistent with this hypothesis—or could even distort the original perception of the sensory data (Guerlain et al. 1999).

10.4.3 TRIGGERING OF INTERNAL REPRESENTATIONS AND RELIANCE ON DATA-DRIVEN PROCESSES AS CAUSES OF HYPOTHESIS FIXATION

The cognitive psychology literature provides strong evidence regarding the impact that an external representation can have upon the corresponding mental model that a person forms and then uses to support problem solving and decision making, as well as the influence of external memory aids (Hutchins 1995; Kotovsky et al. 1985;

Larkin and Simon 1987; Scaife and Rogers 1996). Layton et al. (1994) and Smith et al. (1997) provide additional evidence regarding this type of influence that a decision support system can have on the worker's cognitive processes. Based on their findings, a useful way to model the problem-solving processes of the participating dispatchers and pilots is a characterization based on the triggered problem space (Newell and Simon 1972).

The verbal protocols from a second scenario (see Figure 10.4) used in the studies reported by Layton et al. (1994) and by Smith et al. (1997) illustrate how the task demands can be modeled as the triggering and use of a problem space in which each decision in a sequence of decisions is by itself locally preferred, but that in the end collectively fixates the person on a poor global solution. As an example, Layton et al. describe the performance of one pilot who initially saw the computer's recommended solution, noting that this:

> … illustrates a fascinating example of how particular strategies can lead to poor solutions. His strategy can be characterized as an elimination-by-aspects approach (Tversky 1972), in which the aspects are local decisions about which waypoint to go to next and have been influenced by the route recommended by the software. In particular, in evaluating the path suggested by the computer, this pilot began by saying, "Where should I go next, from PUB [one navigational fix] to TCC [another navigational fix] or from PUB to AMA [yet another navigational fix]?" He selected AMA because it was farther from the storm

FIGURE 10.4 Computer-recommended orange reroute triggering use of an elimination by aspects strategy. (Modified from Layton, C., Smith, P.J. and McCoy, C.E. Design of a cooperative problem-solving system for en-route flight planning: An empirical evaluation, *Human Factors*, 36, Figure 6, p. 93. Copyright 1994 by Sage Publications, reprinted with permission of Sage Publications, Inc.)

west of TCC. He then considered "should I go from AMA to SPS or to ABI or to TCC?" ... He selected SPS. Because of these localized decisions, he never even considered whether this eastern deviation was to be preferred globally to the western route. (p. 367)

As a result, this pilot selected a very questionable route hunting and pecking through a storm. In the debrief afterward, all of the data for that scenario were reviewed, the pilot's response was: "I'd be fired if they knew I had made that choice. How could I have done that?" (I reassured him not only that his data were confidential, but also that we were studying how the design of such software could cause fixation on a poor route—indicating that the software design was to "blame" instead of him.) Further discussion indicated that the pilot had no insights suggestive of how he had arrived at this poor choice.

More abstractly, then, such data suggest a third mechanism through which computers can cause what appears to be fixation or cognitive narrowing, resulting in a poor decision—by influencing the problem space that is constructed through data-driven perceptual processes and which then guides the decision-making process.

Modeling decision making as search through a problem space also highlights another point: The knowledge that an expert has is tacit unless some top-down or bottom-up, data-driven process triggers it. The induced problem space is a powerful determinant in focusing the person's attention, as the decision that needs to be made at each successive node in the search tree strongly influences what data will be viewed and what associated knowledge will be activated.

In short, two heads (the designer of the decision support technology and its user) can only be better than one if they have the right knowledge and it is actually activated and used at the right time.

10.4.4   HYPOTHESES FIXATION—SUMMARY

In Section 10.4, I emphasized that brittle technologies not only can influence worker performance through attentional allocation in multitasking environments but also can produce outcomes that look like hypothesis fixation by influencing perceptual, memory, and problem-solving processes. It needs to be emphasized that these cognitive processes are all consistent with routine human performance when interacting with the world. They are not unique to interactions with technology. However, the designer has to understand how such normal mental processes can be influenced by the design of decision support and process control technologies. The designer must then take this into account when making design decisions in order to help ensure that the worker applies his or her expertise effectively when the software encounters a scenario outside of its competence envelope. Below, I discuss some of the studies completed in efforts to provide guidance for such technology and system design decisions.

10.5   MAKING BRITTLE TECHNOLOGIES USEFUL

As past experience with research studies and fielded systems has overwhelmingly indicated, when designers develop decision support technologies to support complex real-world tasks, these technologies will result in a work system that is brittle. The work system will encounter scenarios that the designers did not anticipate or that are "buggy" because the implemented design did not match the intended design.

Findings to date provide many useful ideas to help improve design decisions, but there is a great deal that we still do not know about how cognitive tools influence operator performance when an anomalous scenario unveils the brittleness of the tool, resulting in an undesirable outcome. Equally important, our understanding of how to use these insights to develop more robust and resilient work systems is still very limited. Below, I discuss various solutions that have been proposed to deal with brittleness, with the caveat that the supporting research is still spotty. Thus, designers need to use their judgment in deciding whether and when to apply such guidance.

10.5.1 Design Concept 1. Human as the Final Arbiter

It is wrong to assume that when a brittle cognitive tool has been introduced, brittle *work system* performance can be avoided by simply assigning responsibility to the system operator or decision maker to make the final decision before an action is executed. Even if the human operator has the knowledge and skill necessary to deal effectively with an anomalous situation and would have applied it if he were working without the support of the technology, the influences of the cognitive tool on the worker's attentional, perceptual, memory, and problem-solving processes may negate effective access to this knowledge and skill (Guerlain et al. 1999; Layton et al. 1994; Smith et al. 2012).

10.5.2 Design Concept 2. Impact of Task Loadings on Reliance

Numerous studies make it clear that the introduction of cognitive tools "under conditions of multiple task load, when manual tasks compete with the automated task for the operator's attention" is likely to result in some level of reliance by the operator, and that such reliance "is found in both naive and expert participants" (Parasuraman and Manzey 2010, p. 381). Whether such performance is simply appropriate reliance given the demands of the multitasking or constitutes "overreliance" or "complacency" depends on a variety of factors that may be difficult to objectively determine.

Thus, task loading is an important design parameter. The technology designer needs to be aware of the potential impact of multitasking on the level of reliance, which, in turn, can affect the performance of the worker in detecting software failures. The designer must therefore carefully consider decisions regarding planned task loadings for the worker across the full range of anticipated use cases. The designer also should consider the need to build in a degree of spare capacity to help ensure resilience (Hollnagel et al. 2006; Smith and Billings 2009) in the face of unanticipated scenarios.

10.5.3 Design Concept 3. Training and Experience as Mitigations

In task environments where multitasking induces some level of reliance that can result in a failure by the person to detect undesirable performance by the technology, such performance can be found in both "naïve and expert participants and cannot be overcome with simple practice" (Parasuraman and Manzey 2010, p. 381). There is, however, some evidence that specific training methods focused on more effective multitasking performance can support more rapid acquisition of dual task skills, which may, in turn, reduce the level of reliance on a cognitive tool (Gopher 1996), but more research is needed on the potential impacts of different training strategies on multitasking and on the detection of technology failures.

Less is known about the potential for training to reduce hypothesis fixation. It is likely that system operators can be trained to more effectively detect and deal with deliberately designed areas of brittleness in a decision support system (such as the deliberate design of a flight planning system that does not consider uncertainty in the weather forecast), assuming appropriately designed information displays are provided. Such training needs to encompass opportunities to actively explore and experience the work system performance across a wide swath of the competence envelope, especially when the work system is being pushed toward the competence envelope boundaries. For example, doctors are taught to develop a differential diagnosis (Richardson 1999) that includes the full range of competing hypotheses in order to reduce hypothesis fixation. To be effective, such training requires effective displays that help the worker understand the performance limitations of the technology.

10.5.4 Design Concept 4. Roles and Responsibilities as Design Parameters

The role assigned to the technology can have a major impact on the likelihood that the human operator will effectively apply his or her knowledge and skill in an anomalous scenario where the software exhibits its limitations or failure modes (Cummings 2004). Alternatives are discussed below.

10.5.4.1 Cognitive Tools as Information Display Systems

Rovira et al. (2007) and Sarter and Schroeder (2001) report findings that software that actively recommends diagnoses or actions is more likely to lead to missed automation failures than software that only supports information display:

> Automated aids that only support processes of information integration and analysis may lead to lower automation bias effects (in terms of commission errors) than aids that provide specific recommendations for certain actions based on an assessment of the available information. (Sarter and Schroeder 2001, p. 396)

An important variation on this is the integration of contextual information displays within the presentation of alerts, instructions, and recommendations. Ockerman and Pritchett (2002), for example, demonstrated that for "the task of planning an emergency descent for a commercial aircraft ... the presence of contextual information in the presentation of an automatically generated emergency descent procedure" helped mitigate "the effects of automation brittleness. By providing pilots with rationale as to the design of the descent procedure, the pilots were better able to correctly determine why a provided procedure was or was not feasible" (p. 382).

10.5.4.2   Cognitive Tools Based on Representation Aiding

The literature on representation aiding and ecological interface design (Burns and Hajdukiewicz 2004; Rasmussen 1983; Smith et al. 2006; Vicente 1999, 2002; Vicente and Rasmussen 1990, 1992) provides a conceptual approach for keeping the worker in the loop. This literature discusses strategies for designing information displays (primarily for process control contexts) that support direct perception to assist in responding to anticipated scenarios, while also providing access to more detailed information to deal with anomalous situations that require knowledge-based processing (Bennett and Flach 1992). (See Chapter 9 for additional details.) An important, underresearched question is the impact of such powerful attention focusing displays on worker performance in anomalous scenarios that were not anticipated in the design of the ecological displays. Jamieson (2002a,b) does provide evidence that a combination of displays based on ecological interface design *along with* task-specific displays based on cognitive task analyses produced better performance than traditional displays for the operation of a petrochemical system for scenarios that were not accommodated during the design process. However, much work needs to be done to understand how such combinations of displays influence an operator's cognitive processes when the design exhibits brittleness.

10.5.4.3   Person as Active Controller versus Passive Monitor

Results by Galster et al. (2001) suggest that the role of the operator as an active controller can also reduce the likelihood that automation failures will be detected. They found that "conflict detection performance was better when [air traffic] controllers were actively involved in conflict monitoring and conflict resolution ('active control') than when they were asked to be passive monitors" (Parasuraman and Manzey 2010, p. 387).

10.5.4.4   Simple Automated Completion of Subtasks

Guerlain et al. (1996) compared abduction/diagnosis (involving the detection of antibodies in a patient's blood on multiple solution problems that could exhibit masking and noisy data) using two different system designs. In one design, the tool automatically completed one of the major subtasks (using certain test results to rule out candidate hypotheses/antibodies). In the other design:

- The medical technologist had responsibility for applying the data to rule out antibodies.
- The software monitored these rule-outs (detected by the software when the user visually marked an antibody on the computer display whenever an antibody was ruled out).
- The software provided the technologist with immediate context-sensitive feedback as soon as it detected a rule-out that it considered suspect.

Both versions reduced slips and mistakes, but they found that, *for those cases where the software had inadequate knowledge and made incorrect rule-outs*, "the system design that automatically completed subtasks for the technologist induced a 29% increase in errors relative to the design that critiqued technologists as they completed the analyses themselves" (Guerlain et al. 1996, p. 101). However, *on cases where the software was competent*, the technologists working alone missed antibodies over twice as often as those supported by the automated rule-out function (11.9% vs. 5.6% error rates).

This highlights a design challenge: Can we develop designs to take advantage of the benefits of more active software support, increasing efficiency and reducing bad outcomes in scenarios where the software is competent, while also reducing the risk of bad outcomes in scenarios that exceed the competence limits of the software?

10.5.4.5 Computer as Critic with Embedded Metaknowledge

Guerlain et al. (1999) and Smith et al. (2012) discuss one possible (partial) solution to this trade-off: Rather than leaving the person on his or her own without active software support, consider designing the computer to act as a coach or critic with embedded metaknowledge to detect cases that might be outside of its range of competence (Hoffman and Ward 2015).

To evaluate this approach, they designed a critiquing system (Fischer et al. 1991; Miller 1986; Silverman 1992) with an interface that provided a convenient tool for running tests and marking intermediate conclusions (see Figure 10.5). As the medical technologist used this interface to work on a patient case, he or she provided informative data that allowed the underlying critiquing functions to compare the technologist's intermediate actions to those prescribed by an expert model in order to

	Donor	D	C	E	c	e	f	V	C^w	M	N	S	s	P1	Le^a	Le^b	Lu^a	Lu^b	K	k	Kp^a	Js^a	Fy^a	Fy^b	Jk^a	Jk^b	Xg^a	Special Type	IS	LISS	IgG	RT	4°	
1	A478	0	0	0	+	+	0	0	0	+	+	0	+	+	0	+	0	+	+	+	0	0	0	+	0	+	+		0	0	0			1
2	B102	0	0	0	+	+	0	0	0	+	+	0	+	+	+	+	0	+	+	+	0	0	+	0	+	+	+		2+	0	2+			2
3	C559	0	0	0	+	+	0	0	0	+	0	+	0	+	0	+	0	+	+	+	0	0	+	0	+	0	+		0	0	2+			3
4	D275	0	+	0	+	+	0	0	0	+	+	+	+	0	0	0	0	+	0	+	0	0	0	+	+	+	+		0	0	0			4
5	E164	0	0	+	+	+	0	0	0	0	+	+	+	+	0	+	0	+	+	+	0	0	0	+	0	+	+		2+	0	0			5
6	F065	+	+	0	0	+	0	0	0	0	+	0	+	+	+	0	+	0	+	+	0	0	+	0	+	+	0		0	0	2+			6
7	G163	+	0	0	+	+	0	0	0	+	+	+	+	0	+	+	0	+	0	+	0	0	0	0	+	+	+		2+	0	0			7
8	H168	+	0	+	+	0	0	0	0	0	+	0	+	+	0	+	0	+	0	+	0	0	+	0	0	+	+		0	0	2+			8
9	R331	+	+	+	0	+	+	0	0	0	0	+	0	0	+	0	+	0	+	+	0	0	+	0	0	+	+		0	0	1+			9
10	A624	+	0	0	+	+	0	0	0	+	+	0	+	+	0	+	+	0	0	+	0	0	0	0	+	0	+		0	0	0			10
	AutoCtrl																												0	0	0			

FIGURE 10.5 Interface design to support interactive critiquing for immunohematologists. (From Smith, P.J., Beatty, R., Hayes, C., Larson, A., Geddes, N. and Dorneich, M. (2012). Human centered design of decisions-support systems, Figure 26.6, p. 609. In J. Jacko (Ed.), *The Human–Computer Interaction Handbook: Fundamentals, Evolving Technologies, and Emerging Applications*, 3rd Edition, 589–621. Copyright 2012 by Taylor & Francis, reprinted with permission of Taylor & Francis.)

provide immediate, context-sensitive feedback when the technologist's actions differed from the expert model (Smith et al. 2012).

In the empirical evaluation of this on cases involving multiple solution problems (some with masking or noisy data) *where the expert system was fully competent,* 33.3% to 62.5% of the practitioners working without the benefit of the critiquing system arrived at incorrect diagnoses on the different cases. The medical technologists receiving the immediate, context-sensitive critiques from the software as they worked through each case had a 0% error rate for all of the cases.

In addition, on the one case that exceeded the competency of the software, 50% of the practitioners working without support from the critiquing system arrived at the wrong diagnosis. In this case, the metaknowledge embedded in the software detected that it was a difficult case that was likely outside of its range of competence because of weak reactions (a weak signal strength) and cautioned the user to proceed carefully by trying to increase the strength of the reaction. The practitioners benefiting from this caution arrived at the wrong diagnosis only 18.75% of the time.

Thus, the available data suggest one approach to deal with brittleness: Designing a critiquing system in which the computer plays the role of a critic observing the person's problem solving rather than the person playing the role of a critic observing the computer's problem solving. If the critiquing system has an unobtrusive interface that enables immediate, context-sensitive feedback, this role can reduce or eliminate bad outcomes (incorrect diagnoses) on cases where the software is competent by detecting and reducing or eliminating the impact of slips and mistakes by the user. It further suggests that the incorporation and use of metaknowledge on cases where the software is incompetent can help to ensure that the user makes active use of his or her own knowledge to make up for the brittleness of the software.

Note that there are some "ifs" in this conclusion, indicating that the details of the functionality and interface design matter. It is not enough to assign the computer the role of critic. The details need to be carefully crafted as well. In addition, it is possible that, if the user is multitasking, he or she could become reliant on the software to catch slips and mistakes, possibly increasing susceptibility to the impacts of any brittleness nested in the software when an anomalous case arises.

10.5.4.6 More Sophisticated Forms of Delegation

The discussion above indicates that within an appropriate design, assigning the computer the role of critic can produce better outcomes even on cases where the critiquing system is incompetent. However, the implication is that, while errors may be reduced, there are little or no efficiency gains since the person needs to manually complete the decision-making task even when supported by the critiquing system. While sometimes getting it right is better than getting it quickly or less expensively, the desire to increase efficiency does need to be considered in design.

Smith et al. (1997) suggest that, for a task like flight planning, the size of the solution space and associated computations is too large for dispatchers to economically complete the task manually, so the question to address is not *whether* we should, but rather, *how* we should embed automated flight planning support in an interaction design that minimizes the potential for hypothesis/plan fixation due to the perceptual, memory, and problem-solving processes discussed earlier.

Smith et al. (1997) suggest the following design solution:

- The dispatcher should specify objectives/constraints for the flight planning software (set with generic defaults that the dispatcher can change for special cases) and delegate the generation of a set of alternative possible flight plans for a flight to the software.
- The dispatcher should view a computer-generated display showing multiple alternative flight plans on a map display showing the weather (to reduce the potential for the automation-induced perceptual framing effect), thus reducing the chances of premature narrowing by driving the dispatcher to look at and evaluate the relevant data and solution space. This set of alternative flight plans should include the best of each viable "class" of solutions (upwind of the storm, downwind of the storm, topping the storm, hunting and pecking through the storm). Routes that are risky (such as a route between two closely spaced storm cells) should be filtered out so that they are not presented in this set of alternatives (Wickens et al. 2015), making the software conservative by filtering out possible solutions that are questionable (applying rules like: Do not recommend a route through a gap in two storm cells for which the actual or forecast spacing is less than some threshold).
- The software should have embedded metaknowledge (Guerlain et al. 1999) that helps it to recognize cases where its competence may be exceeded and to communicate with the dispatcher when such a concern is triggered.
- The broader aviation system should retain the current safety net provided by its design as a distributed work system (Smith et al. 2001, 2007), with dispatchers, airline ATC coordinators, FAA traffic managers, pilots, and controllers all looking at flights from different strategic and tactical perspectives at different look-ahead times to detect and deal with any unsafe conditions that might arise.

This proposed design solution does not eliminate the possibility of a framing effect (i.e., by fixating the dispatcher on those alternatives contained in the recommended "best of class" alternatives), but is hypothesized that it will reduce the likelihood of an automation induced perceptual framing effect while taking advantage of both the efficiency and complementary processing capabilities of the software (Lehner and Zirk 1987). It further provides several human safety nets, including "cooperation" between the designer of the flight planning system and the dispatcher (through the metaknowledge the designer has embedded in the technology).

10.5.4.7 Cognitive Tools That Support Procedural Solutions to Reduce Brittle System Performance: The Collection of Converging Evidence

Guerlain et al. (1999) and Smith et al. (2012) studied a highly experienced human "error detector" whose job was to quickly review the diagnoses made by medical technologists and to decide whether their answers were suspect. This quick judgment included individually considering something akin to the following:

- The perceived prior probability of the answer (Is it an extremely unlikely diagnosis?)
- The perceived likelihood of the available data given the hypothesized diagnosis
- The completeness of the hypothesized answer in explaining all of the available data
- The adequacy of the problem-solving *procedure/process*

This latter assessment was based on a strategy requiring that converging evidence be collected before arriving at a final answer.

10.5.4.8 Cognitive Tools That Estimate the Degree of Uncertainty Associated with a Recommendation

As a final example of approaches to deal with the brittleness of technologies, some researchers have presented data suggesting the positive impact of presenting the confidence level that a decision support system has in its recommendation (McGuirl and Sarter 2006): "Continually updated information about a system's confidence in its own ability to perform its tasks accurately would help operators decide on a case-by-case basis whether to trust the system or to handle the task themselves and thus achieve overall better joint system performance" (p. 663). As with all of the other proposed solutions to deal with brittleness, however, this requires much more extensive research as at face value a number of possible pitfalls for this approach have been suggested, including concerns over brittleness associated with the confidence estimates.

10.6 CONCLUSIONS

I began this chapter with a caution about superficially resonating with misleading labels such as "confirmation bias" and "complacency" and indicated that the field of CSE needs to better understand how technology and system design influences the cognitive processes of the operators or decision makers within a work system. I then outlined the complex influences of alerting and decision support technologies on a whole host of cognitive processes *in the context of different task environments.*

This is the nature of the cognitive triad—the attentional, perceptual, memory, problem-solving, and motor processes of the system operator or decision maker are all influenced by the interaction of the technology design (functionality and interaction design) with the broader system design, which is situated within the task environment and the broader physical, organizational, and cultural environment (Reason 1991, 1997). This broader systems perspective needs to be considered not only for system design but also when generalizing from the results of cognitive experiments. For as Jenkins (1974) pointed out in his tetrahedral model of cognitive experiments, performance in such experiments is similarly influenced by the interactions of task, context, and individual.

The bottom line is that both the design and the integration of new technologies into complex systems need the insights developed by field of CSE. To accomplish this, cognitive engineers need to further contribute to the field of cognitive science—focusing on how system design influences the mental processes and performances of workers. Equally important, cognitive engineers need to design, not just study the impacts of designs developed and implemented by others (Woods and Christoffersen 2002).

REFERENCES

Aaron, B., Nevis, S. and Soto, R. (May, 2015). Multitasking, distraction and cognitive aids. *ASA Monitor*, 79, 30–32.

Bagheri, N. and Jamieson, G.A. (2004). Considering subjective trust and monitoring behavior in assessing automation induced "complacency." In D.A. Vicenzi, M. Mouloua, and O. A. Hancock (Eds.), *Human Performance, Situation Awareness, and Automation: Current Research and Trends* (pp. 54–59). Mahwah, NJ: Erlbaum.

Bahner, E., Huper, A.D. and Manzey, D. (2008). Misuse of automated decision aids: Complacency, automation bias and the impact of training experience. *International Journal of Human–Computer Studies*, 66, 688–699.

Bailey, N. and Scerbo, M.S. (2007). Automation-induced complacency for monitoring highly reliable systems: The role of task complexity, system experience, and operator trust. *Theoretical Issues in Ergonomics Science*, 8, 321–348.

Bennett, K.B. and Flach, J.M. (1992). Graphical displays: Implications for divided attention, focused attention, and problem solving. *Human Factors*, 34(5), 513–533.

Burns, C. and Hajdukiewicz, J. (2004). *Ecological Interface Design*. Boca Raton, FL: CRC Press.

Comstock, J.R. and Arnegard, R.J. (1992). *The Multitask-Attribute Task Battery for Human Operator Workload and Strategic Behavior Research*. Technical Memorandum No. 104174. Hampton VA: NASA Langley Research Center.

Cummings, M.L. (2004, September). Automation bias in intelligent time critical decision support systems. *Proceedings of the American Institute for Aeronautics and Astronautics First Intelligent Systems Technical Conference*, Reston, VA.

Dekker, S.W.A. and Hollnagel, E. (2004). Human factors and folk models. *Cognition, Technology, and Work*, 6, 79–86.

Duley, J.A., Westerman, S., Molloy, R. and Parasuraman, R. (1997). Effects of display superimposition on monitoring of automation. *Proceedings of the 9th International Symposium on Aviation Psychology* (pp. 322–326). Columbus, OH: Association of Aviation Psychology.

Elstein, A.S., Shulman, L.S. and Sprafka, S.A. (1978). *Medical Problem Solving: An Analysis of Clinical Reasoning*. Cambridge MA: Harvard University Press.

Fischer, G., Lemke, A.C., Mastaglio, T. and Morch, A.I. (1991). The role of critiquing in cooperative problem solving. *ACM Transactions on Information Systems*, 9(3), 123–151.

Feltovich, P., Ford, K. and Hoffman, R. (Eds.) (1997). *Expertise in Context*. Cambridge, MA: MIT Press.

Fraser, J.M., Smith, P.J. and Smith, J.W. (1992). A catalog of errors. *International Journal of Man–Machine Studies*, 37, 265–307.

Galster, S. and Parasuraman, R. (2001). Evaluation of countermeasures for performance decrements due to automated-related complacency in IFR-rated general aviation pilots. *Proceedings of the International Symposium on Aviation Psychology* (pp. 245–249). Columbus, OH: Association of Aviation Psychology.

Galster, S., Duley, J.A., Masalonis, A. and Parasuraman, R. (2001). Air traffic controller performance and workload under mature Free Flight: Conflict detection and resolution of aircraft self separation. *International Journal of Aviation Psychology*, 11, 71–93.

Gopher, D. (1996). Attention control: Explorations of the work of an executive controller. *Cognitive Brain Research*, 5, 23–38.

Guerlain, S., Smith, P.J., Obradovich, J., Rudmann, S., Smith, J.W. and Svirbely, J. (1996). Dealing with brittleness in the design of expert systems for immunohematology. *Immunohematology*, 12(5), 101–107.

Guerlain, S., Smith, P.J., Obradovich, J.H., Rudmann, S., Strohm, P. Smith, J.W., Svirbely, J. and Sachs, L. (1999). Interactive critiquing as a form of decision support: An empirical evaluation. *Human Factors*, 41, 72–89.

Hoffman, R.R. and Ward, P. (September/October, 2015). Mentoring: A leverage point for intelligent Systems? *IEEE Intelligent Systems*, 78–84.

Hollnagel, E., Woods, D.D. and Leveson, N.C. (Eds.) (2006). *Resilience Engineering: Concepts and Precepts*. Aldershot, UK: Ashgate.

Hinds, P. and Kiesler, S. (Eds.). (2002). *Distributed Work*. Cambridge MA: MIT Press.

Hollnagel, E. and Woods, D. (2005). *Joint Cognitive Systems: Foundations of Cognitive Systems Engineering*. Boca Raton, FL: Taylor & Francis.

Hutchins, E. (1995). *Cognition in the Wild*. Cambridge MA: MIT Press.

Jamieson, G. (2002a). Empirical evaluation of an industrial application of ecological interface design. *Proceedings of the Human Factors and Ergonomics Society 46th Annual Meeting*. Santa Monica, CA: Human Factors and Ergonomics Society, 536–540.

Jamieson, G. (2002b). *Ecological Interface Design for Petrochemical Process Control: Integrating Task and System-Based Approaches (CEL-02-01)*. Cognitive Engineering Laboratory, University of Toronto.

Jenkins, J.J. (1974). Remember that old theory of memory? Well, forget it! *American Psychologist*, 29, 785–795.

Johnson, P., Duran, A., Hassebrock, F., Moller, J. Prietulla, M., Feltovich, P. and Swanson, D. (1981). Expertise and error in diagnostic reasoning. *Cognitive Science*, 5, 235–283.

Jones, P. and Jacobs, J. (2000). Cooperative problem solving in human–machine systems: Theory, models and intelligent associate systems. *IEEE Transactions on Systems, Man and Cybernetics*, 30(4), 397–407.

Jones, P.M. and Mitchell, C.M. (1995). Human–computer cooperative problem solving: Theory, design, and evaluation of an intelligent associate system. *IEEE Transactions on Systems, Man, and Cybernetics*, 25, 1039–1053.

Klayman, J. and Ha, Y. (1987). Confirmation, disconfirmation and information in hypothesis testing. *Psychological Review*, 94(2) 211–228.

Klein, G.A. (1993). A recognition-primed decision (RPD) model of rapid decision making. In G. Klein, J. Oransanu, R. Calderwood and Zsambok, C. (Eds.), *Decision Making in Action: Models and Method* (pp. 138–147). Norwood NJ: Ablex.

Kolodner, J. (1993). *Case-Based Reasoning*. San Mateo, CA: Morgan Kaufmann.

Kotovsky, K., Hayes, J.R. and Simon, H.A. (1985). Why are some problems hard? Evidence from Tower of Hanoi. *Cognitive Psychology*, 17, 248–294.

Larkin, J.H. and Simon, H.A. (1987). Why a diagram is (sometimes) worth ten thousand words. *Cognitive Science*, 11, 65–99.

Layton, C., Smith, P.J. and McCoy, C.E. (1994). Design of a cooperative problem-solving system for en-route flight planning: An empirical evaluation, *Human Factors*, 36.

Lehner, P.E. and Zirk, D.A. (1987). Cognitive factors in user/expert-system interaction. *Human Factors*, 29(1), 97–109.

Loftus, E. (1975). Leading questions and the eyewitness report. *Cognitive Psychology*, 7, 560–572.

May, P., Molloy, R. and Parasuraman, R. (1983, October). Effects of automation reliability and failure rate on monitoring performance in a multi-task environment. *Proceedings of the 1983 Annual Meeting of the Human Factors Society*, Santa Monica, CA.

McGuirl, J. and Sarter, N. (2006). Dynamic system confidence information supporting trust calibration and the effective use of decision aids by presenting supporting. *Human Factors*, 48(4), 656–665.

Metzger, U. and Parasuraman, R. (2005). Automation in future air traffic management: Effects of decision aid reliability on controller performance and mental workload. *Human Factors*, 47, 35–49.

Miller, G., Galanter, E. and Pribram, K. (1960). *Plans and the Structure of Behavior*. New York: Holt.

Miller, P. (1986). *Expert Critiquing Systems: Practice-Based Medical Consultation by Computer*. New York: Springer-Verlag.

Moray, N. and Inagaki, T. (2000). Attention and complacency. *Theoretical Issues in Ergonomics Science*, 1, 354–365.

Mosier, K.L. and Skitka, L.J. (1996). Human decision makers and automated decision aids: Made for each other? In R. Parasuraman and M. Mouloua (Eds.), *Automation and Human Performance: Theory and Application* (pp. 201–220). Mahwah, NJ: Erlbaum.

Mosier, K.L, Skitka, L.J., Dunbar, M. and McDonnell, L. (2001). Air crews and automation bias: The advantages of teamwork? *International Journal of Aviation Psychology*, 11, 1–14.

Mosier, K.L., Skitka, L.J., Heers, S. and Burdick, M.D. (1998). Automation bias: Decision making and performance in high-tech cockpits. *International Journal of Aviation Psychology*, 8, 47–63.

Muthard, E. and Wickens, C. (2003). Factors that mediate flight plan monitoring and errors in plan revision: Planning under automated and high workload conditions. *Proceedings of the 12th International Symposium on Aviation Psychology*, Dayton, OH.

Mynatt, C., Doherty, M. and Tweney, R. (1978). Consequences of confirmation and disconfirmation in a simulated research environment. *Quarterly Journal of Experimental Psychology*, 30(3) 395–406.

Newell, A. and Simon, H. (1972). *Human Problem Solving*. Englewood Cliffs, NJ: Prentice Hall.

Ockerman, J. and Pritchett, A. (2002). Impact of contextual information on automation brittleness. *Proceedings of the 2002 Annual Meeting of the Human Factors and Ergonomics Society Annual Meeting*, 46(3), 382–386.

Parasuraman, R. and Manzey, D. (2010). Complacency and bias in human use of automation: An attentional integration. *Human Factors*, 52(3), 381–410.

Parasuraman, R., Molloy, R. and Singh, I.L. (1993). Performance consequences of automation-induced "complacency." *International Journal of Aviation Psychology*, 3, 1–23.

Rasmussen, J. (1983). Skills, rules and knowledge: Signals, signs, symbols and other distinctions in human performance models. *IEEE Transactions on Systems Man and Cybernetics*, 13(3), 257–266.

Reason, J. (1991). *Human Error*. New York: Cambridge Press.

Reason, J. (1997). *Managing the Risks of Organizational Accidents*. Aldershot, UK: Ashgate.

Richardson, W.S. (1999). Users' guides to the medical literature: XV. How to use an article about disease probability for differential diagnosis. *JAMA*, 281(13) 1214–1219.

Roth, E.M., Bennett, K.B. and Woods, D.D. (1987). Human interaction with an 'intelligent' machine. *International Journal of Man–Machine Studies*, 27, 479–525.

Rovira, E., McGarry, K. and Parasuraman, R. (2007). Effects of imperfect automation on decision making in a simulated command and control task. *Human Factors*, 49, 76–87.

Sarter, N.B. and Schroeder, B. (2001). Supporting decision making and action selection under time pressure and uncertainty: The case of in-flight icing. *Human Factors*, 43, 573–583.

Scaife, M. and Rogers, Y. (1996). External cognition: How do graphical representations work? *International Journal of Human–Computer Studies*, 45, 185–213.

Sheridan, T.B. (2002). *Humans and Automation: System Design and Research Issues*. New York: John Wiley & Sons.

Silverman, B.G. (1992). Survey of expert critiquing systems: Practical and theoretical frontiers. *Communications of the ACM*, 35(4), 106–128.

Skitka, L., Mosier, K. and Burdick, M. (1999). Does automation bias decision making? *International Journal of Human–Computer Systems*, 51, 991–1006.

Simon, H.A. (1957). *Models of Man: Social and Rational*. New York: John Wiley and Sons.

Smith, P.J., Beatty, R., Hayes, C., Larson, A., Geddes, N. and Dorneich, M. (2012). Human centered design of decisions-support systems. In J. Jacko (Ed.), *The Human–Computer Interaction Handbook: Fundamentals, Evolving Technologies, and Emerging Applications*, 3rd Edition (pp. 589–621). Boca Raton: CRC Press.

Smith, P.J., Bennett, K. and Stone, B. (2006). Representation aiding to support performance on problem solving tasks. In Williges, R. (Ed.), *Reviews of Human Factors and Ergonomics, 2*. Santa Monica CA: Human Factors and Ergonomics Society.

Smith, P.J. and Billings, C. (2009). Layered resilience. In C. Nemeth, E. Hollnagel and S. Dekker (Eds.), *Resilience Engineering Perspectives, Volume Two* (pp. 413–430). Aldershot, UK: Ashgate.

Smith, P.J., Giffin, W., Rockwell, T. and Thomas, M. (1986). Modeling fault diagnosis as the activation and use of a frame system. *Human Factors*, 28(6), 703–716.

Smith, P.J., McCoy, E. and Layton, C. (1997). Brittleness in the design of cooperative problem-solving systems: The effects on user performance. *IEEE Transactions on Systems, Man, and Cybernetics*, 27(3), 360–371.

Smith, P.J., McCoy, E., Layton, C. and Bihari, T. *Design Concepts for the Development of Cooperative Problem-Solving Systems*. OSU Tech. Report, CSEL-1992-07, 1992.

Smith, P.J., McCoy, E. and Orasanu, J. (2001). Distributed cooperative problem-solving in the air traffic management system. In G. Klein and E. Salas (Eds.), *Naturalistic Decision Making*, (pp. 369–384). Mahwah NJ: Lawrence Erlbaum.

Smith, P.J., Spencer, A.L. and Billings, C. (2007). Strategies for designing distributed systems: Case studies in the design of an air traffic management system. *Cognition, Technology and Work*, 9(1), 39–49.

Thomas, L. and Wickens, C.D. (2006). Effects of battlefield display frames of reference on navigation tasks, spatial judgments, and change detection. *Ergonomics*, 49, 1154–1173.

Tversky, A. (1972). Elimination by aspects: A theory of choice. *Psychological Review*, 79, 281–299.

Vicente, K. (1999). *Cognitive Work Analysis: Toward Safe, Productive, and Healthy Computer-Based Work*. Mahwah NJ: Lawrence Erlbaum.

Vicente, K. (2002). Ecological interface design: Progress and challenges. *Human Factors*, 44 (1), 62–78.

Vicente, K. and Rasmussen, J. (1990). The ecology of human–machine systems II: Mediating "direct perception" in complex work domains. *Ecological Psychology*, 2, 207–249.

Vicente, K. and Rasmussen, J. (1992). Ecological interface design: Theoretical foundations. *IEEE Transactions on Systems, Man and Cybernetics*, 22, 589–606.

Wason, P. (1960). On the failure to eliminate hypotheses in a conceptual task. *Quarterly Journal of Experimental Psychology*, 12, 129–140.

Wickens, C.D., Gempler, K. and Morphew, M.E. (2000). Workload and reliability of traffic displays in aircraft traffic avoidance. *Transportation Human Factors Journal*, 2, 99–126.

Wickens, C.D., Sebok, A., Li, H., Sarter, N. and Gacy, A.M. (2015). Using modeling and simulation to predict operator performance and automation-induced complacency with robotic automation. *Human Factors*, 57(6), 959–975.

Wiener, E.L. (1981). Complacency: Is the term useful for air safety? *Proceedings of the 26th Corporate Aviation Safety Seminar* (pp. 116–125). Denver, CO: Flight Safety Foundation.

Woods, D.D. (1996). Decomposing automation: Apparent simplicity, real complexity. In R. Parasuraman and M. Mouloua (Eds.), *Automation and Human Performance* (pp. 3–18). Mahwah, NJ: Erlbaum.

Woods, D.D. and Christoffersen, K. (2002). Balancing practice-centered research and design. In: M. McNeese and M.A., Vidulich (Eds.), *Cognitive Systems Engineering in Military Aviation Domains* (pp. 121–136). Wright-Patterson AFB, OH: Human Systems Information Analysis Center.

Woods, D.D. and Hollnagel, E. (2006). *Joint Cognitive Systems: Patterns in Cognitive Systems Engineering*. Boca Raton, FL: Taylor & Francis.

Woods, D.D., Patterson, E. and Roth, E.M. (2002). Can we ever escape from data overload? A cognitive systems diagnosis. *Cognition Technology and Work*, 4, 22–36.

Woods, D.D. and Roth, E.M. (1988) Cognitive engineering: Human problem solving with tools. *Human Factors*, 30(4), 415–430.

11 Design-Induced Error and Error-Informed Design: A Two-Way Street

Nadine Sarter

CONTENTS

11.1 INTRODUCTION

Human error and its relation to system safety and performance have been a major concern and topic of interest in the cognitive systems engineering community, as well as an important part of the roots of psychology and human factors for well over a century (James 1890). Much of the early work focused on the classification and prevention of erroneous actions and assessments. Later, once the inevitability of human error was acknowledged, the emphasis shifted to error prediction, error tolerance, and error management, that is, the detection, explanation, and recovery from errors.

One reason for the considerable interest in human error was the widely held belief that erroneous actions are the "root cause" of incidents and accidents and need to be eliminated to increase system safety. In fact, even current accident statistics still cite human error as being responsible for 70%–80% of all mishaps. "*Runaway Train Blamed on Human Error*" or "*Pilot Error to Blame for Airplane Crash*"—those headlines continue to dominate media reports immediately following an accident.

This tendency to blame the operator and his or her actions, referred to as the "Bad Apple Theory" by Dekker (2006), has been replaced by the "New View" of human error as an indication of deeper underlying "system" problems. Today, most researchers in this field of investigation share this broader systems view. They consider human error a symptom, not a cause—it does not provide, but rather requires an explanation.

One contributor to human and joint system error is poor design of technological artifacts, which can enhance human expertise or degrade it—"make us smart" or "make us dumb" (Norman 1993). Design-induced errors—the focus of this chapter—were discussed as early as the beginning of the twentieth century (see Hoffman and Militello 2008) and saw a surge of interest during the years following World War II. At that time, much of the work on human error was conducted in the context of aviation, and in the 1940s, Alphonse Chapanis suggested that many aspects of "pilot error" were actually "designer error." In 1967, the National Transportation Safety Board published an entire report entitled "Aircraft Design-Induced Pilot Error." And in 1996, the Federal Aviation Administration (FAA) issued a report on pilot-automation interfaces that included as one of its 51 recommendations that "the FAA should require the evaluation of flight deck designs for susceptibility to design-induced flightcrew errors and the consequences of those errors as part of the type certification process" (Abbott et al. 1996, p. 9). Design-induced errors have been defined as errors that are "carried out by a human where the primary reason for the inappropriate human behaviour can be shown to be a fault or faults in [the design of; the author] a piece of electronic or mechanical equipment" (Day et al. 2011, p. 1).

While errors can be a consequence of poor design, they also represent an important opportunity for learning and for informing and improving design—error and design as a two-way street. Errors can lead to design changes in the aftermath of incidents or accidents, as a result of consumer complaints or based on findings from usability studies and, more generally, empirical ergonomics research. They help inform design guidelines and the certification of safety-critical equipment. As Woods and Sarter (2000, p. 10) stated: "Human Factors began and has always been concerned with the identification of design-induced error (ways in which things make us dumb) as one of its fundamental contributions to improved system design."

The goal of this chapter is to examine and illustrate this two-way street: the role of design (as opposed to other factors such as training or procedures) in inducing erroneous actions and assessments as well as the contribution of the latter to design changes and the development of design guidance. To this end, I will first discuss the changing view of human error from the root cause of an accident to, at most, a contributing factor that can be explained by poor design. Next, various ways in which design can induce, or increase the likelihood of errors at various stages of information processing will be described. Examples of design-induced errors will be presented using both simple everyday tools and devices and more complex technologies. Finally, a system-oriented perspective of human error will be highlighted that views erroneous actions as the result of multiple interacting factors—not design alone—associated with human agents, artifacts, and the environment in which this triad interacts to produce system performance.

11.2 WHO'S TO BLAME?

In the aftermath of an accident, the determination of culpability (the degree to which a person involved should be held responsible) is often emphasized over trying to understand why the mishap occurred and how it can be prevented from happening again. In most cases, the finger is pointed at the operator and his or her erroneous actions, especially early on in an investigation. In aviation, for example, between 70% and 80% of accidents are blamed on pilot error (Wiegmann and Shappell 2003). This tendency can be explained, in part, by looking at a particular engineering approach to system safety: defenses in depth.

In complex high-risk systems, multiple layers of defenses, such as safety features of the system or standard operating procedures (SOPs), are introduced to reduce the likelihood of catastrophic events. However, each layer may be weakened at times by dynamic gaps or holes, such as violations of SOPs or a system that is inoperative on a given day because of maintenance or failure. When these holes appear at the same time and line up in sequence, an accident trajectory is no longer stopped, and a catastrophic event can occur (see Figure 11.1). Importantly, one of these layers, and often the last line of defense, is human operators—their selection, training, and capabilities as influenced by the designed and natural environment in which they are performing.

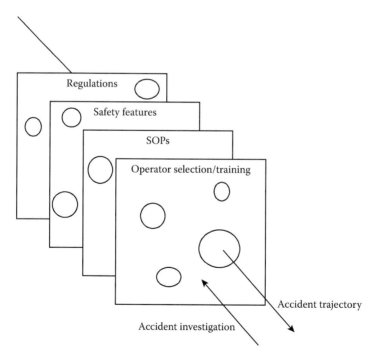

FIGURE 11.1 The concept of defense in depth—multiple layers of defenses, each with its own weaknesses that need to line up for an accident to occur. (Adapted from Reason, J. (2000). Human error: Models and management. *BMJ*, 320, 768–770.)

After an accident, investigators tend to move backward through the chain of defenses in search of a "root" cause. The first layer they encounter often will be the human operator(s). If any shortcomings or weaknesses are identified for this layer, such as erroneous actions or assessments, this can trigger the stopping rule—the end of the search for a culprit because human error is considered an acceptable and familiar cause of accidents (e.g., Leveson 2011). Instead of considering the error a symptom and one of many possible contributors to the accident, human error is labeled the "root cause" of the mishap.

While this conclusion is neither useful nor appropriate in most cases, care must be taken to avoid dismissing the possibility of willful or negligent acts too quickly. To this end, a fault tree analysis can be employed (e.g., Reason 1997). As we move from the left to the right side of the decision tree shown in Figure 11.2, sabotage, substance abuse, and reckless behavior are eliminated. The culpability of an individual diminishes while systemic and organizational reasons move to the forefront.

The dashed vertical line in the above figure (added by the author) indicates the transition point from a possibly negligent operator to a system- or design-induced error. This determination is based, in part, on Johnston's Substitution Test (Johnston 1995). For this test, the person involved in an accident is substituted for one or more

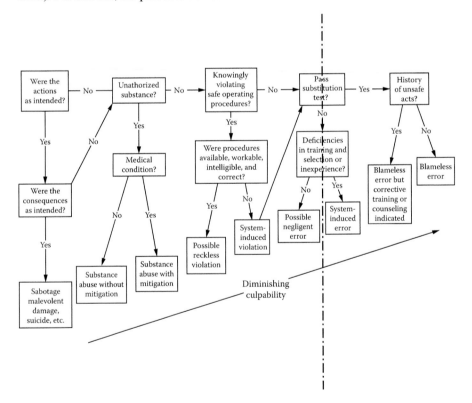

FIGURE 11.2 Reason's (1997) decision tree for determining the culpability of unsafe acts. (From Reason, J. (1997). *Managing the Risks of Organisational Accidents*. Burlington, VT: Ashgate, Figure 9.4, p. 209, reprinted with permission of Ashgate Publishing.)

well-motivated, equally competent, and similarly qualified operators. The question whether it is likely that the other operators would have acted the same way is then asked. If the answer to this question is "yes," then blaming the person is unacceptable. Instead, one has to consider the possibility of system-induced deficiencies related to the person's capabilities, training, selection, or experience. The operator should be assumed to have acted in a locally rational manner (Simon 1991); that is, they performed as well as their peers and as well as possible given tendencies and limitations of human cognition, as well as the limited amount of information and time they may have had to make a decision or take an action. In addition to factors such as inadequate training, their behavior may be explained by poor equipment and feedback design—a factor that will be discussed in more detail in Section 11.3.

11.3 THE ROLE OF DESIGN IN HUMAN ERROR

A range of design-related shortcomings can invite human error, including inadequate control arrangements, poor system feedback, improper hazard advisories and warnings, inadequate system instructions and documentation, inappropriate levels of automation for a given task or circumstance, and high system coupling and complexity (see Day et al. 2011). Wilde (1982) mentions risk homeostasis as another contributing factor (see also Rasmussen 1983). Risk homeostasis refers to the situation where designers try to reduce safety risks by increasing the number of defenses in a system. This approach can have the opposite, unintended effect that users assume the changes made to the system allow them to increase productivity by moving closer to the safety boundaries, thus defeating the designer's purpose and actually increasing the risk of an incident or accident.

One way to categorize and illustrate design contributions to human error is by relating them to the "gulfs of execution and evaluation" (Norman 1986). The gulf of execution refers to difficulties with identifying and implementing the possible and required steps to perform an intended action or achieve a goal with a given artifact. The gulf of evaluation, on the other hand, refers to the effort required to assess the state and/or behavior of a technology and to determine whether an action has had the intended outcome. The role of design is to minimize or bridge both gulfs.

One effective means of overcoming the gulf of execution is, for example, the introduction of direct manipulation interfaces (DMIs; Shneiderman 1998) where the desired action is performed directly on the interface instead of being described in the machine's language (as in command language interfaces). A DMI shows all objects and actions of interests and allows the user to manipulate those virtual objects using actions that correspond, to an extent, to the manipulation of corresponding physical objects. Bridging the gulf of evaluation, on the other hand, requires that feedback is designed such that it provides attention guidance to the user, supports easy integration and interpretation of information, and incurs low information access costs.

11.4 GULF OF EXECUTION

The gulf of execution can be related to two stages of information processing described by Parasuraman et al. (2000): action selection and action implementation.

11.4.1 Action Selection

Action selection, the choice of "what to do next," can be affected by design in various ways. One problem occurs when a design does not clearly indicate the affordances of a system (Norman 1988). In other words, the interface does not help the user identify the range of actions that the artifact can be used for, given the capabilities, goals, and past experiences of this person. Instead, it relies on the user to infer, bring to mind, and compare all possible options in a cognitively demanding fashion.

Another way in which design can affect action selection is by biasing a user toward one or few (possibly incorrect) actions. This can be illustrated using the example of decision support systems—computer-based technologies that help users gather and integrate information with the goal of supporting a choice among various courses of action. Decision aids can lead to poor choices when they suggest an action to the user without explaining the (possibly limited) data and (possibly flawed) reasoning that formed the basis for the recommendation (e.g., Guerlain et al. 1996; Layton et al. 1994; Smith et al. 1997).

For example, Sarter and Schroeder (2001) conducted a full-mission simulator study where pilots who were flying in potential inflight icing conditions were assisted by icing-monitoring automation. The system indicated either the type of icing (wing or tail plane) or the required responses to the particular icing condition. When presenting accurate information, the automation significantly improved performance (measured as the percentage of icing-induced stalls); however, incorrect advice by the system in 25% of the trials resulted in a nearly 100% increase in the stall rate, compared to baseline performance without the automation. Here, pilots relied on the decision support system because they lacked information about its trustworthiness across different icing conditions. They took the incorrect system-recommended action even though information about the actual icing condition (in the form of airframe and yoke buffet) was available.

In a subsequent study, where information about the confidence of the icing system in its own judgment was provided to pilots, trust calibration and performance improved significantly. Pilots adopted system recommendations in case of high system confidence; however, when system confidence was low, they based their action selection on the available icing-related cues from the aircraft (McGuirl and Sarter 2006).

11.4.2 Action Implementation

Action implementation also relies on perceived affordances. In this case, the user tries to determine how the device can be used to execute an intended action. One simple example of a display that hinders action implementation is the design of automobile fuel gauges that do not indicate whether the gas tank opening is located on the right or left side of the vehicle. These gauges leave the driver guessing and sometimes taking the wrong step (moving the vehicle to the wrong side of the pump), especially in case of a rented or unfamiliar vehicle.

Another often-cited example of poor design that affects action implementation is the design of some doors. The design logic is to install flat push-bar handles on the

FIGURE 11.3 Illustration of affordances suggested by the design of door handles. (Adapted from http://blogs.adobe.com/interactiondesign/2011/06/what-makes-a-good-user-experience/.)

sides of doors that can only be pushed; in contrast, pull-type handles are appropriate when the sides of the doors are meant to be pulled. This logic, illustrated in Figure 11.3, is sometimes violated, resulting in the person trying to perform the intended action—opening the door—incorrectly and unsuccessfully.

Problems with action implementation can occur also when no reminders are provided to ensure that a sequence of actions is completed—a violation of the principle of closure (Shneiderman 1998). One example is the design of most copying machines that do not remind the user—either through a message on the control panel when the user logs out or possibly by using a transparent lid—to remove the original after he or she is done with copying. Another example in the context of aviation operations is the Control Display Unit of the Flight Management System. In order for pilot entries to take effect and be executed by the automation, the pilot needs to press the EXECUTE button as the last step in a sequence of actions. However, no reminder is provided and pilots sometimes forget this step only to be surprised that the automation fails to take expected actions.

11.5 GULF OF EVALUATION

The gulf of evaluation encompasses two stages of information processing: information acquisition and information integration. Design deficiencies can lead to errors at both stages.

11.5.1 Information Acquisition

Problems with information acquisition are often encountered with designs that are characterized by data availability—the mere presence of some data in some location—rather than observability (Woods et al. 1999). Observability is achieved by cleaning a set of observations ("raw data") of errors, reducing sources of unreliability, making the data relevant to the user, and organizing the data in ways that support understanding.

Data availability creates problems by affecting both bottom-up and top-down attention allocation. Bottom-up attention allocation, that is, the management of attentional resources driven by objects or cues in our environment, suffers when (1) unnecessary or irrelevant data are presented and capture attention in an involuntary fashion, thus distracting from more important information, or (2) display elements fail to attract attention in a data-driven fashion, when necessary, through timely and salient indications. Top-down attention allocation refers to the voluntary allocation of attention to objects or regions in our environment, based on prior knowledge or expectations. This form of attention is affected by data availability when, for example, a user engages in visual search but cannot locate the target due to display clutter.

Display clutter can be defined as "the presence of performance and attentional costs that result from the interaction between high data density, poor display organization, and abundance of irrelevant information" (Moacdieh and Sarter 2014, p. 5). These costs include a degradation of monitoring and signal/change detection (Schons and Wickens 1993) and delays in visual search (Neider and Zelinsky 2011). For example, a well-known problem with the detection of relevant changes due to data overload and clutter has been reported on modern flight decks where pilots sometimes fail to notice changes in the status and/or behavior of their automated flight deck systems—a breakdown in mode awareness (e.g., Sarter and Woods 1994, 2000). This problem can be explained, in part, by the fact that annunciations associated with these changes are not sufficiently salient to capture attention as they are embedded in a data-rich and highly dynamic display, the Primary Flight Display (e.g., Nikolic et al. 2004).

A loss of mode awareness, and associated accidents, can also occur as a result of a high degree of similarity of feedback across different conditions. In one such case, while vectoring an aircraft for an approach, the air traffic controller provided radar guidance to a waypoint that the aircraft was supposed to cross at an altitude of 5000 ft. After that waypoint, the approach profile called for a 5.5% slope, or 3.3° angle of descent. While entering the angle of descent, "−3.3," into the Flight Control Unit (FCU), the crew did not notice that the automation was operating in the "heading/ vertical speed" (HDG/V/S) mode. In this mode, the entry "−3.3" is interpreted by the system as a descent rate of 3300 ft/minute, whereas in the track/flight path angle mode (TRK/FPA)—the intended mode—the same entry would have resulted in the (correct) −3.3° descent angle, the equivalent of an 800 ft/minute rate of descent. As a result of this confusion, the aircraft descended rapidly, struck trees and impacted a mountain ridge about 3 nm from the waypoint. Figure 11.4 shows the very similar feedback associated with the two automation modes, HDS/V/S and TRK/FPA, when the pilots entered "−3.3" into the FCU of the aircraft.

11.5.2 INFORMATION INTEGRATION

Information integration can be hindered by poor design when, for example, an interface fails to comply with the proximity compatibility principle. The proximity compatibility principle states that if sources of information need to be mentally integrated for performing a given task, they should be presented in close spatial proximity, their integration should be supported through the use of common features such as color or shape (Wickens and Carswell 1995), or they should be presented in some integrated display that provides an effective representation of the combined impact of these sources or raw data (Smith et al. 2006). On the other hand, when

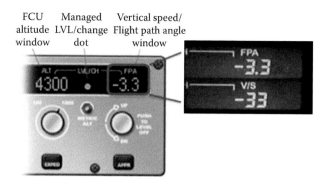

FIGURE 11.4 Indications associated with a flight path angle (FPA) of −3.3 (resulting in an 800 ft/minute descent; top right) and a vertical speed (V/S) of −3300 ft/minute (lower right) on the FCU. (Adapted from http://lessonslearned.faa.gov/ll_main.cfm?TabID=2andLLID =57andLLTypeID=2.)

focused attention on one piece of information is required and other information should not be considered or integrated, that information should be clearly separated. Failure to comply with this principle incurs mismatches between task and display proximity and leads to errors and performance decrements (Rothrock et al. 2006).

Problems with information integration can also occur when a design does not support visual momentum, which is critical for supporting a user in transitioning between multiple data views and information-seeking activities. Visual momentum refers to "the impact of a transition from one view to another on the cognitive processes of the observer, in particular on the observer's ability to extract task-relevant information" (Woods 1984). A high degree of visual momentum can be achieved using a range of techniques such as longshots (in display design: a summary or overview display) or landmarks (features that are visible at a glance and provide information about location and orientation). Low visual momentum can result in difficulties with locating important data, becoming lost, and increased mental workload, which, in turn, can lead to tunnel vision and errors. Thus, both the proximity–compatibility principle and visual momentum are ways to overcome the problem of forcing serial access to highly related data that need to be integrated (Woods et al. 1994).

11.6 ERROR-INFORMED DESIGN

So far, I have focused on how design can induce erroneous actions and lead to poor performance. However, the relationship between error and design is more appropriately characterized as a two-way street. The occurrence of errors has triggered design improvements and the development of design guidance in the interest of error prevention. For example, in aviation, accidents resulting from the failure of the flight crew to extend the flaps and slats before takeoff in the absence of reminders have led to the introduction and requirement of takeoff configuration warning systems on flight decks.

For a number of reasons, such design changes are more likely to be adopted and tend to occur more frequently in the context of simpler devices, such as consumer products, than with highly sophisticated technologies that are employed in high-risk environments. Modifying the latter is very costly because it often requires mandatory retrofitting of a large number of units, retraining of operators, and certification of the modified equipment. Still, considerable and increasing efforts are currently under way in these domains by those involved in the certification of safety-critical equipment. This is especially the case for learning from past experience with the goal to prevent errors and their potentially catastrophic consequences.

Various techniques and approaches for identifying likely and actual design-induced errors have been developed and tested. For example, HET (human error template; Stanton et al. 2006) is an error identification technique that was designed specifically for the detection of design-induced errors on flight decks. It takes the form of a checklist that focuses on 12 error modes (e.g., task execution incomplete, wrong task execute). Other techniques, such as SHERPA (systematic human error reduction and prediction approach; Harris et al. 2005) and HEIST (human error in systems tool), can be employed for the same purpose but are not domain-specific.

Another approach for anticipating and avoiding errors in design was advocated by Sebok et al. (2012) who developed ADAT (The Automation Design Advisor Tool), a research and software automation design tool for flight deck technology. This tool combines empirical data and models of human performance to identify perception-, cognition-, and action-related problems in the design of interfaces for current and future automated cockpit systems. It helps evaluate and compare designs and identify possible human–automation interaction concerns, such as mode errors. ADAT provides research-based guidance for resolving a wide range of design issues related to display layout, noticing of changes/events, meaningfulness and confusability of information, system complexity, and procedures.

The development of much needed error prediction tools is one reason why the complete elimination of errors may not be desirable; it would prevent both designers and operators from learning from experience. Instead, inducing or replicating and studying human error in "safe" environments, such as simulators, is critical for assessing the potential for a design to increase the likelihood of erroneous actions and assessments and to determine what types of errors can be expected.

11.7   DESIGN-INDUCED—REALLY?

Throughout this chapter, I have used the phrase "design-induced error." However, one of the hallmarks of cognitive ergonomics (or cognitive systems engineering) is the systems perspective advocated by these disciplines. This perspective emphasizes that the performance of a joint cognitive system can be explained only by considering the contributions and interactions of all its elements—human agents, artifacts/representations, and environmental factors—in other words, the cognitive triad (Woods and Roth 1988; see Figure 11.4). Errors or, more generally, breakdowns in performance should be considered symptoms of a mismatch between at least two components of the system. Thus, "design-induced error" appears to be an oversimplification. It merely shifts blame from the human operator "at the sharp end" to the artifact (or, rather, the creator of the artifact "at the blunt end").

For example, some have blamed the earlier cited problem of mode errors on modern flight decks on poor design. However, a more thorough analysis and research

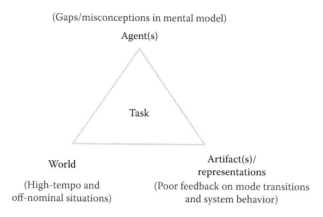

(Gaps/misconceptions in mental model)

Agent(s)

Task

World

Artifact(s)/
representations

(High-tempo and
off-nominal situations)

(Poor feedback on mode transitions
and system behavior)

FIGURE 11.5 The cognitive triad (the text in parentheses refers to the contributions of all three elements to the problem of mode awareness).

into this issue have highlighted that mode awareness tends to be lost when poor feedback design combines with gaps and misconceptions in operators' model of the system. Poor mental models lead to less than optimal monitoring of the technology in a top-down fashion, while poor feedback can result in a failure to capture attention in case of unexpected changes in the status/behavior of the automation. In addition, the combination of these two factors will most likely result in mode errors in the context of high-tempo and/or off-nominal conditions (Sarter and Woods 1994, 2000). Thus, all three work system elements contribute to the problem: the human agents (poor mental models), the artifact (poor feedback design), and the world (time constraints and rare circumstances), as illustrated in Figure 11.5.

REFERENCES

Abbott, K., Slotte, S., Stimson, D., Bollin, E., Hecht, S. and Imrich, T. (1996). *The Interfaces between Flightcrews and Modern Flight Deck Systems* (Federal Aviation Administration Human Factors Team Report). Washington, DC: Federal Aviation Administration.

Day, R., Toft, Y. and Kift, R. (2011). Error-proofing the design process to prevent design-induced errors in safety-critical situations. *Proceedings of the HFESA 47th Annual Conference.* Ergonomics Australia—Special Edition.

Dekker, S. (2006). *The Field Guide to Understanding Human Error.* Burlington, VT: Ashgate.

Guerlain, S., Smith, P.J., Obradovich, J.H., Rudmann, S., Strohm, P., Smith, J.W., Svirbely, J. (1996). Dealing with brittleness in the design of expert systems for immunohematology. *Immunohematology*, 12(3), 101–107.

Harris, D., Stanton, N., Marshall, A., Young, M.S., Demagalski, J. and Salmon, P. (2005). Using SHERPA to predict design-induced error on the flight deck. *Aerospace Science and Technology Journal*, 9, 525–532.

Hoffman, R.R. and Militello, L.G. (2008). *Perspectives on Cognitive Task Analysis: Historical Origins and Modern Communities of Practice.* Boca Raton, FL: CRC Press/Taylor & Francis.

James, W. (1890). *The Principles of Psychology (Volume 1).* New York: Henry Holt.

Johnston, N. (1995). Do blame and punishment have a role in organizational risk management? *Flight Deck*, 33–36.

Layton, C., Smith, P.J. and McCoy, E. (1994). Design of a cooperative problem-solving system for en-route flight planning: An empirical evaluation. *Human Factors*, 36, 94–119.

Leveson, N. (2011). *Engineering a Safer World: Systems Thinking Applied to Safety*. Cambridge, MA: MIT Press.

McGuirl, J. and Sarter, N. (2006). Supporting trust calibration and the effective use of decision aids by presenting dynamic system confidence information. *Human Factors*, 48(4), 656–665.

Moacdieh, N. and Sarter, N. (2014). Display clutter: A review of definitions and measurement techniques. *Human Factors*. doi: 0018720814541145.

Neider, M.B. and Zelinsky, G.J. (2011). Cutting through the clutter: Searching for targets in evolving complex scenes. *Journal of Vision*, 11(14), 1–16.

Nikolic, M., Orr, J. and Sarter, N. (2004). Why pilots miss the green box: How display context undermines attention capture. *International Journal of Aviation Psychology*, 14(1), 39–52.

Norman, D.A. (1986). Cognitive engineering. In D.A. Norman and S.W. Draper (Eds.), *User-Centered System Design: New Perspectives on Human–Computer Interaction*. Hillsdale, NJ: Lawrence Erlbaum Associates.

Norman, D.A. (1988). *The Design of Everyday Things*. New York: Doubleday.

Norman, D.A. (1993). Cognition in the head and in the world. *Cognitive Science*, 17, 1–6.

Parasuraman, R., Sheridan, T.B. and Wickens, C.D. (2000). A model of types and levels of human interaction with automation. *IEEE Transactions on Systems, Man, and Cybernetics—Part A: Systems and Humans*, 30, 286–297.

Rasmussen, J. (1983). Skills, rules, and knowledge; Signals, signs, and symbols, and other distinctions in human performance models. *IEEE Transactions on Systems, Man, and Cybernetics, SMC*-13, 257–266.

Reason, J. (1997). *Managing the Risks of Organisational Accidents*. Burlington, VT: Ashgate.

Reason, J. (2000). Human error: Models and management. *BMJ*, 320, 768–770.

Rothrock, L., Barron, K., Simpson, T.W., Frecker, M., Ligetti, C. and Barton, R.R. (2006). Applying the proximity compatibility and the control-display compatibility principles to engineering design interfaces. *Human Factors and Ergonomics in Manufacturing & Service Industries*, 16, 61–81.

Sarter, N. and Schroeder, B. (2001). Supporting decision making and action selection under time pressure and uncertainty: The case of in-flight icing. *Human Factors*, 43, 573–583.

Sarter, N. and Woods, D.D. (1994). Pilot interaction with cockpit automation: II. An experimental study of pilots' models and awareness of the flight management system. *International Journal of Aviation Psychology*, 4, 1–28.

Sarter, N. and Woods, D.D. (2000). Team play with a powerful and independent agent: A full-mission simulation study. *Human Factors*, 42(3), 390–402.

Schons, V.W. and Wickens, C.D. (1993). *Visual separation and information access in aircraft display layout* (Technical Report ARL-93-7/NASA-A 3I-93-1). Savoy, IL: University of Illinois, Aviation Research Laboratory.

Sebok, A., Wickens, C., Sarter, N., Quesada, S., Socash, C. and Anthony, B. (2012). The automation design advisor tool (ADAT): Development and validation of a model-based tool to support flight deck automation design for NextGen operations. *Human Factors and Ergonomics in Manufacturing and Service Industries*, 22(5), 378–394.

Shneiderman, B. (1998). *Designing the User Interface: Strategies for Effective Human-Computer Interaction* (1st edition). Addison Wesley Longman.

Simon, H.A. (1991). Bounded rationality and organizational learning. *Organization Science*, 2(1), 125–134.

Smith, P.J., Bennett, K. and Stone, B. (2006). Representation aiding to support performance on problem solving tasks. In Williges, R. (Ed.), *Reviews of Human Factors and Ergonomics, Volume 2*. Santa Monica CA: Human Factors and Ergonomics Society.

Smith, P.J., McCoy, E. and Layton, C. (1997). Brittleness in the design of cooperative problem-solving systems: The effects on user performance. *IEEE Transactions on Systems, Man and Cybernetics*, 27, 360–371.

Stanton, N., Harris, D., Salmon, P.M., Demagalski, J.M., Marshall, A., Young, M.S., Dekker, S.W.A. and Waldmann, T. (2006). Predicting design-induced pilot error using HET (Human Error Template)—A new formal human error identification method for flight decks. *Journal of Aeronautical Sciences*, 107–115.

Wickens, C.D. and Carswell, C.M. (1995). The proximity compatibility principle: Its psychological foundation and relevance to display design. *Human Factors*, 37(3), 473–494.

Wiegmann, D.A. and Shappell, S.A. (2003). *A Human Error Approach to Aviation Accident Analysis*. Burlington, VT: Ashgate.

Wilde, G.J.S. (1982). The theory of risk homeostasis: Implications for safety and health. *Risk Analysis*, 2, 209–225.

Woods, D.D. (1984). Visual momentum: A concept to improve the cognitive coupling of person and computer. *International Journal of Man–Machine Studies*, 21(3), 229–244.

Woods, D.D. (1991). The cognitive engineering of problem representations. In G.R.S. Weir and J.L. Alty (Eds.), *Human–Computer Interaction and Complex Systems*. Academic Press, London.

Woods, D.D., Johannesen, L., Cook, R.I. and Sarter, N. (1994). *Behind Human Error: Cognitive Systems, Computers, and Hindsight*. Crew Systems Ergonomic Information and Analysis Center (CSERIAC), Dayton, OH (State of the Art Report).

Woods, D.D., Patterson, E.S., Roth, E.M. and Christoffersen, K. (1999). Can we ever escape from data overload? A cognitive systems diagnosis. In *Proceedings of the Human Factors and Ergonomics Society 43rd Annual Meeting* (pp.174–178). Santa Monica, CA: Human Factors and Ergonomics Society.

Woods, D.D. and Roth, E.M. (1988). Cognitive systems engineering. In M. Helander (Ed.), *Handbook of Human–Computer Interaction*. North-Holland: Elsevier Science B. V.

Woods, D.D. and Sarter, N. (2000). Learning from automation surprises and 'going sour' accidents. In N.B. Sarter and R. Amalberti (Eds.), *Cognitive Engineering in the Aviation Domain* (pp. 327–353). Hillsdale, NJ: LEA.

12 Speaking for the Second Victim

Sidney W. A. Dekker

CONTENTS

12.1 INTRODUCTION

"What you're doing, Sidney," Jim Reason said, looking at me intently, "is trying to crawl into the skull of a dead man." He turned to gaze at the teacup in his hand, shook his head, and grumbled. "How is that possible?"

The way he looked, it was not a question. We were in a side discussion during a conference more than a decade ago. The topic was process-tracing methods from cognitive systems engineering (Woods 1993), which I had attempted to popularize in the first version of the *Field Guide* (Dekker 2002). How far can we take these methods to understand why it made sense for practitioners to do what they did? Is it possible to "reconstruct the mind-set" as I had called it at the time? When things have gone terribly wrong, and an incident or accident has happened, reconstructing the mind-set can become a matter of "speaking for the second victim." Second victims are practitioners involved in an incident or accident that (potentially) harms or kills other people (passengers, patients, co-workers), and for which they feel personally responsible (Dekker 2013). Speaking for them is no longer just about technical authenticity, but about learning, and about ethics and justice. Of course, some accidents take the lives of the practitioners as well. This raises the challenge of speaking for practitioners who no longer are around to provide us with any first hand insight into their assessments and actions.

A broad issue in cognitive systems engineering is at stake here. If we want to learn from practitioners' interactions with each other and technology, we need to study their practice. Process-tracing methods are part of a larger family of cognitive task analysis, but aim specifically to analyze how people's understanding evolved in parallel with the situation unfolding around them during a particular problem-solving episode. Process-tracing methods are extremely useful, if not indispensable, in the investigation of incidents and accidents. A scenario that leads to these develops both autonomously and as a result of practitioners' management of the situation, which gives it particular directions toward an outcome.

Typically, process tracing builds two parallel accounts of the problem-solving episode: one in the context-specific language of the domain, and one in concept-dependent terms. These two interlink: It is in and through the latter that we can discover or recognize (even from other domains) regularities in the particular performance of practitioners during the episode (Hutchins 1995; Woods et al. 2010). What we look at is not a meaningless mass of raw data, but a set of patterns of human performance. The context-specific or domain-dependent particulars make sense through the concepts applied to them. What looks like a flurry of changing display indications and confused questions about what the technology is doing, for example, can be made sense of by reading into them the conceptual regularities of automation surprise (Sarter et al. 1997). At the same time, the conceptual can only be seen in the particulars of practitioners' performance; its regularities begin to stand out across different instances of it.

The question is: Does an authentic representation of the process require us to engage directly with the practitioners who were there? Can process tracing speak for the second victim, particularly if they are no longer around? Despite Jim Reason's remark, I have always been encouraged by those who use a process-tracing method and who are careful about the reach of the claims, and who are made about this method willing to expose and acknowledge the constraints that make the claims possible. A dismissal about "crawling into a dead man's skull" has long felt unrepresentative of what we try to do. Let's turn to an example, and cognitive systems engineering attempts to generate ideas about what could have gone on "in the skull of a dead man." I will then finish with the ethical imperative to at least try, on occasion, to speak for the second victim.

12.2 THE COGNITIVE ENGINEERING OF SPEAKING FOR OTHERS

In one project, I had been asked by another country's investigation board to supply a human factors report for a fatal accident that happened to an aircraft of the type I was flying at the time. This plane had had a flight crew of three—all of whom perished, in addition to six other occupants. There had been an initially inexplicable speed decay for more than a minute, late on the approach, with the airplane stalling close to the ground about a mile short of the runway. It plowed into a muddy field and broke apart. As it turned out, this airplane type is equipped with automation where one part is responsible for tracking the vertical and horizontal approach path to the runway (the autopilot) and the other part is responsible for maintaining (through engine autothrottle) the correct airspeed to do so. The latter part was getting readings from a faulty (but not flagged as "failed") radar altimeter on the captain's side that suggested that the aircraft had already landed. The crew knew about the radar altimeter problem. The training package for this aircraft had implied that if the co-pilot's autopilot does the flying, it also uses the radar altimeter on that side to maintain correct thrust and thus airspeed. So, the crew had set up an approach in which the co-pilot's side was doing all the flying. All good, cleared to land.

Yet, the crash showed that it is, in fact, always the radar altimeter from the captain's side that supplies height information to the autothrottle system (the one that takes care of airspeed on an approach such as this). This fact about the radar altimeter could be found in none of the manuals and none of the training for this aircraft. It probably resulted from legacy issues (This jet has, in one form or another, been flying since the 1960s, starting off with many fewer systems and much less automation and complexity.) Let's limit ourselves to the pilots here (not the designers from way back). The kind of archeology that generates insights about what they might have understood or not understood is not about crawling into the skull of a dead man. It is in part about conscientious investigative work. If no pilot could have known about this because it was never published or trained in any manual they ever saw, it is pretty safe to conclude that the pilots involved in this accident did not know it either. The global pilot community of this aircraft type could not have known it on the basis of what it was told and trained on. "What other booby traps does this airplane have in store for us?" was how the highly experienced and knowledgeable technical pilot of my own airline at the time put it.

Because of its intermittent radar altimeter fault (essentially believing the jet was already on the ground), that radar altimeter was telling the engines to go to idle. This while the crew believed the autothrottle was getting faithful information from the co-pilot's radar altimeter, as that is what they thought they'd set up. The co-pilot's autopilot, in other words, was diligently flying the approach track to the runway (or trying to keep the nose on it at least) while the autothrottle system had already decided it was time to land. As a result, speed bled away. The aircraft had been kept high by air traffic control and had had to go down and slow down at the same time—this is very difficult with newer jets, as they have highly efficient wings. Going down and slowing down at least requires the engines to go to idle and remain there for a while. In other words, the action of the autothrottle system (while actually the result of a misinformed radar altimeter) was entirely consistent with crew intentions and expectations. It was only during the last seconds that the crew would have had to notice that the engines remained idle, rather than slowly bringing some life back so as to help them stay on speed. In other words, the crew would have had to suddenly notice the *absence* of change. That is not what human perceptual systems have evolved to be good at.

Speaking for the others requires us to understand the messy details of the world that killed them. Was this all the result of a rushed approach? Allusions to this were made at the time. With hindsight, now knowing the landing checklist was completed below the altitude of 1000 ft at which it "should have been," these guys must have been rushed, getting hot and high. Typical, typical. There is not a chief pilot in the world who would not experience this as one of the more vexing operational risks. But were they rushed? I approached the investigators again and asked how many track miles the crew knew they had to go at various time fixes during the approach. To understand the most basic thing about workload, after all, it'd be good to plot time available against tasks they knew they still had to complete to get the jet ready for landing. The investigators came back and presented me with a little table that detailed the radial distance to the airport at various times during the approach.

It was useless. Why? Imagine the jet on a downwind leg, passing the airfield. The radial distance to the field decreases as it nears the field, and the jet passes the field at a closest point of a couple of miles. Then it continues on, and the radial distance will increase again, up to, say, 12 miles. Then the jet might make, roughly, a 180° turn back onto final approach. During that turn, radial distance will essentially stay the same. Only on final approach will radial distance to the field count down in a way that is consistent with how track miles are gobbled up. My pedagogy must be lousy, as it took three attempts to persuade the investigators that radial distance was not what I needed. Granted, it was early in the investigation, and an accurate radar plot of exactly where the jet was in relation to the airport had not been recovered yet. Once I got the plot, I did the geometry myself.

Even then, it was not trivial. The track that the jet ended up flying (and from which we could predict how much time there was to touchdown at any point during the approach) was not necessarily the track the crew knew they would fly and would have had represented in their flight management systems and on their moving map displays. Approach control gives corner shortcuts and leg extensions and speed restrictions to individual jets so as to merge approaching traffic as efficiently as possible. The track miles a crew believes they have to go at any moment in the approach, then, might become less, or more. It can be hard to predict. In this case, it required us to match the clearances from air traffic control given during the approach, and see how this updated both the track and the crew's evolving understanding of how much time was left. That the jet was vectored onto final approach *above* the approach path for the runway threw an ultimate wrench into this understanding, as it suddenly and radically redefined "being rushed" or "hot and high." In all of this, it was always sobering to realize that the time between getting the approach clearance and them dying in the mud short of the runway would be less than what it takes you to read this paragraph.

In a press conference upon the release of the final report, the chair of the investigation board was asked how it was possible that the crew had "failed" to perceive the final decay in their airspeed. The chair looked down, and sighed: "For this, we do not have an explanation." Yet on his desk had been a 130-page report full of cognitive engineering explanations for precisely that decay. Perhaps, as one of the investigators told me later, it was too early for this board and this chairman to see the usefulness of process tracing. They had not been ready for it. Perhaps it indeed amounted to crawling into the skulls of a dead crew—witchcraft, sorcery, black magic. The kind of thing that was far removed, in any case, from the image of "science" and engineering in accident investigation, a carefully crafted image, whose assurances of objectivity and positivism supposedly lend investigations the credibility they need (Galison 2000). Years later, I asked the investigator whether the insights of process tracing could have been incorporated better. Yes, they probably would have been, the investigator assured me. They probably would have been.

12.3 THE ETHICS OF THE COGNITIVE ENGINEERING OF SPEAKING FOR OTHERS

It would have been nice. In this case, as in many others, the alternative of speaking *for* the dead, after all, is to speak *about* the dead. Lots of people do this in the aftermath of accidents—from media to colleagues to investigation boards.

Speaking about the dead is apparently easy; much easier than speaking for the dead. And it can quickly turn into speaking badly about the dead. Listen to the voice of another investigation board, involved because the aircraft had been built in their country. It concluded that the crash resulted from a poorly managed, nonstabilized approach that resulted in a high crew workload during a critical phase of flight; the crew's subsequent failure to detect and respond to the decreasing airspeed and impending onset of the stick shaker during the final approach despite numerous indications; the crew's failure to abort the nonstabilized approach and initiate a go-around; and after stick shaker onset, the captain delayed the thrust increase that then resulted in a fully developed aerodynamic stall and ground impact.

Linearity and hindsight offer those who speak about the second victim plenty to go on. They can start from the smoking hole in the ground and look back to expose a series of mistakes; they can line these mistakes up, and show the oh-so obvious way that all this would have led to the bad outcome. The vocabulary that gets deployed when they do this is organized around presumed human deficits. It is about their shortcomings, their delays and failures, their poor management and poor decisions. All it sees, from the human remains, is what they did not do. Three human lives, rich with experiences, aspirations, intentions, hopes, and dreams (and the assumption that they were going to turn around the airplane and fly back, just like on any other work day), are formally reduced to a short paragraph of deficiencies. This is speaking badly about the dead.

There are at least two strong ethical reasons why we should deploy all that our field has to offer to counter these sorts of characterizations. The first reason is that the dead second victims are not just dead. They were—and in a sense always will be— colleagues, mentors, students of ours. And that's just their professional relationships. The second reason is that when we speak for the second victim, we do not just speak for the second victim or their loved ones. Cognitive systems engineering speaks for practitioners—dead *or* alive—because it is able to illuminate the conditions of their work in ways that few other fields can: show the constraints and obligations of their daily existence, the patterns of cognition and performance, the regularities of their interaction with devices, systems, organizations. If what the second victims did made sense to them at the time, it will likely make sense to others like them as well. That means that the ethical commitment to speak for the second victim irradiates beyond the second victim, to all practitioners involved in that activity, and to those responsible for the operation, regulation, and design of their work.

And cognitive systems engineering applies to the future at least as much as it might to the past. Learning from the past—for example, through process tracing—is a matter of abstracting away from its contextual mess, ambiguities, and indeterminacies, to begin to see conceptual regularities. These we can take into the future. Automation surprises, error intolerance—such concepts allow us to explain, to predict, to help guide design, to perhaps prevent. Indeed, the very strength of cognitive systems engineering lies in its ability to conceptualize (Woods and Hollnagel 2006), to see regularities at levels of resolution where others see only "human error." Speaking for the second victim speaks as loudly for them as it does for everyone who could have been in their shoes. It speaks for the past as much as for the future.

Of course, we might throw up our hands and admit defeat; say that we cannot crawl into the skull of a dead man; concede that time is irreversible and that reconstructing somebody's mind-set is impossible. But that does not mean that there is nothing systematic about what people do in interaction with complex systems, about why they do what they do. It does not mean that everything about their mind and the world in which its understanding unfolded is unintelligible, inexplicable. It does not mean that there is nothing predictable about how things go right or wrong. And it does not release us from the ethical responsibility to at least try. If all there is in the past is dead people, whose intentions and machinations are forever closed off to our analysis and attempts at understanding, then we have little to say about the future too.

REFERENCES

Dekker, S.W.A. (2002). *The Field Guide to Human Error Investigations*. Bedford, UK: Cranfield University Press.

Dekker, S.W.A. (2013). *Second Victim: Error, Guilt, Trauma and Resilience*. Boca Raton, FL: CRC Press/Taylor & Francis.

Galison, P. (2000). An accident of history. In P. Galison and A. Roland (Eds.), *Atmospheric Flight in the Twentieth Century* (pp. 3–44). Dordrecht, Netherlands: Kluwer Academic.

Hutchins, E.L. (1995). *Cognition in the Wild*. Cambridge, MA: MIT Press.

Sarter, N.B., Woods, D.D., and Billings, C. (1997). Automation surprises. In G. Salvendy (Ed.), *Handbook of Human Factors/Ergonomics*. New York: Wiley.

Woods, D.D. (1993). Process-tracing methods for the study of cognition outside of the experimental laboratory. In G.A. Klein, J.M. Orasanu, R. Calderwood, and C.E. Zsambok (Eds.), *Decision Making in Action: Models and Methods* (pp. 228–251). Norwood, NJ: Ablex.

Woods, D.D., Dekker, S.W.A., Cook, R.I., Johannesen, L.J. and Sarter, N.B. (2010). *Behind Human Error*. Aldershot, UK: Ashgate.

Woods, D.D. and Hollnagel, E. (2006). *Joint Cognitive Systems: Patterns in Cognitive Systems Engineering*. Boca Raton, FL: CRC Press/Taylor & Francis.

13 Work, and the Expertise of Workers

Amy R. Pritchett

CONTENTS

13.1 INTRODUCTION

The goal of this chapter is to review the concept of *work*, and the expertise it demands of its workers. This concept reflects a unique and beautiful interplay in which workers understand the environment well enough to mirror it, patterning and adapting their activities to drive the dynamics around them to a desired end. In the process, experts construct artifacts around them to reflect their intentions, plans, and memory such that the cognitive system extends "outside the head." This dynamic becomes even richer with teamwork.

To illustrate these concepts, this chapter examines two work domains: first, perhaps more accessible to many readers, the kitchen; and, second, aviation. The four key points raised in this examination are then reiterated in a discussion of their implications for design of cognitive systems. I indulgently conclude with a discussion of work as a source of joy and beauty that cognitive systems engineering can foster and support.

Throughout, I shall assume a definition of work as "activity, cognitive and physical, to achieve a purpose within one's environment," strategically paraphrasing that found in common dictionaries. Thus, by definition, work has a purpose, although sometimes this purpose may itself evolve with circumstances: as a pilot, for example, my purpose is generally a pleasant flight for my passengers until something breaks on the aircraft, at which point safety takes priority. The work may be cognitive, or physical, or both: as a visual thinker who needs paper to sketch and outline in order to form a thought, I am not certain a division between the two types of work is useful anyways.

Finally, work is conducted within an environment or domain. This profoundly shapes what the worker needs to do to achieve his or her purpose, the subject of our next section.

13.2 WORKERS: ANTS OR EXPERTS?

To describe how an agent capable of executing only a few simple rules might create complex behaviors, Simon (1996) described the path of an ant on the beach:

> We watch an ant make his laborious way across a wind- and wave-molded beach... I sketch the path on a piece of paper. It is a sequence of irregular, angular segments not quite a random walk, for it has an underlying sense of direction, of aiming toward a goal.
>   ...Viewed as a geometric figure, the ant's path is irregular, complex, hard to describe. But its complexity is really a complexity in the surface of the beach, not a complexity in the ant. (p. 51)

Through time, this analogy has both helped and hindered my own understanding of workers and work. It helped through its emphasis on the role of the environment: the complexity of work is so driven by the environment that, without understanding the topology of the domain, the activities of the workers are hard to describe, predict, or explain. Conversely, this analogy has hindered my understanding of the expertise of workers by promoting the easy interpretation that they may be viewed as "worker ants" (or perhaps "worker bees"?) needing only some simple stimulus–response capabilities to work within their domain. In some domains, the ability to execute a few simple skills and rules may allow the novice to function in carefully scaffolded situations—but with expertise, the worker can go far beyond carefully constrained situations to fluidly and creatively drive the environment toward his or her goals. Let's examine this interplay between worker and work domain in Section 13.3 using a familiar example: the kitchen.

13.3 BAKING A CAKE: WORK AS A MIRROR OF THE DOMAIN

I might approach baking a cake with particular goals for its taste, texture, and nutrition. The cake itself demands certain ingredients—the substrate of the cake, the leavening agents, the flavoring, any gluten or binding agent, and so on—and equipment that affects changes upon the batter until the cake can stand alone—the whisk, the oven, and the pan, for example.

But work in the kitchen in general—and baking a cake in particular—is governed by laws of physics and chemistry. As illustrated in Figure 13.1, these may be represented as generalized functions that occur within the cake batter if both of the right physical elements are combined at the right times by the actions of the worker. These generalized functions mediate the choice of physical elements relative to the goals. A particular goal for taste, for example, may require certain flavorings, but these choices then also affect which leaven and binding agent are appropriate. A change in leaven then may require different chemical changes with heat—does the heat serve to

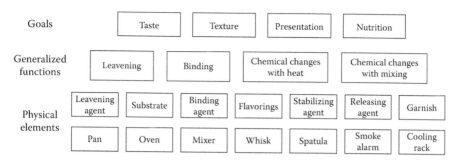

FIGURE 13.1 An abstraction hierarchy of baking.

enlarge the air bubbles within a foam of egg whites, or is the leavening more a by-product of an acid–base reaction between buttermilk and baking soda, with the heat of the oven then binding the matter together? The temperature and its duration will affect the degree of caramelization of any sugars in the ingredients, which will affect taste. Thus, any choice here will then affect other general functions, and so on.

A tasty cake, then, requires its worker to execute an effective path through this space of options. Selecting this path involves the cognitive effort of planning and decision making, but then the baker is also physically involved in the processes—beating the egg whites to soft peaks, for example—with his or her motions becoming an intimate part of the dynamic.

A novice baker may choose to proceduralize his or her behavior through a recipe and automate some functions with electrical appliances. This does not lessen the complexity of the work domain. Instead, a good recipe lessens the complexity of the worker's interaction with the work domain to a degree commensurate to his or her expertise.

Further, one measure of a recipe is whether it is inherently robust; that is, it represents a path through space of options that minimizes the sensitivity of the outcome to variations in the worker's activities or in the ingredients themselves. Even an experienced cook may refer to a tried-and-true recipe for a birthday cake for a first grader or other equally stressful situation demanding a robust outcome.

But, within this work domain, expertise allows the baker to explore the space of options creatively and fluidly. This is when the magic occurs, not just in terms of outcome (yum!) but in terms of process: the worker and the domain work together to create. Perhaps the baker brings a special ingredient to the kitchen one day—let's say, locally grown first-of-the-season peaches—and the environment and the process serve to generate new ideas and new tastes. Another expert, observing, can generally predict what the baker will do next based on the state of the batter and the heat of the oven, even in the face of specific surprises such as the addition of balsamic or basil to the cake batter.

This brings us to the first of four points in this chapter:

Point 1: *Expert workers understand the dynamics of their work domain and capitalize upon them to the extent that their own activity is a mirror reflection of their environment.*

Cognitive work analysis has some formal terms relevant here (see, e.g., Chapter 6). The kitchen is itself a source of *affordances*: the "eaters" in the house might simply open the refrigerator. For example, look at those lovely Georgia peaches, and think of the flavors that they may ultimately afford in tonight's dessert, while the more experienced cook sees many more intermediate affordances in terms of how the kitchen's physical space and tools might be applied when making the said dessert. Likewise, the dynamics of the kitchen is full of "constraints": some ingredients can only be heated so far, or beaten to a limited extent, before adverse dynamics emerge, such as whipped cream turning to butter.

Finally, many possible paths can be taken through the space in Figure 13.1, where each represents a different *strategy*. Strategies may be limited by available resources or by the workers' expertise: the novice may not realize anything is possible beyond the graphic procedure on the back of the cake mix box, but to the expert, the choice of strategy can be a creative response to the moment. (As an extreme example, one might think here of various cooking shows that challenge chefs to make-do with bizarre ingredients in a limited space of time.)

Indeed, such creative responses to the moment have been described by Woods and Hollnagel (2006, p. 20) by the "Law of Fluency: [Expert] work occurs with a facility that belies the difficulty of the demands resolved and the dilemmas balanced." Section 13.4 then takes this fluent interplay with the environment further, to the point that the expert not only responds to the environment but also constructs it.

13.4 CONSTRUCTING THE WORK DOMAIN TO SUPPORT ONE'S WORK

Cooking an American Thanksgiving dinner has always been a challenge to my foreigner self with its many dishes, most of which compete for attention in the half-hour before serving time. Over the years, I have built up the physical resources of my kitchen, but this has only highlighted where my own human limitations have become the limiting constraints. That last half-hour before serving, well, I can't fit it all in—and worse yet is the impact of my poor working memory, with turkey overcooked when forgotten in the oven and with missing silverware on the table.

Resolving such problems requires workers to develop another aspect to their expertise. Beyond providing fluid responses to the current state of the environment, expert human workers can plan significant aspects of their work and then construct the environment. This stage requires a switch from viewing work as a reaction to the environment, to viewing work as also constructing the environment.

In my own case, every year I dust off a Gantt chart for the Thanksgiving dinner to tailor it to the dinner time, special requests for dishes, and whether the turkey is still frozen. The ideal timeline is evaluated for workload spikes, and as much work as possible is moved earlier in the schedule: the table can be set once the turkey is in the oven, then the pie crusts mixed and rolled and the green beans trimmed into the steamer insert. I believe the American idiom here is staying "ahead of the game," such that my human limitations are off-weighed by forethought.

I also consciously apply alerting systems and reminders. Some are already built into my kitchen, such as the singing kettle, the timer on the oven, and the timer on the microwave, and I have developed the habit of using them. Others are more overtly created. Early in the day, I lay out all the serving dishes in a row, the more generic with labels, some represented by surrogates like the hot plate to go under the sweet potato casserole. During the last mad rush, when I cannot even recall my guests' names let alone which dishes are still in the oven, each serving dish is filled and taken out to the table such that a quick glance reminds me of what remains to be done: when the last serving dish is carried out, dinner's on! I no longer believe that, to be an expert, I must remember everything in my head, or even think it possible—instead, I consciously construct the environment to protect against my feeble memory.

This construction of the environment has been noted in other food-and-drink venues, as noted by Kirlik (1998). He observed, for example, a particularly efficient short-order cook who divided his grill into regions associated with rare, medium, and well done (p. 708). The initial placement of meat on the grill corresponded to the desired final state, with the orders for the more-cooked placed further to the right. With each pass of the grill, the cook moved all the meats left. Kirlik then described the impact on the cook's work created by this construct: "In this way, each piece of meat ... would signal its own completion as it reached the left barrier of the grill surface."

This brings us to the second point of this chapter:

Point 2: *Expert workers not only respond to their work domain—they schedule their activities and create artifacts within it that reflect their intentions, plans, and memory to themselves and others.*

Examples of such expert manipulation of the work environment are not limited to culinary work domains. Beyer and Holtzblatt's method for observing work in situ, Contextual Inquiry, includes a specific tasking to capture the artifacts workers construct to support memory and distribute workload, and to analyze them for the insights they bring about work in that domain (Beyer and Holtzblatt 1997).

Likewise, Hutchins (1995) examined how pilots set indicators of various speeds important to lowering the flaps and flying the approach. This practice is institutionalized such that moveable "speed bugs" are built into airspeed indicators; these speed bugs are normally set some 25 to 30 minutes before landing, when the aircraft is close enough to landing to estimate aircraft weight and wind speed during approach yet before crew workload is increased with all the tasks associated with landing. Indeed, the speed bugs are also used in other phases of flight—before starting a flight, for example, they are set to the "V1" and "V2" speeds reflecting go/no-go conditions should an engine fail during takeoff.

13.5 ADAPTING TO THE DYNAMICS OF THE WORK DOMAIN

The preceding sections discussed how workers respond to their work domain and, further, may construct artifacts into the environment embodying their plans and memory. Now, let's continue the transition from examining work in the context of the everyday environment of a home kitchen to examining work in the context of a more complex environment: aviation. The important transition here is that the desired outcome is, in engineering parlance, the output of a closed feedback loop in which the worker needs to constantly regulate the environment to maintain straight-and-level flight.

Some of the first studies of piloting examined how pilots control the aircraft, specifically referring to the task of moving their control stick, rudder pedals, and throttle to steer the aircraft to the correct altitude and thrust for their desired flight condition. Methods for analyzing the aircraft's dynamics had been formalized between the 1920s and the 1950s, using differential equations to relate the output of the aircraft (dynamic state) to its inputs (the positions of the control surfaces, i.e., its rudder, aileron, elevator, and throttle setting). By this logical engineering paradigm, all that remained was to identify the corresponding differential equation relating the output of the human pilot (movement of the controls) to his or her input (perception of aircraft dynamic state). If the human pilot would just conform to a fixed input–output relationship, then the linkage of the pilot's output to the aircraft's input, and the aircraft's output to the pilot's input, could be made as a continuous feedback loop, and total system behavior could thence be mathematically predicted.

However, the human pilot did not accommodate this engineering paradigm. To be sure, when asked to control one aircraft (or dynamic system in a laboratory), the pilot's behavior could be reliably fit to a differential equation. The problem was that, when asked to then control a different aircraft, the same pilot's behavior was described by a different differential equation. Change the aircraft, and the pilot's control behavior changes. This inconsistency in behavior between aircraft was found to happen consistently within the pilot population—trained pilots' behavior changes between aircraft.

So how to model the pilot then? A seminal paper by McRuer and Jex (1967) captured the growing realization that the pilot cannot be modeled independent of the controlled system. Their Crossover Model highlighted what is constant: the combined pilot–aircraft model. Put mathematically, if the dynamics of the controlled element are described as "K_C" and those of the pilot are described as "K_P," then the product of these two functions is always described by a simple relationship around the important Crossover Frequency: the combined time delays of the aircraft and pilot and a simple integrator or time lag that provides a smooth response toward zero error between desired and actual state. This combined relationship is the invariant, and the aircraft dynamics, as a simple, cold mechanical system, are what they are: the implication is that the pilot adapts his or her behavior, such that K_P can be solved for once K_C is known.

McRuer and Jex stress that this Crossover Model is a fit to observed pilot behaviors. With one exception, it cannot be explained by looking at the innards of the pilot's perceptual or physiological mechanisms, but instead by the goals of any good pilot: smooth tracking performance and zero error. The exception is the time delay added by the pilot, which includes basic perceptual and cognitive delays; however, even this fundamental aspect of the human can be reduced when the pilot can anticipate the future motion of the aircraft. (At higher frequencies than the Crossover Frequency, some structural aspects of human physiology may also come into play, such as resonant modes within the muscular–skeletal system.)

This leads us to the third point of this chapter:

Point 3: *Within the feedback loops of expert workers and environmental dynamics, expert workers adapt their behavior to achieve their desired outcomes.*

This adaptability only starts to break down when the situation becomes unmanageable for a given combination of task and environment. In the case of aircraft control, for example, the handling qualities of an aircraft are judged principally not by the pilot's performance in controlling the aircraft (this performance may be very similar over a range of aircraft), but instead by the effort required by the pilot to adapt to the aircraft (which can vary widely between aircraft, and is the main subject of the categories of the original Cooper–Harper scale as developed to evaluate the fly-ability of aircraft). The Cooper–Harper ratings for a specific aircraft can vary between different flight tasks: a fighter aircraft, for example, will hopefully be judged as easy to maneuver in a dogfight even as this same maneuverability may be unmanageable over a long cruise flight.

This adaptability is described mathematically by the Crossover Model for the specific tasks of pursuit and compensatory control. More recently, other mathematical models of invariants in human-system behavior have been identified in a similar vein by Mettler and Kong (2013) examining the observed behaviors of human pilots of small, remotely piloted UAVs in a guidance task around obstacles. They identified a few simple maneuver types and descriptors of the situations when each maneuver would be selected; these behaviors are reasonable and near-optimal. With these descriptors as a framework, pilots' flights could be explained with little residual variance. Further, this residual variance was significantly smaller for expert pilots compared to the less experienced pilots, that is, the ability to adaptively create the desired outcome becomes more reliable and less variable with expertise.

Likewise, I believe that we can explain many other tasks this way. Some may also be derived mathematically from some analysis of task and environment. For example, a pilot also chooses where to direct his or her visual attention, and when. Given that foveal vision is remarkably narrow (roughly the central 2° of the eyes' field of view), this choice must carefully reflect the dynamics of the environment. Figures 13.2 and 13.3 represent the three parts of this choice. First, Figure 13.2 represents the task of directional control and guidance as a set of three nested control loops, each with its own inherent bandwidth. At the center is the fastest control loop: controlling the bank of the aircraft, which can diverge (and thus requires attention) within a second or two. The middle control loop relates the bank of the aircraft to its heading: during straight-and-level with a near-zero bank, the heading will only diverge slowly, while during a turn, the heading will change at a standard rate (usually 3° per second) such that a 30° turn can be initiated and then checked just before 10 seconds, and then checked more frequently only as the new heading is acquired. Finally, the outer control loop relates heading to track over the ground; as the aircraft reaches a new waypoint (on a time scale of tens of minutes in cruise, and minutes during approach), a new target heading is set toward the next target. Thus, the task itself has an inherent structure and specific timing requirements that need to be adapted to the dynamics of each aircraft.

Second, Figure 13.3a shows the "Standard T" layout of the flight deck displays, essentially a world standard since before World War II. The fastest control loop

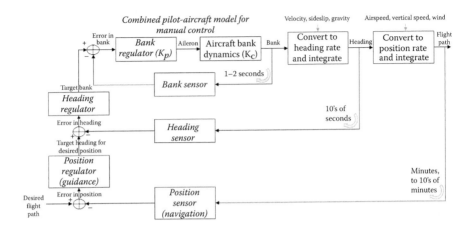

FIGURE 13.2 The systems dynamics viewpoint of directional control and guidance as a set of nested control loops.

concerns altitude, which is shown in the altitude indicator in the top-center and is mounted directly in front of the pilot. The middle control loop for heading relates the bank on the altitude indicator to the directional gyro shown immediately below it; when banking, the turn-and-bank indicator on the bottom left can also be referenced. Finally, the least-frequent outer control loop identifies a target heading to fly to a desired course or waypoint, which would start with the directional gyro but then reference navigation displays and charts located further away. Thus, the displays needing to be referenced the most often are placed front-and-center, and the rest arrayed progressively around it.

Finally, Figure 13.3b demonstrates how this aspect of the pilot's work—scanning the flight instruments—carefully capitalizes upon the Standard T layout to support aircraft control. To receive an instrument flight rating, a pilot must demonstrate a scan that constantly starts at the altitude indicator and then, importantly, returns to it after scanning any one other display. Thus, following the thicker arrows, the pilot's scan may go from altitude indicator to directional gyro and back to the altitude indicator, then check airspeed on the top left and back to the altitude indicator, and then to check altitude on the top right and back to the altitude indicator. With this scan, the fastest-changing information—altitude—will have been sampled three times in the scan listed above, and then the slower-changing heading, airspeed, and altitude each only sampled one time. Further, the scan can be adapted to the task. As shown with the thinner arrows, the pilot may also look to the turn-and-bank indicator in the bottom left when seeking a standard-rate turn, and may also look to the vertical speed indicator in the bottom right during a climb or descent. Thus, the expert pilot's scan is adapted to the environment and task in a manner that can be predicted mathematically given knowledge of the aircraft dynamics and the maneuver that is immediately sought.

For other tasks, the pilot's adaptation may be described qualitatively. Similar to my Thanksgiving Gantt chart, an expert pilot will reschedule his or her activities around circumstances. While waiting for a delayed pre-departure clearance, the pilot

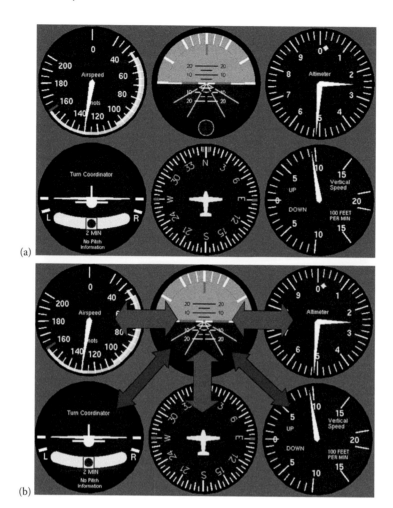

FIGURE 13.3 The flight deck viewpoint of directional control and guidance. (a) The Standard T layout of displays. (b) The pilot scan, always returning to the fastest dynamic (altitude indicator in the center).

may perform much of the before-takeoff checklist as possible; on the other hand, with an expedited pre-departure clearance, the pilot may instead perform the checklist during taxi-out to the departure threshold. Told at top of descent to expect a particular landing approach, the pilot may pull out that approach chart and pin it to the control column and, during low-tempo periods, start to enter in its various speeds and decision altitudes. With an engine running hot, the pilot will check it more often. And so on.

Put another way, to predict what the expert *will do*, examine what the expert *will need to do* for a given task and situation. A first step here is the general intent of work domain analysis (WDA, see Chapter 6): understand what is happening in the environment, including the broader dynamics within the environment and the specific affordances for (and constraints on) possible actions by the expert. (As an engineer, I have also argued that the qualitative representation of the work domain provided by WDA often can, and should, be analyzed for its system dynamics such as the control bandwidths noted earlier, and also computationally simulated [Pritchett et al. 2014]).

Thus, relating the work domain to the expert's tasks requires winnowing down a large set of possible actions (or action sequences, or strategies) into the activities that would represent the best trade-off between effort and resulting performance. In some cases, such analysis might identify a clear best plan of action that an expert would be likely to exhibit. A complex environment, however, likely has several different patterns of activity that reflect different measures of performance (e.g., passenger comfort vs. accurate tracking of the flight path) and that reflect different trade-offs between effort and performance (see, e.g., Hollnagel's [2009] discussion of the Efficiency–Thoroughness Trade-Off or ETTO Principle). Thus, an expert pilot may choose to apply an effort-ful strategy capable of precise tracking of the final approach, but during cruise, where such a level of effort would be unsustainable and unnecessary, a pilot may apply a slacker, less onerous strategy.

The expert's capability to select strategies balancing performance and effort also entails multitasking. For example, when a pilot chooses to perform, say, the pre-descent checklist, he or she must necessarily choose a strategy for controlling the aircraft that does not require his or her full attention. Indeed, this drives not just how he or she performs multiple tasks, but when: hopefully the pilot chooses to perform the checklist when the performance goals for control do not require a high-effort strategy!

Further, these strategies can also include the choice of when and how to balance performance and effort by assigning work to other agents, automated or human. Thus, in Section 13.6, we move from a small training aircraft—single pilot and no autopilot—to the more realistic domain of the multi-crew, flexibly automated flight deck, interacting with air traffic control.

13.6 FEIGH'S SECOND LAW OF COGNITIVE ENGINEERING

Flying with another expert pilot is a joy. Standard roles have evolved for the "pilot flying" and the "pilot managing" in which each knows what the other is doing, and how to support it. For example, Hutchins et al. (2013) describe the lovely interplay between two pilots during the takeoff roll, where verbal callouts and even their hand movements carefully reflect their independent-yet-coordinated contributions. It is important to note, however, that their coordination may divvy up the essential taskwork of flying the aircraft between them, requiring less taskwork effort from each, yet this coordination also adds new teamwork activities that serve to inform each other and to interleave their actions.

If the goal of aviation is to enable the safe transit of aircraft, air traffic controllers are vital to ensure safe separation and, in busy airspace, maintain safe traffic flows. But, do they contribute any taskwork? They do not physically move the aircraft. Instead, they serve as the consummate team workers, gathering information about each of the aircraft by radar and voice communications, and communicating a safe target flight path for each flight crew to control toward. This teamwork shapes the environment of the pilots—instead of being able to fly anywhere, air traffic regulations and clearances constrain their allowable actions—and creates numerous new actions for the pilots around communication and reporting to air traffic control. Likewise, the teamwork provided by the air traffic controller is so incredibly specialized and complex that this division of labor requires a whole separate role with finely honed selection and training requirements.

This brings us to the fourth point of this chapter. Woods and Hollnagel (2006), when discussing the "Laws of Coordination" in joint cognitive systems, noted that "Achieving coordinated activity across agents consumes resources and requires ongoing investments and practice." In this chapter, which is specifically examining work, let us further examine the impact of teamwork using a phrase coined by my colleague Karen Feigh as a form of entropy:

Point 4: *Feigh's Second Law of Cognitive Engineering: Divvying up the taskwork within a team creates more work—teamwork.*

This second law highlights that the addition of team members is not a facile matter, for it can increase the total work of the team. If worker A is saturated, then a second team member, worker B, can be a good addition if he or she can assume half the taskwork. Now, each may only need to do 50% of the minimum required taskwork (or may opt to choose more effortful strategies to increase performance), but their total work will be greater to the degree that teamwork is added.

How much teamwork is added, and its particulars, is a rich topic that can easily expand into numerous related areas including the social and cognitive aspects of human–human teamwork. This perspective on work does afford some basic distinctions. First, teamwork will necessarily be increased when different team members' tasks are interleaved and need to be coordinated. (Imagine, for example, one person pouring melted butter into the mixing bowl one tablespoon at a time, which the other person then beats it into the sauce after each pour: the start of one requires the stop of the other, and the amount of butter poured in may then drive how much beating needs to follow. In such situations, the activities of both are driven by the other, and must be carefully coordinated.) Conversely, activities that are largely distinct can be split up with minimal added coordination. (Imagine instead a division of taskwork where one person makes the batter while the other butters the cake pan, requiring minimal interaction.) Thus, the choice of division within the teamwork is important. Second, designating one team member as "responsible" for the team's outcome can establish the need for monitoring and supervision of other team members' work, which, depending on the circumstances, can be itself a significant tasking. Third, the addition of a team member may be motivated for error checking capability as much as for divvying up the taskwork; in this case, the monitoring may be performed both ways, which each team member checking the results of each other. In the flight deck, for example, some key actions by the pilot flying must be cross-checked by the pilot managing, such as the selection of autopilot modes and target altitudes, headings, and speeds.

For example, the lack of such a habitual callout and cross-check of autopilot modes within an airline is attributed by the National Transportation Safety Board (NTSB) as one of several contributing factors that placed a Boeing 777 in the wrong autopilot mode during the final approach of Asiana Flight 214 in July 6, 2013, and its subsequent stall and crash short of the runway at San Francisco airport (NTSB 2014). Interviews with the flight crew within the airline about general practices found that that the pilot flying may or may not call out commanding a new autopilot mode, and the pilot managing may or may not verbally confirm the mode. Thus, a review of the cockpit voice recorder for this particular flight found the trainee captain, high and fast on the final approach, did not call out the selection of the Flight Level Change mode (which switched the throttle to essentially idle instead of tracking airspeed) and the instructor pilot in the right seat, apparently busy with setting the flaps, did not appear to notice the change.

This teamwork extends also to automation: allocating work functions to automation creates monitoring and cross-checking duties for the flight crew. An autopilot, for example, can assume the taskwork of actively controlling the aircraft, but, unless the automation is certified and trusted to be completely autonomous (a rare situation, i.e., on the commercial flight deck, this is only partially true for only the auto-land function), then human–automation teamwork is created, including programming or configuring the automation, monitoring the automation, or responding to automated alerts or requests for inputs.

Most categorizations of levels of automation focus on the taskwork delegated to the automation, without accounting for the resulting teamwork demanded of the pilot. Unfortunately, a so-called "high level of automation" may appear to transfer a significant amount of taskwork to the automation but in operation be unmanageable with the teamwork it imposes on the pilot. This effect has been noted in aviation with the use of autoflight systems that can take over all aspects of control and guidance, but may trigger workload spikes for the flight crew in terms of sudden reprogramming when a new air traffic clearance needs to be laboriously typed into the flight management computer. In a more generalizable laboratory task, Kirlik demonstrated that the choice of delegating a function to automation (or not) is itself a strategy that balances performance gains with the effort to program and monitor the automation, to the point that sometimes "an aid can (and should) go unused" (Kirlik 1993).

Conversely, a low level of automation, such as an alerting system, may allow a worker to relinquish a substantial degree of monitoring or other low-level activities. This can be very helpful in some contexts, such as a meat thermometer that alerts when the Thanksgiving turkey is ready, freeing the cook for other duties or a glass of wine. However, in other cases, such delegation may be problematic. The introduction of flight deck alerting systems, for example, created fears of a primary-backup inversion in which the pilots may relinquish their role as the primary monitor to the backup alerting system (Wiener and Curry 1980). As some of these so-called alerting systems became more capable and also commanded avoidance maneuvers, others noted situations in which pilots were "cognitively railroaded" into following the commands because they had insufficient time and capacity to verify or refute the automation's output (Pritchett 2001).

The introduction of team members also adds artifacts and dynamics to the work domain, to the extent that team members help create each other's environment. The flight deck and the air traffic controller station both contain voice radio sets for communication, and the controller additionally has radar. Their teamwork actions employ carefully designed structures, including standard phraseology and published routes, so that any pilot can talk to any controller. These structures should be included in the analysis of each worker's activities, but have not been historically an integral part of many work domain analyses (see Burns et al. 2004 for a discussion of how different analysts have approached such structures). While some may term them "intentional" dynamics or describe them as "soft" constraints on behavior, I would argue that (in aviation at least) these structures do not derive from the intentions of the worker, and operationally are just as "hard" as many physical aspects of the environment. A pilot may view an air traffic instruction as an exact target (particularly when he or she then overhears how it carefully steers them between other aircraft above and below), while some so-called physical limitations such as maneuvering speed, comfortable bank limits, or maximum landing weight can be overridden in an emergency.

Thus, the teamwork aspects of the work domain can be consciously constructed to create a work environment for each worker that balances the demands of taskwork and teamwork, and that provides structures for easy, effective teamwork—even among flight crew, or between pilots and controllers who have never interacted before. Such designs may include concepts of operation, or those aspects of standard operating procedures that exist to establish standard team interactions and crew coordination.

To follow up on this notion of designing work environments, the next section starts by reviewing this chapter's four key points about work. With each point, Section 13.7 also discusses its implication for designing work environments.

13.7   FOUR TAKEAWAYS, AND IMPLICATIONS FOR DESIGN

This chapter has, by this point, identified a sizeable list of capabilities of an expert worker: an expert understands the dynamics of the environment; an expert can construct the environment to support work; an expert further knows which strategies he or she can enact to drive the environment toward his or her purpose; and ultimately an expert adapts which strategies he or she chooses to apply to reflect his or her goals for performance relative to the effort required. This work includes both taskwork and teamwork to the point that workers construct each other's environment.

In their book on joint cognitive systems (2006), Woods and Hollnagel argue that the cognitive systems engineering community needs not only to understand these attributes of work—we also need a design perspective that turns these understandings into improved design:

> if we are to enhance the performance of operational systems, we need conceptual looking glasses that enable us to see past the unending variety of technology and particular domain (Woods and Sarter 1993)... But the challenge of stimulating innovation goes further. A second strand of processes is needed that links this tentative understanding to the process of discovering what would be useful. Success occurs when "reusable" (that is, tangible but relevant to multiple settings) design concepts and techniques are created to "seed" the systems development cycle. (Woods and Christoffersen 2002, p. 3)

First, let us examine what such seeds we can find in each of the points raised thus far about work. Point 1 stated that *Expert workers understand the dynamics of their work domain and capitalize upon them to the extent that their own activity is a mirror reflection of their environment.* For a comparative novice, a recipe or standard operating procedure should be designed to provide a robust, effective path through the many constraints and affordances of the domain. For a comparative expert, on the other hand, the design challenge is to allow for flexible, fluent responses to the domain. This latter point is especially hard in a highly automated environment: it is too easy to design automation that operates only in one way and in doing so provides the human operator with only one pattern of activity to mirror its actions.

Similarly, Point 2 stated that *Expert workers not only respond to their work domain—they schedule their activities and create artifacts within it that reflect their intentions, plans, and memory to themselves and others.* Designs can conflict with, or support, several aspects of this point. The first aspect is the scheduling of activity. A design should not presuppose—or worse yet enforce—a single schedule. For example, an electronic checklist in the cockpit can be built upon not only knowledge of the nominal order of a checklist but also underlying "procedure context" information that recognizes when steps can be performed in different orders or skipped altogether in response to immediate circumstances (Ockerman and Pritchett 2000).

The second aspect of this point is the construction (and placement) of artifacts. Often, designs seek to be "clean" and "uncluttered." This principle can be good when it causes the designer to review and reflect on the core attributes that are really needed by the worker, but this principle can also be taken too far. Specifically, the design should not obstruct the worker's efforts to create artifacts, and otherwise construct and manipulate their environment. Instead, the designer should be thoughtful but also humble enough to build in mechanisms that allow the user to "finish the design" (Rasmussen and Goodstein 1987). In a flight deck, this can include leaving space for paper reminders in the physical space, for entry of alternate potential flight routes in the flight management system so that they can be invoked quickly at a later, higher-tempo time if necessary, for example.

Point 3 discussed how *Expert workers adapt their behavior to the dynamics of the domain to achieve their desired outcomes.* An interesting aspect for designers here may focus on the measures by which we evaluate a design. Given that an expert worker will "do what it takes" to achieve work goals, their ability to achieve satisfactory performance does not necessarily reflect a good design, but instead may only signal that the design is not so bad as to obstruct performance. Instead, measures of design should examine how much it requires the worker to adapt to the design in nominal conditions—and how much it supports the worker in adapting to unusual conditions.

Finally, Point 4 noted that *Divvying up the work within a team creates more work—teamwork.* Thus, designers need to be mindful of several points. The first is that adding any team member to the work environment—whether human or automated—will not necessarily reduce the total work demanded of each team member. This is particularly true when the allocation of taskwork between team members is interleaved or, worse yet, does not create coherent breakdown of work. For example, Feigh and Pritchett (2014) noted how nonlinear effects in each team member's workload can arise when they are each assigned different high-level tasks that rely on awareness of the same information or on interdependent judgments of similar phenomena; in such a case, each team member may need to perform the same underlying activities around information gathering and judgment, only minimally reducing the task load of each and increasing the combined task load of the combined team.

Likewise, Feigh and Pritchett (2014) argued that designers should seek to establish coherent roles within a team of workers:

> One attribute of a coherent function allocation can be viewed from the bottom up: its functions share (and build upon) obvious, common constructs underlying all their activities… Another attribute can be viewed from the top down: [each team member's] functions … contribute towards work goals in a manner that is not only apparent …, but can be purposefully coordinated and adapted in response to context.

The value of coherent roles is particularly important when designing automation and integrating it into operations. As noted by Dekker and Woods (2002), the design of automation inherently creates a function allocation that is described by what the automation will do, leaving the human to pick up whatever tasks are "left over" as well as the teamwork demanded of the (human) worker to inform, command, and monitor the automation. When these leftovers are collectively an incoherent set, it is difficult for human workers, no matter how expert, to coordinate their activities and adapt to the demands of the work environment. By this measure, the concept of "level of automation" is only useful to describe the capability of the machine, but because it does not describe what the rest of the team will need to do to coordinate with the automation, it is not useful to the designer of the overall operation.

Thus, while Woods and Hollnagel called for the "seeding" of design, the "sprouting" of new designs is not easy: just understanding the work doesn't by itself ignite the spark of creation. Indeed, this chapter's approach so far has been to isolate the construct of "work" from the other constructs in the triad of joint cognitive systems (namely, people and technology) and then to further decompose work into four points. But the insights just provided by decomposing work into these four points then need to be applied in the reverse manner, reassembling the notion of designing work and the work environment back into a holistic picture in which each aspect of a design must support not one aspect but instead all aspects of effective work. And thus, beyond the separate points here, we may again refer to Woods and Hollnagel (2006, p. 8) for the separate perspective of envisioning the patterns inherent to effective work: patterns in coordinated activity, patterns in resilience (anticipating and adapting to surprise and error), and patterns in affordance (supporting the worker's natural ability to express forms of expertise in the work environment).

13.8   THE JOY OF A JOB WELL DONE

True happiness comes from the joy of deeds well done, the zest of creating things new.

Antoine de Saint-Exupery

Work, when done well, can transcend the mere mechanics of standard operating procedures into mindfulness and joy. Indeed, this joy is not readily apparent in an abstraction hierarchy, with its focus on the work environment (as important as this is). I posit that the joy comes in the melding of the worker's activity with the dynamics of the environment, until together they do something that could occur with neither alone. Indeed, the act of creation is itself work.

While the word *work* denotes "purposeful activity"—a good thing when the purpose is worthy and the activity is healthy—in modern parlance, work is often connected in the negative. Merriam-Webster's discussion of work first lists as synonyms *travail, toil, drudgery*, and *grind*, all evoking exertion that does not reflect any intrinsic motivation within the worker, or reward the worker by any but the most calculating means such as a salary. With this exhausting perspective on work comes the common assertion that we need to reduce one's workload and effort required. Indeed, the NASA Task Load Index (TLX) method invites workers to self-assess their workload on six scales, where four of these scales describe "Effort" or various aspects of "Demand."

One needs to dig deeper to find the positive in work. Further down its list of synonyms of work, Merriam-Webster also lists more positive terms: *calling, pursuit*, and *métier*. All imply some fit between the activity inherent to the work and a capability or gift inherent to the worker. Perhaps this transcends to the other TLX scales: "Performance" allows the worker to self-assess if his or her purpose was achieved, and "Frustration" allows the worker to express those obstructions that discouraged, irritated, or annoyed them as he or she attempted to achieve his or her purpose. My own experience with pilots has given me a deal of affection for this last rating scale, given the number of times pilots have reported markedly higher frustration scores with automated systems that did not support their work; indeed, given that pilots typically "do what it takes" to maintain their desired level of performance, reports of frustration can be an excellent diagnostic of automated "aids" that do not support their work.

So, perhaps then our designs can be analyzed for whether they frustrate or foster the joy of work and the expert worker. Doing so requires an improved language and understanding of work: indeed, it should be analyzed as rigorously as people and technology, the other components with which it intersects in Woods' discussions of innovations in joint cognitive systems. This chapter has reviewed my own reflections on the key points underlying work, but there is more to be done in envisioning how these observations about work then reliably translate into better design.

REFERENCES

Beyer, H., and Holtzblatt, K. (1997). *Contextual Design: Defining Customer-Centered Systems*. London: Elsevier.

Burns, C.M., Bisantz, A.M., and Roth, E.M. (2004). Lessons from a comparison of work domain models: Representational choices and their implications. *Human Factors*, 46, 711–727.

Dekker, S.W., and Woods, D.D. (2002). MABA-MABA or abracadabra? Progress on human–automation co-ordination. *Cognition, Technology and Work*, 4, 240–244.

Feigh, K.M., and Pritchett, A.R. (2014). Requirements for effective function allocation: A critical review. *Journal of Cognitive Engineering and Decision Making*, 8, 23–32.

Hollnagel, E. (2009). *The ETTO Principle: Efficiency–Thoroughness Trade-Off*. Burlington, VT: Ashgate.

Hutchins, E. (1995). How a cockpit remembers its speeds. *Cognitive Science*, 19, 265–288.

Hutchins, E., Weibel, N., Emmenegger, C., Fouse, A., and Holder, B. (2013). An integrative approach to understanding flight crew activity. *Journal of Cognitive Engineering and Decision Making*, 7, 353–376.

Kirlik, A. (1993). Modeling strategic behavior in human–automation interaction: Why an "aid" can (and should) go unused. *Human Factors*, 35, 221–242.

Kirlik, A. (1998). Everyday life environments. In W. Bechtel and G. Graham (Eds.), *A Companion to Cognitive Science* (pp. 702–712). Malden, MA: Blackwell.

McRuer, D.T., and Jex, H.R. (1967). A review of quasi-linear pilot models. *IEEE Transactions on Human Factors in Electronics*, 3, 231–249.

Mettler, B., and Kong, Z. (2013). Mapping and analysis of human guidance performance from trajectory ensembles. *IEEE Transactions on Human–Machine Systems*, 43, 32–45.

National Transportation Safety Board (NTSB) (2014). *Accident Report for Asiana 214*. Downloaded March 15, 2015 at http://dms.ntsb.gov/pubdms/search/document.cfm?docID=419822&docketID=55433&mkey=87395.

Ockerman, J.J., and Pritchett, A.R. (2000). A review and reappraisal of task guidance: Aiding workers in procedure following. *International Journal of Cognitive Ergonomics*, 4, 191–212.

Pritchett, A.R. (2001). Reviewing the role of cockpit alerting systems. *Human Factors and Aerospace Safety*, 1, 5–38.

Pritchett, A.R., Feigh, K.M., Kim, S.Y., and Kannan, S.K. (2014). Work models that compute to describe multiagent concepts of operation: Part 1. *Journal of Aerospace Information Systems*, 11, 610–622.

Rasmussen, J., and Goodstein, L.P. (1987). Decision support in supervisory control of high-risk industrial systems. *Automatica*, 23(5), 663–671.

Simon, H.A. (1996). *The Sciences of the Artificial* (3rd ed.). Cambridge, MA: MIT Press.

Wiener, E.L., and Curry, R.E. (1980). Flight-deck automation: Promises and problems. *Ergonomics*, 23, 995–1011.

Woods, D.D., and Christoffersen, K. (2002). Balancing practice-centered research and design. In M. McNeese and M.A. Vidulich (Eds.), *Cognitive Systems Engineering in Military Aviation Domains* (pp. 121–136). Wright–Patterson AFB, OH: Human Systems Information Analysis Center.

Woods, D.D., and Hollnagel, E. (2006). *Joint Cognitive Systems: Patterns in Cognitive Systems Engineering*. Boca Raton, FL: CRC Press.

Woods, D.D., and Sarter, N.B. (1993). Evaluating the impact of new technology on human–machine cooperation. In *Verification and Validation of Complex Systems: Human Factors Issues (NATO ASI Series)* (pp. 133–158). Springer: Berlin Heidelberg.

14 Designing Collaborative Planning Systems
Putting Joint Cognitive Systems Principles to Practice

Emilie M. Roth, Elizabeth P. DePass,
Ronald Scott, Robert Truxler,
Stephen F. Smith, and Jeffrey L. Wampler

CONTENTS

14.1   INTRODUCTION

The challenges associated with designing effective automation are well documented (Bradshaw et al. 2013; Christoffersen and Woods 2002; Roth et al. 1997). Clumsy automation can result in brittle performance, miscalibration of trust, deskilling of the users of the automation, and lack of user acceptance (Lee 2008; Lee and See 2004; Parasuraman and Riley 1997). Designing effective decision aids is particularly challenging when the underlying problem-solving technology is "opaque" in the sense that, while the general principles by which the algorithm generates solutions may be understandable, it is not readily apparent exactly why a particular solution "bubbled to the top" in any specific situation (Cummings and Bruni 2009; Roth et al. 2004). Examples of opaque technologies include some types of mathematical optimization algorithms, complex mathematical simulations, and fusion algorithms.

247

In this chapter, we review three decision aids that we developed for military transport organizations where the solution generation technology underlying the decision aid was opaque (DePass et al. 2011; Scott et al. 2009; Truxler et al. 2012). Each decision aid addressed a different aspect of transportation planning and was based on a different automated planning technology. What they shared was the goal to create collaborative planning systems that enabled the human planners to understand, direct, and contribute to the problem solution process.

The chapter describes these three collaborative planning systems with the aim of illustrating the visualization and user interaction design principles that we developed to foster more effective collaborative performance between the technology and the people that together constitute the *joint cognitive system* (Woods and Hollnagel 2006). These design principles provide concrete illustrations of the more abstract-level principles for effective joint cognitive systems that were pioneered by David Woods and his colleagues (Christoffersen and Woods 2002; Hollnagel and Woods 1983; Klein et al. 2004; Roth et al. 1987; Sarter et al. 1997; Woods 1986; Woods and Hollnagel 2006; Woods and Roth 1988). The goal of the chapter is to contribute to the literature on reusable concepts and techniques (or *design seeds*) for the design of joint cognitive systems (Woods and Christoffersen 2002).

14.2   THE NEED FOR A JOINT COGNITIVE SYSTEMS APPROACH

One of the canonical paradigms for decision support assumes that algorithms will generate a solution working on their own. The user's role is to merely provide it the information it needs, and then act on its "recommendation." This has been called the "Greek Oracle" paradigm (Miller and Masarie 1990). Decision support applications built on the Greek Oracle paradigm continue to be created to this day. Roth et al. (1997) documented some of the most serious deficiencies that have been observed with this approach. These negative consequences include the following:

- *Brittleness in the face of unanticipated variability.* Automated problem solvers perform well for the set of cases around which they are designed, usually routine or nominal ones, but performance quickly breaks down when confronted with situations that had not been anticipated by system designers.
- *Deskilling/irony of automation.* Reliance on the automation for routine cases reduces opportunity for users to exercise and extend their skills (*deskilling*). Paradoxically, users are expected to detect and deal with the most challenging cases beyond the capability of the automated problem solver. Bainbridge (1983) coined the term "irony of automation" to describe a parallel dilemma that was noted in the introduction of industrial automation.
- *Biasing human decision process.* Another serious concern is that the Greek Oracle paradigm can, in some cases, lead to *worse* performance than if the human was working unaided. Automated problem solvers can bias people's information-gathering activities, narrow the set of hypotheses they consider, and increase the likelihood that they will fail to come up with a correct solution when the automated problem solver fails (Guerlain et al. 1996; Layton et al. 1994; Smith et al. 1997; Smith, this volume).

Experience with decision aids built on the Greek Oracle paradigm highlight the need for an alternative decision-aiding paradigm—one that treats the people and technology as collaborative partners that jointly contribute to the problem-solving endeavor—a joint cognitive system. People on the scene have access to real-time information and common-sense knowledge not available to the automated problem-solver. Their contribution to successful joint performance, particularly in unantici-pated situations, needs to be explicitly recognized and fostered in the design of the software—in terms of both functionality and interaction design.

14.3 PRINCIPLES FOR DESIGN OF EFFECTIVE JOINT COGNITIVE SYSTEMS

Over the last 30 years, David D. Woods and his collaborators have developed design principles for effective joint cognitive systems. Woods and Hollnagel (2006) provide a synthesis of these design principles that include the following:

Observability: A foundational principle for effective joint cognitive systems is that the assumptions, objectives, findings, and actions of the partners in the problem solving process—the people and the technologies—be jointly observable. This can be best achieved by providing a *shared representation* of the problem to be solved that team members can inspect and contribute to. This establishes a "common ground" that makes coordinated activity and collabo-rative problem-solving possible (Clark and Brenna 1991; Klein et al. 2005). It enables actions and intentions of each of the parties to be mutually understood and allows divergence of goals or incorrect assumptions to be rapidly detected and resolved. A by-product and closely related concept to observability is *understandability*—the ability of each partner to understand the others' con-tribution to a proposed problem solution.

Directability: A second important characteristic of effective joint cognitive systems is the ability of the person to direct and redirect resources, activities, and priorities as situations change. This implies that the technology should be directable. Software mechanisms ("levers") are needed to enable the person(s) on the scene to define limits on the automation, modify default assumptions, and guide problem solution. As has been pointed out by Ackoff: "The optimal solution of a model is not an optimal solution of a problem unless the model is a perfect representation of the problem, which it never is" (Ackoff 1979, p. 97, as quoted in Woods and Hollnagel 2006). The people on the scene will inevitably know about affordances and constraints associated with the particular situation that the automated problem solver will have no knowledge of. Problems of brittleness can be mitigated by enabling persons on the scene to redirect the software based on their situated knowledge.

Directing Attention: A related feature of effective joint cognitive systems is the ability of each actor (people or technology) to orient the attention of the other(s) to critical problem features, as well as reorient attention focus as new infor-mation becomes available or changes occur in the world. This implies the ability of the collaborating actors to flag key factors to be considered in problem solutions and new findings that may signal a need to revise priorities, solution considerations, and constraints.

Shifting Perspectives: This refers to the power of joint cognitive systems to leverage multiple views, knowledge, and stances in pursuit of problem solution. All of the actors in joint cognitive systems (people and technology) need to be able to broaden and winnow the solution space through seeding (suggesting directions or providing candidate solutions), reminding, and critiquing.

The four high-level functions of collaborative systems sketched above are intended to be in the service of two fundamental objectives of joint cognitive systems:

Broadening: This is the ability of the joint cognitive system to perform better than any actor (person or technology) working alone. This is achieved by broadening the set of candidate solutions explored and the range of factors considered in evaluating these solutions relative to what would be likely (or, in some cases, possible) by either the person(s) or automated problem solver working on their own.

Adaptability: This is the ability of the joint cognitive system to more effectively handle unanticipated characteristics of the situation and recognize and adapt fluidly to dynamic changes for more resilient performance.

Our research team has had the opportunity to explore the above ideas and principles in the design of joint cognitive systems for military transportation organizations as part of a broader program to design work-centered support systems (Eggleston 2003; Scott et al. 2005). Here, we describe three of the collaborative planning systems we developed and some of the concrete techniques that we employed in each case to enable transportation planners to be active partners in the problem-solving process.

14.4 EMBODYING JOINT COGNITIVE SYSTEMS PRINCIPLES

In each of the three cases, the existing problem-solving technology was opaque, requiring us to come up with different strategies for allowing transportation planners to understand and contribute to the problem-solving process. The first system leveraged simulation technology to enable transportation planners to rapidly develop and investigate multiple courses of action alternatives for moving people and cargo to meet mission objectives (DePass et al. 2011). The second system leveraged mathematical optimization algorithms to allow personnel to revise mission plans during execution in order to respond to dynamic changes (Scott et al. 2009). The third system also leveraged a (different) mathematical optimization algorithm, this time to support allocation of airlift resources to meet transportation requirements under conditions of limited resources, and fluidly changing requirements and priorities.

14.5 COURSE OF ACTION EXPLORATION TOOL

The Course of Action (COA) exploration tool was developed for the U.S. Transportation Command (USTRANSCOM), which is charged with directing and executing transportation needs for movement of personnel and distribution of goods (DePass et al. 2011). One of USTRANSCOM's responsibilities is to identify potential ways a transportation requirement might be satisfied—the modes of movement (air, sea, or ground), the ports to pass through, and the mix of vehicles to be used—referred to as a transportation COA.

A transportation planner may be tasked with very quickly producing multiple COA options in response to either actual or projected needs. In generating and comparing options, both speed of travel and carrying capacity need to be considered. For example, air movements are faster than sea movements, but the capacity of a typical ship is many times the capacity of a plane. As a consequence, while an airplane might get the first plane load to its destination faster, it may take less time to complete delivery of all items (referred to as closure day) when going by sea. Other considerations include port capacity—airports and seaports can have severe limits on space or material-handling capability, limitations in available aircraft or ships, and overall cost, among many others.

One noteworthy characteristic of this domain is that the planner is often required to generate a "rough COA" even before the problem is fully specified. For example, planners may start the planning process to send humanitarian relief to a part of the world where an earthquake had just hit before having a clear idea of what or how much is to be moved. They can begin to think through what ports could be used to quickly get supplies into the affected area without having a full specification of movement requirement. This is reminiscent of the "skeletal plans" that experts in other domains have been observed to develop that serve to guide early decision making, while preserving options to enable adaptation as a situation unfolds (e.g., Suchman 1987).

At the time that we started this project, there were no software tools specifically intended to support this type of rapid COA creation. The procedures in place relied on planner experience and informal "back-of-the-envelope" computations. There was a mathematical simulation model for large-scale strategic air and sea movement analyses, but those are orders of magnitude more complicated than the COA problems of our planners. This simulation tool, while powerful, was opaque in that it included multiple default assumptions and constraints that were not apparent through the user interface. Further, it required detailed domain and modeling expertise to use, needed extensive precise data inputs (e.g., exact details on the type and amount of cargo), and took on the order of hours to run. These factors made it ill-suited for rapid COA development. However, we were able to leverage this simulation technology to create a more collaborative COA exploration tool better suited to the needs of our planners.

Our design objective was to approximate the ease and speed of generating a "back of envelope" solution, while at the same time leveraging the power of sophisticated transportation simulation tools for generating COAs. We wanted to allow individuals with minimal expertise to come up with reasonable "ballpark" answers quickly, while at the same time allowing for the problem definition to be refined over time and the solution precision to be correspondingly improved as more information on the transportation problem became available. Most importantly, we wanted to provide the opportunity for the transportation planner to rapidly develop multiple COAs so as to broaden the set of solutions considered as well as the set of factors used in evaluating alternative solutions (e.g., cost, impact on other ongoing missions).

To accomplish these goals, we needed to modify the simulation software so that

- It ran more quickly, generating a solution in approximately 1 second, making it possible to run multiple alternatives in a short period of time.
- It could be run without requiring detailed problem definition inputs.
- Default simulation assumptions and constraints were apparent to and changeable by the transportation planner.
- "What if" exploration as well as solution comparison along multiple factors (e.g., cost, robustness, impact on other missions) was supported so as to broaden the range of solutions generated as well as the factors used to evaluate alternatives.

The resulting tool was called the Rapid Course of Action Analysis Tool (RCAT). Figure 14.1 shows a screenshot of a map visualization that serves as shared representation for defining the transportation problem and exploring elements of a solution. Transportation planners can explore characteristics of airports and seaports as well as draw candidate transportation routes and obtain rapid feedback on route characteristics such as cycle time and throughput (i.e., short tons/day). The map display and interaction mechanisms provide direct support for observability and directability.

Observability and directability are further supported via assumptions editors that enable transportation planners to see and modify default assumptions underlying computations, such as cargo onload and offload times, and the number of maximum-on-ground (MOG) slots for aircraft to park at an airfield at once. Figure 14.1 shows a

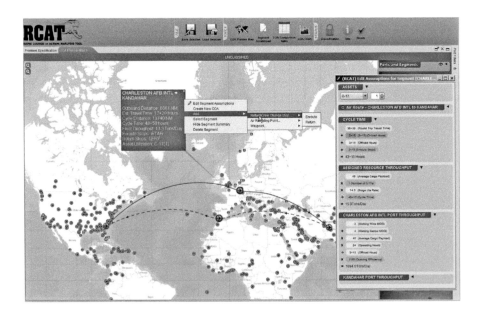

FIGURE 14.1 A screenshot of a map in the RCAT display illustrating the ability to define transportation routes and quickly see results such as cycle time and throughput.

number of default values that planners can review and modify. In the example given in Figure 14.1, the planner has modified the onload hours from a default value of 3 hours 15 minutes to 2 hours 15 minutes. Note that the default value remains visible (it is shown in brackets) along with the user-entered value as a reminder that the default value has been overridden.

In addition, RCAT will alert transportation planners if a routing they specified violates constraints such as the maximum number of hours a crew can fly without a rest stop, or the maximum distance an aircraft can fly without refueling. This illustrates directing attention to factors that need to be considered in problem solution.

A major challenge in developing RCAT was the need to bridge the gap between the detailed problem specification input that is required by the simulation model and the fact that our planners would be unlikely to know the problem at this level of detail at the time problems are first posed to them. To bridge this gap, we developed utilities that convert high-level specifications of the sort that the transportation planners are likely to have into lower-level specifications that the model needs to run. For example, transportation planners can input minimal information about the cargo to be moved (e.g., total short tons) if this is the only information available to them. From this minimal information, RCAT makes (reasonable) default assumptions about the likely characteristics of the cargo that transportation planners can then view and modify if they choose. An example is shown in the load details section in Figure 14.2. The transportation planner entered 1000 short tons as the load. From this, RCAT generated default values for load characteristics that the planner can view and modify.

This capability supports adaptability, accommodating problem specifications at different levels of granularity as details emerge over time. It also fosters observability

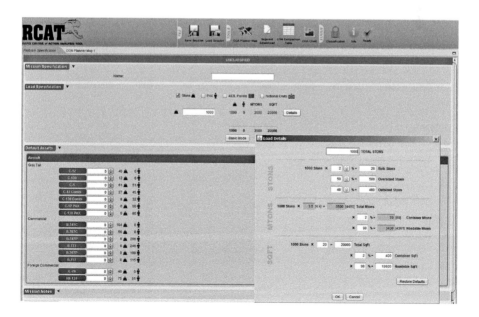

FIGURE 14.2 A screenshot of a transportation problem specification RCAT display.

by revealing to transportation planners the default assumptions that RCAT makes, and directability by allowing them to override those assumptions.

RCAT was also designed to enable planners to rapidly generate and compare alternative solutions. Multiple solutions can be displayed simultaneously on a map view, as well as in a tabular view that allows transportation planners to compare solutions on dimensions of interest including the number and type of assets utilized, cost, and closure time (the time by which all movement requirements are completed). These features in combination supported rapid "what if" exploration of the solution space, broadening the set of factors considered and solutions explored, encouraging shifting perspectives and broadening of the solution space.

We conducted a user evaluation of RCAT. Current transportation planners utilized RCAT on representative problems and then filled out a feedback questionnaire. Planner feedback indicated that RCAT would enable them to consider a broader set of solutions because of the ease and speed of exploring alternatives. The tool has since been adopted by the organization and is routinely used in support of transportation COA generation and comparison.

RCAT illustrates several concrete techniques for more effective joint cognitive systems. Specifically, RCAT

- Utilizes visualizations that enable a shared representation of the problem space: Observability
- Gives planners ways to view and modify model assumptions: Observability, directability, adaptability
- Flags constraints and violations: directing attention
- Accommodates problem specifications at different levels of granularity (as details emerge): Adaptability
- Speeds up solution time to afford more iteration/exploration (~1 seconds): Broadening
- Gives planners ways to intuitively and rapidly compare/evaluate distinct solutions: Shifting perspectives, broadening
- Supports "What If" exploration: Broadening

14.6 DYNAMIC REPLANNING TOOL

A second collaborative planning system that we developed that illustrates joint cognitive systems principles is the Global Response Synchronization (GRS) tool. GRS utilizes a mathematical optimization algorithm to support dynamic replanning of airlift missions during mission execution (Scott et al. 2009). It was developed for the Air Mobility Command, which is the air transportation component of USTRANSCOM responsible for detailed planning, scheduling, and tracking airlift missions worldwide. Missions are initially planned by mission planners. Twenty-four hours before a planned mission launch, responsibility is transferred to the execution floor where Duty Officers are responsible for handling last minute problems (e.g., due to mission delays). This requires balancing competing airlift demands and ensuring that multiple mission constraints are met, including airfield operating hours and airfield MOG limits, cargo, aircrew, and refueling constraints.

GRS leveraged a coordinated suite of timeline visualizations that we had built and tested previously (Roth et al. 2009). The timeline visualizations allowed Duty Officers to assess the impact of a change in one mission (e.g., a delay) upon other missions (e.g., creating a MOG situation at an airfield that impacts multiple missions). Duty Officers could manually drag individual missions along a timeline and visualize the impact this had, including on other missions in a "what if" mode.

Figure 14.3 is a screenshot of a multi-mission timeline view from this earlier prototype. In this example, a delay in mission 301YT causes it to stay on the ground at OKAS airfield longer than it was initially scheduled to do. As a consequence, three missions (301YT, 311YT, and 711YT) are now scheduled to be on the ground at OKAS at the same time but OKAS has only room for two (MOG is 2). This results in visual alerts on the mission timeline for each of the three missions at the point in time where the MOG conflict occurs (depicted in the prototype as red dots) as well as an alert on the OKAS Parking MOG time bar (depicted in the prototype as red highlighting of the portion of the time bar during which OKAS is scheduled to exceed MOG). The stacked vertical arrows in Figure 14.3 point to where these visual alerts appear. Another problem that results from the delay of mission 301YT is that the crew on that mission will exceed their crew duty day in the middle of the last leg of the flight (depicted as a red dot on the mission timeline and red highlighting on the

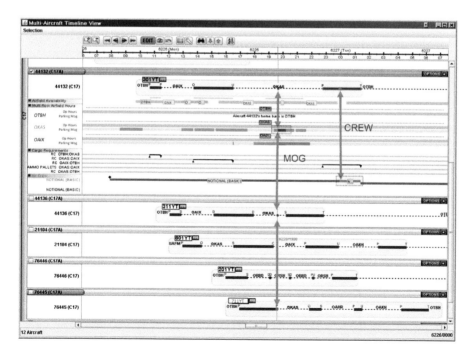

FIGURE 14.3 A screenshot of a multi-mission timeline showing several planning constraint violations.

aircrew time bar corresponding to the duration during which the crew is expected to be in violation of the crew duty day limit). This is shown in Figure 14.3 by the vertical arrow between the point on the mission timeline when the crew duty day limit is exceeded and the corresponding location on the aircrew time bar. Note that the arrows in Figure 14.3 are not part of the prototype display. They have been added to the figure to facilitate explanation of how the display works.

With this prior prototype, the Duty Officers could attempt to fix the problem by manually dragging missions along the timelines until the MOG situations and crew duty day issues were resolved and all constraints across all missions were satisfied. While fixing problems was possible, it was a cumbersome trial-and-error process that could clearly benefit from more automated support. This motivated the need for a tool that could automatically generate solutions that satisfied the across-mission constraints. GRS was designed to meet this need while preserving a joint cognitive systems approach.

Key domain characteristics guided our design of GRS:

- *The time horizon for replanning decisions is on the order of minutes.* In most cases, taking up to tens of minutes to find and implement a replanning solution is generally acceptable—particularly when the missions involved are high profile. This meant that it was possible to include the Duty Officer as an active participant in exploring the solution space.

- This is an environment with *missing and imperfect information.* Cargo and passenger load and relative priority of missions are generally not available in the databases or are not kept up to date. Duty Officers have learned to keep track of this information via informal means (e.g., phone calls, private notes). This meant that Duty Officers would be aware of important considerations that the automated planner would have no way of knowing about, and would thus need mechanisms to constrain and direct the automated planner.

- *There is no fixed scoring function.* There are multiple measures of goodness for a plan—completing all the missions as quickly as possible, having a robust schedule, getting aircrews home at appropriate times, getting high-priority cargo delivered first, minimizing flight hours or fuel usage, and minimizing time spent doing time-consuming human re-coordination as schedules change. Any of these factors might take on highest importance for different people or even the same person in different contexts. Thus, any fixed scoring function is bound to yield a brittle solution. There is a need for the Duty Officer to help guide and evaluate solutions along dimensions that may not be considered or properly weighted by the optimization algorithm.

- The *high dimensionality of the search space* makes trial-and-error search unworkable, but the Duty Officer can (if given appropriate visualization tools) effectively evaluate the goodness of one or more solutions proposed by an automated planner.

- The goal of Duty Officers is *satisficing rather than optimizing.* There is no single, unique, best solution; rather, there can be multiple acceptable solutions that may trade-off across problem dimensions. One solution may entail delay of a high-priority mission, whereas another solution may entail delay of several, less high-priority missions. This meant that it would be important to provide the Duty Officer with multiple alternative options to compare and select from.

Because the dynamic replanning task was not time critical, it was possible to design a collaborative planning system where the Duty Officer could iterate on a problem solution. Further, because there can be multiple acceptable solutions in this domain, we persuaded the optimization algorithm designers to have the automated planner output multiple solutions that all met the problem constraints but that varied on a number of dimensions (e.g., how many and which missions were changed) rather than outputting a single "best solution" based on some predefined scoring function.

Observability was achieved by having all of the interaction between the Duty Officer and the automated planner occur in the context of the previously developed suite of timeline visualizations. This allowed the Duty Officer to understand the initial problem state passed to the automated planner, the constraints within which the planner will be operating, and the solution that is passed back.

Directability was achieved by providing mechanisms to allow the Duty Officer to add to or modify constraints in the initial state passed to the automated planner, so as to direct and constrain the problem solution set. A Duty Officer can "lock" an entire mission—declaring that it cannot be moved by the automated planner. They can also indicate that a mission should not be cancelled—the automated planner may reschedule it, but may not cancel it. In addition to placing constraints on the set of missions to be rescheduled, the Duty Officer may directly specify constraints on the upper or lower bounds of the times of individual events (e.g., takeoffs and landings). If the Duty Officer knows that a mission needs to take off from its initial airfield by 0400, he can specify that as an upper bound on takeoff to the automated planner.

Figure 14.4 shows an example of the GRS timeline visualization displaying multiple missions requiring changes during execution to meet all within- and across-mission constraints. Missions with an open lock icon above (e.g., missions 805N1, 125PG, and 8C919) are free to be rescheduled or even cancelled by the automated planner; missions with an open lock but also an arrow below a takeoff or landing point (e.g., Mission 124PQ) mean that there is an upper or lower temporal bound placed on those specific events. Missions with a struck out C icon (e.g., Mission 32101) mean that the mission can be rescheduled but not cancelled. Finally, missions with a closed lock (e.g., Mission 127PQ) mean that they are locked in place and cannot be changed in any way by the automated planner. Thus, the Duty Officer has visibility into the problem state that the automated planner is trying to solve, satisfying the high-level principle of observability, and has software mechanisms that can be used to express constraints that the automated planner may not otherwise be aware of, satisfying the high-level principle of directability.

Once the Duty Officer has set up the problem, the automated planner can be called up to generate solutions that satisfy those constraints. When the automated planner has completed its computations (generally within 30 seconds), a results display is presented, which shows multiple alternative solutions that the Duty Officer can inspect and choose among. While it is not possible to meaningfully trace why or how

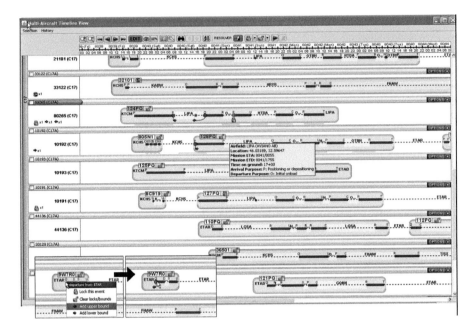

FIGURE 14.4 A screenshot of the GRS multi-mission timeline that Duty Officers can use to define and constrain the problem sent to the automated planner for solution.

a particular solution is generated by an optimization algorithm (there is typically a random search component), it is possible to provide the Duty Officer with tools to inspect and evaluate the appropriateness of a given solution.

Figure 14.5 provides a sample GRS results display. Each table column represents a possible solution to the replanning problem generated by the automated planner. The

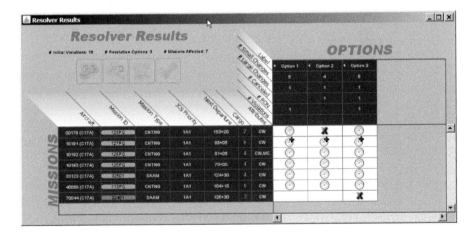

FIGURE 14.5 A screenshot of the GRS results table.

rows of the table represent individual missions that are rescheduled by at least one of the solutions. By scanning the columns of the table, the Duty Officer can quickly get a sense of which and how many missions are changed by each solution option and the nature of the change (e.g., small or large time delay, switch in rest location, cancelled). In this example, Option 1 results in small delays in five missions and a large delay for one mission. In contrast, Options 2 and 3 result in one mission being cancelled (depicted by an X in a cell), as well as missions with delays. Note that Options 2 and 3 differ in which mission is cancelled.

A key design challenge in developing the automated planner was to ensure that the multiple alternative options presented to Duty Officers for consideration were operationally distinct (Downey et al. 1990). It would be straightforward but ineffective to have the automated planner output multiple options that differ in trivial ways (e.g., small differences in takeoff and landing times). Significant software development effort was expended to ensure that the solution alternatives differed in meaningful ways (e.g., different number of mission changes, different particular missions changed, different crew rest locations).

Our display also provided Duty Officers with visibility into important factors in evaluating a solution that the automated planner was not able to take into account, such as priority of the mission, priority of the cargo, time until the next takeoff of this mission (a mission scheduled to take off soon may be essentially untouchable, due to lack of time to re-coordinate), and user-provided mission annotations. Thus, the joint cognitive system was able to take important factors into account in evaluating alternative candidate solutions that the automated planner by itself had no knowledge of. By directing attention to relevant information that was not taken into consideration by the automated planner, we hoped to reduce the decision bias observed with other automated decision support systems (see Smith, this volume, for a review of the impact of brittle technology on user cognition).

The Duty Officer can also view a solution in the original multi-mission timeline view and modify it manually. Thus, the Duty Officer is an active partner in solution generation by (1) defining the initial state passed to the automated planner and imposing constraints on valid solutions; (2) being given multiple, operationally distinct, alternative solutions from which to choose; (3) having their attention directed to key information about missions that may be relevant to solution selection that the automation was not able to take into account; (4) having the capability to further fine-tune the resulting plan manually. These features allow Duty Officers to actively partner with the technology in generating a solution that best meets the local conditions and constraints.

This combination of capabilities allowed the Duty Officers and the automated planner to bring multiple perspectives in generating and evaluating solutions, broaden the set of options considered (beyond those either the user or the automated planner might have otherwise generated), and adapt the solution to local conditions and constraints that the automated planner could not be made aware of.

A user evaluation of replanning performance using GRS was conducted. It compared performance when using the full GRS capabilities (including the automated planner capability) versus when using the GRS visualizations only (without automated planner capability). Twelve Air Mobility Command personnel (8 mission planners and 4 Duty Officers) participated in the evaluation. A within-subject, counterbalanced design was used. Half the participants first solved two multi-mission replanning problems using the full GRS capabilities (automation condition) followed by two (different but comparable) problems using the GRS visualizations only (manual condition). The other half of the participants experienced the conditions in the reverse order. The four problems were also counterbalanced across conditions so that any given problem occurred equally often in the manual and automation conditions.

Participants were run one at a time and experienced approximately an hour and a half of training before performing the evaluation problems, completing the evaluation session in 5 or less hours. While some of the participants had participated in earlier knowledge acquisition sessions, the majority had no prior experience with GRS.

Multiple objective performance measures were collected, including time to solution and quality of solution. The automated aid (automation condition) enabled participants to generate solutions significantly faster than in the manual condition (mean of 4 vs. 8 minutes, $F = 30.25$, df $= 1, 18$, $p < 0.001$). Further, it enabled them to identify better solutions that resulted in significantly fewer missions changed (3.4 vs. 3.7, $t = 3.04$, $p < 0.05$), fewer sorties changed (5.8 vs. 12.7, $t = 3.26$, $p < 0.01$), and less overall mission delays (0.4 vs. 16.6 hours, $t = 2.35$, $p < 0.05$) when compared with the manual condition. There were no statistically significant differences between conditions in participant self-assessed situation awareness or workload as measured by NASA TLX self-report ratings.

Participant feedback collected via the post-test questionnaire reinforced the conclusions from objective performance data. Participants provided high scores (mean 6.5 or greater on an 8-point scale) for usability, usefulness, learnability, impact on own work, and impact on the overall mission of AMC. In addition, participants gave high scores to ability to understand, evaluate, and redirect solutions, as well as high scores to overall "trust" in the automated solutions. They also gave high scores to "ability to generate better solutions with the automated aid than currently possible (in the same amount of time)."

The Air Mobility Command has since decided to implement the GRS capabilities.

In summary, GRS embodies the core principles of joint cognitive systems and extends the set of concrete techniques for achieving key functions of effective joint cognitive systems. Specifically, GRS

- Provides a visual representation of the problem that needs to be solved and the constraints that need to be met: Observability
- Enables users to control the initial state passed to the automation by specifying constraints on valid solutions: Directability, adaptability
- Generates multiple, operationally distinct, alternative solutions that the user can inspect and compare: Shifting perspectives, broadening
- Provides additional information relevant to solution evaluation that the automation could not take into account but that the user can leverage in winnowing the solution set: Directing attention; shifting perspective

14.7 DYNAMIC RESOURCE ALLOCATION TOOL

The third collaborative planning system we developed focused on the initial stages of mission planning that occur at USTRANSCOM. USTRANSCOM receives transportation movement requests from customer military organizations distributed across the world called COCOMs (Combatant Commands). Before handing off an air movement request to the Air Mobility Command for detailed mission scheduling and execution, USTRANSCOM transportation planners must decide whether the movement requirement can be supported, what type of aircraft to use, and whether the requested delivery date can be met. A key challenge is that the set of movement requirements typically exceeds the available assets, requiring decisions to be made about which movement requirements will be supported as requested, and which will need to be delayed. We developed a prototype tool called Adaptive Transportation Loads for Airlift Scheduling (ATLAS) to support this initial airlift asset allocation decision.

A common challenge in environments where demand exceeds available resources is how to define and communicate priorities to guide effective allocation and reallocation. This challenge is accentuated in environments such as USTRANSCOM where (1) there are multiple, organizationally distinct customer groups (COCOMs) vying for the same resources, so that relative priority across groups is difficult to gauge; (2) the individuals who are assigned the responsibility of allocating resources (USTRANSCOM transportation planners) are organizationally distinct from the customer groups, and must rely on input from the customers to assess relative priority; (3) there are gradations of priorities that are not captured by the formal, prioritization schemes; (4) demands for resources come in over time and are difficult to anticipate; and (5) priorities dynamically shift within and across groups with changing circumstances whereas the formal priorities remain fixed.

Airlift scheduling is particularly challenging because movement requirements and available airlift capacity both change over time in ways that are largely unpredictable. While USTRANSCOM normally requests 21 days notice of upcoming airlift requirements, new rapid-response, high-priority requirements often emerge. When crises arise anywhere in the world, whether military or civil, USTRANSCOM will often be tasked to transport personnel and equipment on very short notice.

The airlift allocation problem is further complicated by the fact that requirements vary in priority, but those priorities are often not fully articulated. While USTRANSCOM has a formal priority scheme in place, it is relatively coarse. Most movement requirements have the same formal priority, reducing its usefulness for guiding resource allocation within and across COCOMs. For example, a requirement by one COCOM to move water filtration equipment may have the same formal priority as a requirement by another COCOM to move ship maintenance equipment. This makes it difficult for USTRANSCOM transportation planners to decide which requirement to meet and which to delay in cases where aircraft asset limitations make it impossible to meet the stated required delivery date for both requirements. In practice, there is a richer set of informal priorities that are communicated verbally between customers and USTRANSCOM personnel. However, the relative priorities of requirements may not be fully uncovered and communicated by the COCOMs until they are informed that not all their transportation requirements will be able to be met by their required delivery date. For example, a COCOM may request a transportation move to bring in helicopters and another move to bring in ship maintenance equipment. It is only when they are informed that USTRANSCOM cannot meet both movement requirements as stated that the COCOM customer will go back to the movement request initiators to determine which requirement in fact has the higher priority.

Further, relative priorities within and across COCOMs can change dynamically. The COCOM may at first indicate that the helicopter movement has the higher priority, but revise their priorities if a ship malfunction arises that increases the urgency of receiving the ship maintenance equipment. Additionally, an entirely new requirement may emerge, such as a need for humanitarian disaster relief, which requires reevaluation of priorities across requirements. Consequently, there is a need to capture and track these informal operational priorities as new requirements arrive, and operational conditions change, so that they can be effectively used to guide airlift allocation.

A final source of complication is that USTRANSCOM staff has limited visibility into the total projected requirements and capacity. Typically, transportation planners may be aware of the movement requirements for the particular COCOM they deal with, but they have no visibility into total projected requirements across COCOMs or total available air assets at different points in time. Similarly, personnel who are responsible for managing airlift assets have limited visibility into total projected movement requirements.

Our design challenge was to create a collaborative planning system that would enable transportation planners to capture finely graded priorities within and across organizationally distinct customer groups so as to guide the automated planner, as well as provide the planners with mechanisms to enable them to modify priorities so as to agilely respond to dynamic changes in movement requirements, priorities, and availability of assets.

The resulting prototype, ATLAS, allocates airlift assets to requirements in a progressive manner, enabling adaptation in the face of changing requirements and priorities. ATLAS uses an automated scheduler based on incremental, constraint-based search and optimization procedures (Smith et al. 2004). The automated scheduler and visualizations enable transportation planners to (a) incrementally commit transportation resources, reserving degrees of freedom to accommodate late changes in requirements and priorities; (b) incorporate informal priorities and changes in those priorities; and (c) consider and make trade-offs in priorities across organizational boundaries.

ATLAS includes a set of coordinated visualizations that are integrated with an automated scheduler. Views include a Notional Mission Timeline View, a Capacity View, and a Requirement Tag View. The ATLAS visualizations enable all user groups involved in asset allocation decisions to have a shared view of total available assets and the demand on those assets over time, as well as relative movement priorities. The scheduler is called up in the context of those visualizations, with software mechanisms provided to transportation planners to enable them to constrain and direct the automated scheduler behavior.

Figure 14.6 shows an ATLAS screenshot of a display that provides transportation planners an overview of mission requirements per day against projected capacity. The top pane provides a Notional Timeline view that displays projected movement requirements along a timeline. Each row in the visualization represents a notional mission that satisfies one or more movement requirement. Data can be directly read from a mission's visualization, such as its origin, destination, its temporal constraints (earliest departure, earliest arrival, latest arrival, etc.), and its cargo contents.

The bottom pane in Figure 14.6 shows the Capacity View. This is a histogram of the number of aircraft required per day to support the missions shown in the top pane. The horizontal bar toward the top of the Capacity View shows the projected total number of aircraft available per day (i.e., the capacity limit). Thus, the transportation planner can directly perceive time periods where requirements are approaching or exceeding capacity as well as time periods where additional requirements could be fit in.

One of the unique aspects of ATLAS is that it includes a mechanism for transportation planners to communicate to the software fine-grained informal priorities to use in making airlift allocation decisions—particularly when movement requirements exceed available assets. This is achieved via Tag Views. Figure 14.7 provides an

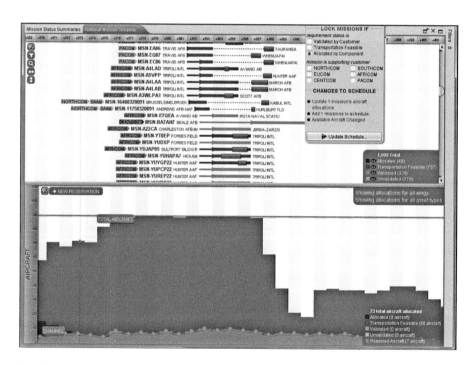

FIGURE 14.6 An ATLAS screenshot showing a mission timeline view (top pane) and a capacity view (bottom pane).

FIGURE 14.7 An ATLAS screenshot of a Requirements Tag View that can be used to specify and change relative mission priorities.

example screenshot of a Tag View. A transportation planner can assign meaningful tags to movement requirements and indicate relative priorities across tags. Tags at the same horizontal level in the display are treated as equal priority, whereas tags above are treated as higher priority. In the example shown in Figure 14.7, a mission moving water filtration equipment has been assigned the same priority as a 23rd wing support mission, and both have higher priority than all the mission types shown in rows underneath them. The transportation planner can prioritize and reprioritize missions as events evolve, by simply dragging their associated tag labels up and down in this view. In this way, transportation planners are able to express fine-grained informal priorities and changes in priorities that they uncover through close communication with the customer representatives at the COCOM to guide asset allocation that the automation would otherwise have no way of knowing about.

Multiple software mechanisms are provided to support directability. One means for directing the scheduler are the modifiable tag priorities that have already been discussed. Transportation planners can also specify the scope of missions to be included in an automated reschedule. For example, in normal operations, ATLAS can be directed to schedule new missions with an emphasis on minimizing disruption to previously scheduled missions. In this mode, it considers a narrow aperture of possible missions (e.g., just the new missions) from among which it will make priority-based trade-offs. Other situations may require more encompassing reprioritization across missions. In those cases, transportation planners can broaden the ATLAS rescheduling aperture to include already scheduled missions.

Transportation planners also have control of airlift capacity made available to the scheduler. On any given day, the number of aircraft available for allocation can change for various operational causes. Transportation planners can modify airlift capacity for a given time interval by dragging the capacity limit bar representing the total aircraft available for allocation up or down for that time period (see Figure 14.6). All subsequent updates to the schedule will respect this new capacity limit. As in the case of the other systems we have described, ATLAS provides transportation planners the opportunity to develop and compare multiple alternative schedules in a "what if" mode, enabling them to be an active partner in deciding which schedule best balances the multiple requirements and constraints (not all of which are knowable by the scheduler).

A user evaluation of ATLAS was conducted. Active transportation planners were given the opportunity to exercise a dynamic prototype of ATLAS on representative scenarios. They were then asked to provide feedback via a written questionnaire. Participant feedback, as reflected in both verbal comments and questionnaire ratings, indicated that the prototype was usable, useful, and would positively affect operations. Participant ratings indicated that ATLAS improves visibility of requirements and capacity, would improve users' ability to utilize priorities in allocating airlift assets, and would enable them to more rapidly reallocate airlift assets to accommodate changing requirements and priorities.

In summary, ATLAS illustrates a number of specific techniques intended to enable the joint cognitive system to be more adaptive in the face of changing circumstances—particularly when confronted with changes in requirements and priorities that inevitably arise. Specifically, ATLAS

- Provides shared representations of the problem to be solved and constraints that need to be respected: Observability
- Provides mechanisms by which users can contribute to and revise the problem specification: Directability
- Provides ways to express (informal) priorities and changes in priorities that the automation would otherwise not know about: Directability, adaptability
- Enables planners to control rescheduling scope, to minimize plan disruption when possible while accommodating situations that require broader scope replanning: Perspective shift, adaptability
- Allows planners to rapidly generate and compare multiple solutions: Broadening

14.8 DISCUSSION

Woods and colleagues (e.g., Woods and Hollnagel 2006, Chapter 12) proposed high-level design principles for effective joint cognitive systems. In our work, we have developed a number of work systems intended to embody those principles, resulting in demonstrably usable and useful collaborative planning systems. At the core of the planning systems is a set of visualizations that provide a shared representation of the problem space, providing common ground for the multiple actors (people and technology) who are collectively working on the problem. The goals and assumptions of the automated planner, the additional constraints on the problem solution imposed by the person on the scene, and the candidate solutions are all made externally visible and changeable within the context of these visualizations. Equally fundamental is providing the person on the scene with multiple software mechanisms ("levers") for constraining and guiding problem solutions. In our case, this included software mechanisms for changing model assumptions, adding constraints, and articulating informal, changing priorities. Ultimately, the objective of these principles and techniques for effective joint cognitive systems is to foster broadening—enabling the joint system to identify and explore more options than would be possible by either the person or the technology working alone, as well as to foster adaptability in the face of unanticipated variability—enabling the joint cognitive system to more agilely adapt to changing circumstances.

Ultimately, the objectives of our projects were not only to develop concrete prototypes in support of particular application problems but also to contribute to the Cognitive Engineering corpus of reusable techniques for fostering more effective joint cognitive systems. It is our hope that our work has served to extend the range of available techniques for fostering collaborative systems, empowering the people on the scene to actively partner in generating solutions that best meet the local conditions and constraints.

ACKNOWLEDGMENTS

We would like to acknowledge the earlier contributions of Robert Eggleston and Randall Whitaker on work-centered design that helped inform our thinking. Bob and Randy participated with us on prior work-centered design systems for the Air Mobility Command and were instrumental in forging our interdisciplinary team approach to work-centered design. This research was funded by the Air Force Research Laboratory (711 Human Performance Wing and Information Directorate) and the United Stated Transportation Command. Distribution A: Approved for public release; distribution unlimited. 88ABW Cleared 09/23/2014; 88ABW-2014-4498.

REFERENCES

Ackoff, R.L. (1979). The future of operational research is past. *Journal of the Operational Research Society*, 30, 93–104.

Bainbridge, L. (1983). Ironies of automation. *Automatica*, 19, 775–779.

Bradshaw, J.M., Hoffman, R.R., Johnson, M., and Woods, D.D. (2013). The seven deadly myths of 'autonomous systems'. *IEEE Intelligent Systems*, 28, 54–61.

Christoffersen, K., and Woods, D.D. (2002). How to make automated systems team players. In E. Salas (Eds.), *Advances in Human Performance and Cognitive Engineering Research*, Volume 2 (pp. 1–12). St. Louis, MO, Elsevier Science.

Clark, H.H., and Brenna, S.E. (1991). Grounding in communication. In L.B. Resnick, J.M. Levine, and S.D., Teasley (Eds.), *Perspectives on Socially Shared Cognition* (pp. 222–233). Washington DC: American Psychological Association.

Cummings, M.L., and Bruni, S. (2009). Collaborative human–computer decision making. In S. Nof (Ed.), *Handbook of Automation* (pp. 437–447). Berlin: Springer-Verlag.

DePass, B., Roth, E.M., Scott, R., Wampler, J.L., Truxler, R., and Guin, C. (2011). Designing for collaborative automation: A course of action exploration tool for transportation planning. In *Proceedings of the 10th International Conference on Naturalistic Decision Making* (pp. 95–100). May 31–June 3, 2011, Orlando, FL.

Downey, E., Brill, J.R., Flach, J.M., Hopkins, L.D., and Ranjithan, S. (1990). MGA: A decision support system for complex, incompletely defined problems. *IEEE Transactions on Systems, Man, and Cybernetics*, 20, 745–757.

Eggleston, R.G. (2003). Work-centered design: A cognitive engineering approach to system design. In *Proceedings of the Human Factors and Ergonomics Society 47th Annual Meeting* (pp. 263–267). Santa Monica, CA: Human Factors and Ergonomics Society.

Guerlain, S., Smith, P.J., Obradovich, J.H., Rudmann, S., Strohm, P., Smith, J.W., and Svirbely, J. (1996). Dealing with brittleness in the design of expert systems for immunohematology. *Immunohematology*, 12, 101–107.

Hollnagel, E., and Woods, D.D. (1983). Cognitive systems engineering: New wine in new bottles. *International Journal of Man–Machine Studies*, 18, 583–600.

Klein, G., Feltovich, P., Bradshaw, J.M., and Woods, D.D. (2005). Common ground and coordination in joint activity. In W. Rouse and K. Boff (Eds.), *Organizational Simulation* (pp. 139–184). New York: Wiley.

Klein, G., Woods, D.D., Bradshaw, J., Hoffman, R.R., and Feltovich, P.J. (2004). Ten challenges for making automation a "team player" in joint human–agent activity. *IEEE Intelligent Systems*, November/December, 91–95.

Layton, C., Smith, P.J., and McCoy, E. (1994). Design of a cooperative problem-solving system for en-route flight planning: An empirical evaluation. *Human Factors*, 36, 94–119.

Lee, J.D. (2008). Review of pivotal human factors article: "Humans and Automation: Use, Misuse, Disuse, Abuse." *Human Factors*, 50, 404–410.

Lee, J.D., and See, K.A. (2004). Trust in automation: Designing for appropriate reliance. *Human Factors*, 46, 50–80.

Miller, R.A., and Masarie, F.E. Jr. (1990). The demise of the "Greek oracle" model for medical diagnostic systems. *Methods of Information in Medicine*, 29, 1–2.

Parasuraman, R., and Riley, V. (1997). Human and automation: Use, misuse, disuse, abuse. *Human Factors*, 39, 230–253.

Roth, E.M., Bennett, K., and Woods, D.D. (1987). Human interaction with an 'intelligent' machine. *International Journal of Man–Machine Studies*, 27, 479–525.

Roth, E.M., Hanson, M.L., Hopkins, C., Mancuso, V., and Zacharias, G.L. (2004). Human in the loop evaluation of a mixed-initiative system for planning and control of multiple UAV teams. In *Proceedings of the Human Factors and Ergonomics Society 48th Annual Meeting* (pp. 280–284). Santa Monica, CA: Human Factors and Ergonomics Society.

Roth, E.M., Malin, J.T., and Schreckenghost, D.L. (1997). Paradigms for intelligent interface design. In M. Helander, T. Landauer and P. Prabhu (Eds.), *Handbook of Human–Computer Interaction* (2nd ed., pp. 1177–1201). Amsterdam: North-Holland.

Roth, E., Scott, R., Whitaker, R., Kazmierczak, T., Truxler, R., Ostwald, J., and Wampler, J. (2009). Designing work-centered support for dynamic multi-mission synchronization. In *Proceedings of the 2009 International Symposium on Aviation Psychology* (pp. 32–37). Dayton, Ohio: Wright State University.

Sarter, N.B., Woods, D.D., and Billings, C.E. (1997). Automation surprises. In G. Salvendy (Ed.), *Handbook of Human Factors/Ergonomics* (2nd Ed., pp 1926–1943). New York: Wiley.

Scott, R., Roth, E.M., Deutsch, S.E., Malchiodi, E., Kazmierczak, T., Eggleston, R., Kuper, S.R., and Whitaker, R. (2005). Work-centered support systems: A human-centered approach to intelligent system design. *IEEE Intelligent Systems*, 20, 73–81.

Scott, R., Roth, E.M., Truxler, R., Ostwald, J., and Wampler, J. (2009) Techniques for effective collaborative automation for air mission replanning. In *Proceedings of the Human Factors and Ergonomics Society 53rd Annual Meeting* (pp. 202–206). Santa Monica, CA: HFES.

Smith, P.J., McCoy, E., and Layton, C. (1997). Brittleness in the design of cooperative problem-solving systems: The effects on user performance. *IEEE Transactions on Systems, Man and Cybernetics*, 27, 360–371.

Smith, S.F., Becker, M., and Kramer, L.W. (2004). Continuous management of airlift and tanker resources: A constraint-based approach. *Mathematical and Computer Modeling*, 39, 581–598.

Suchman, L.A. (1987). *Plans and Situated Actions: The Problem of Human Machine Communication*. New York: Cambridge University Press.

Truxler, R., Roth, E., Scott, R., Smith, S., and Wampler, J. (2012). Designing collaborative automated planners for agile adaptation to dynamic change. In *Proceedings of the Human Factors and Ergonomics Society 56th Annual Meeting* (pp. 223–227). Santa Monica, CA: HFES.

Woods, D.D. (1986). Cognitive technologies: The design of joint human-machine cognitive systems. *AI Magazine*, 6, 86–92.

Woods, D.D., and Christoffersen, K. (2002). Balancing practice-centered research and design. In M. McNeese and M.A. Vidulich (Eds.), *Cognitive Systems Engineering in Military Aviation Domains*. Wright–Patterson AFB, OH: Human Systems Information Analysis Center.

Woods, D.D., and Hollnagel, E. (2006). *Joint Cognitive Systems: Patterns in Systems Engineering*. Boca Raton, FL: Taylor & Francis.

Woods, D.D., and Roth, E.M. (1988). Cognitive systems engineering. In M. Helander (Ed.), *Handbook of Human–Computer Interaction* (1st ed., pp. 3–43). New York: North Holland (reprinted in N. Moray, Ed., *Ergonomics: Major Writings*. Taylor & Francis, 2004).

15 The FireFox Fallacy: Why Intent Should Be an Explicit Part of the External World in Human Automation Interaction

Christopher A. Miller

CONTENTS

15.1 INTRODUCTION

I began my career as a neophyte human factors engineer working for Honeywell's research and development laboratories. I was hired to work on a feeder project to DARPA (Defense Advanced Research Projects Agency) and the U.S. Air Force's Pilot's Associate (Banks and Lizza 1991)—a project that was, for its time, both very large and very, very innovative at the forefront of thinking in human–automation integration efforts.

Beginning in the early 1980s, the U.S. Air Force had initiated the development of a human-adaptive, information and automation management technology that came to be known as the "Pilot's Associate" (PA). PA, and all of the subsequent "associate" systems, consisted of an integrated suite of intelligent subsystems that were designed to share (among themselves and with the pilot) a common understanding of the mission, the current state of the world, the aircraft, and the pilot himself or herself. Associate systems then used that shared knowledge to plan and suggest courses of action and to adapt cockpit information displays and the behavior of aircraft automation to better serve the inferred pilot intent and needs.

The stated goal of the PA program was "to explore the potential of intelligent systems applications to improve the effectiveness and survivability of post-1995 fighter aircraft" (Banks and Lizza 1991). But some of the inspirations for the program were, semi-seriously, R2D2 from the *Star Wars* movies and an earlier fictional "automated assistant"—the "FireFox" from a 1977 novel by Craig Thomas, subsequently made into a film starring Clint Eastwood in 1982. In *FireFox*, a prototype neural interface helmet directly links the pilot's thoughts to the aircraft systems, allowing him (or, hypothetically, her) an "edge" in dogfighting—which came in handy when challenging advanced Soviet MiGs.

While neither of the two PA programs (one led by McDonnell-Douglas and the other by Lockheed) seriously considered neural interfaces, they did their best to adopt a technology that was intended as a rough functional equivalent: intent inferencing. Intent inferencing (Hoshstrasser and Geddes 1989), now more commonly referred to in Artificial Intelligence as "Plan Recognition" (Carberry 2001), was designed to make reasonable guesses about what the pilot was intending to do based on observations of his or her actions, augmented by knowledge of the current context and the initial mission plan.

For example, if the pilot had begun descending, and especially if he or she was near the end of the mission plan and in the vicinity of the planned destination airfield, the system might infer that he or she intended to land. Once knowledge of the pilot's intent could be tied to a known task or activity in this fashion, then various forms of automated aiding appropriate to that task and intent context could be provided—for example, approach templates could automatically be shown on displays, warnings and alarms could be tuned to the landing plan, systems health status could be checked for the ability to support landing, and gear could be automatically lowered at appropriate elevations, and so on. By making automation and information systems aware of the specific task(s) the pilot intended, they could be more accurately adapted to provide more appropriate support.

As a junior engineer, I worked on a DARPA and U.S. Air Force program peripherally related to Lockheed's PA program, developing machine learning techniques to acquire novel tactical plans and information management plans to be tied to the outputs of the intent inferencing algorithms and, thereby, to control automation and display behavior when those new plans were inferred as active. Therefore, it should be noted that my access to the core PA program was limited. What I will report about PA and the use of intent inferencing are my personal observations and anecdotes, based on the access I had available to me. In most cases, I do not have quantitative data to back up these observations and impressions.

As a technological development, intent inferencing was not invented during the PA program, but it was certainly enhanced and applied in novel ways. In fact, one could reasonably argue that it was given an early and important practical trial in that effort, which improved understanding of how and when to apply it. In my opinion, however, it was not used in a fashion in keeping with the tenants of cognitive engineering—and this led to a suboptimal interaction between human and automation. Although this was not apparent to me, or presumably to others, during the project itself, this realization has formed the basis of much of my work since that time. Unfortunately, I continue to see others attempt the same maladaptive pattern of human–machine interaction to this day.

15.2 A TALE OF TWO ASSOCIATES

Woods and Roth (1988) defined cognitive systems engineering as a focus on a triad of mutually constraining, interrelated factors: the world to be acted upon, the agent(s) who act on the world, and the "external representations through which the agent experiences the world" (p. 67). They say "…computerization creates a larger and larger world of cognitive tasks to be performed. More and more we create or design cognitive environments" (p. 60).

In my opinion, an important element was missing from the early PA design and its use of intent inferencing to guide and control automation behavior: the focus on the "external representation through which the agent experiences the world." This is not to say that PA "blinded" the pilot to what the world and his aircraft was doing. The PA cockpit had more than the then-current set of controls and displays—and one of the explicit innovations of the program was the application of intent inferencing to the task of dynamically configuring those cockpit elements to more precisely serve the needs of the pilot. Instead, PA's use of intent inferencing largely omitted any mechanism through which the pilot could "effectively experience" an important and novel aspect of his world when using the associate: the associate system itself.

One PA program had developed a set of "10 Commandments" for effective associate system design. One of the most important of these was that the effort the pilot required to control the associate must be less than the effort saved by the associate. While this is a fine principle in general, in one regard I feel that it led the development effort astray.

The PA program was intended to provide an automated "associate" to the *single* pilot of an advanced fighter aircraft. This future pilot didn't exist yet, of course, but the representative pilots of then-current advanced fighters (and, to the best of my limited knowledge, all the subject matter experts interviewed in the course of the design effort) were primarily familiar with single pilot operations. They were used to communicating and coordinating with a wingman, but not with a co-pilot. They relied on their own expertise and on then-prevalent automation and were, perhaps, not used to performing the "teamwork" tasks (Morgan et al. 1993) that are required in multi-agent work environments. In some regards, then, I fear, giving such a person an assistant was like giving any solo worker an assistant: it can readily seem like more work to tell the assistant how to do the job than just doing it one's self. Of course, it may actually *be* more work if the assistant is ignorant of the task domain or incompetent in performing tasks, but the perception can be more worrisome than the reality proves to be—at least in successful supervisor/subordinate or peer-to-peer teamwork relationships.

My impression is that the combination of the exciting new technology of intent inferencing and the fear of overworking the pilot (especially via coordination tasks that were not a part of existing fighter operations) together produced a pronounced aversion to the use of explicit communication about intent—either from the pilot to the associate or the associate to the pilot. Instead, we strove to build something functionally close to a FireFox—which always "just knew" what the pilot needed and tried to provide it. There were, to the best of my knowledge, no explicit displays of inferred intent and no ability to tell the PA directly what the pilot wanted it to do. (A "levels of authority" scheme enabled some coarse-grained, a priori functional authorization for PA behaviors, but this was designed to be employed primarily pre-mission and restricted the pilot to authorizing or de-authorizing broad classes of functionality.) After all, wouldn't it be better for the system to "just know" what the pilot intended and then do it? Clearly, that would save workload and thereby adhere to the design commandment… right?

Of course, that sort of "just knowing" what is needed happens extremely rarely in human–human teams outside of rigorously scripted domains. If and when it is possible, it undoubtedly saves workload, but is it *plausible*? In fact, it is occasionally a source of humor: for example, in the relationship between Radar O'Reilly, assistant to Colonel Blake in the M*A*S*H television series. Radar seemed "to have extrasensory perception, appearing at his commander's side before being called and finishing his sentences" ("List of M*A*S*H Characters," n.d.) and regularly provided forms, ordered supplies, and took actions before his bumbling boss had gotten around to thinking they were necessary. The fact that this dynamic is funny to us is testament to the fact that it is uncommon, if not unreal.

This fact was brought home to me when I later began work on the U.S. Army's Rotorcraft Pilots' Associate (RPA)—a program whose goal was to develop an associate system to aid the *two* pilots of an advanced attack/scout helicopter. The RPA program ran from 1995 through 2001 and was led by Boeing Helicopters. RPA was a 5-year, $80-million program that culminated in an extended series of flight tests in the 2001–2002 time frame. I led an effort as part of a Honeywell Laboratories subcontract to design the RPA's Cockpit Information Manager (CIM)—a subsystem that, in conjunction with a dynamically tracked intent inferencing system, would manage the two crew members' controls and displays to improve performance and reduce workload (Miller 1999; Miller and Hannen 1999). As such, I was intimately involved in initial design and prototyping phases and was peripherally involved in later implementation and testing.

In early design phases, when we were familiarizing ourselves with the tasks and behavioral patterns of current Apache crewmen, one thing became obvious: these two crew members were spending a large portion—nearly a third by one informal estimate—of their time in intercrew communication and crew coordination behaviors. That is, they were *talking* to each other, explicitly, about what they were doing, what they intended to do, what they would like the other person to do, and so on. This was very unlike the operations of single seat fighter aircraft where there was no one else to talk to. There was an explicit hierarchical relationship between the pilot and co-pilot/gunner (who was generally the senior crewman and commander), yet these two pilots managed most of their interactions verbally and explicitly. There was little or no attempt for the subordinate pilot to "just know" through inference what the commander intended and do it automatically. Instead, they talked about it.

Of course they did. This is what humans *do*—using the powerful tools of natural language and shared mental models to coordinate behavior for efficiency and resiliency. They use it to establish shared knowledge of intent at multiple levels—so that reactions that have to be made very quickly can at least be informed by shared knowledge (Shattuck and Woods 1997). They also use it to trap errors of misaligned perception and intention within the team and correct them dynamically before they can produce large mismatches in behavior that take substantially longer to sort out and repair and can, sometimes, lead to devastating outcomes. (It should be noted, however, that natural language, while extraordinarily expressive, is not without its own flaws of misinterpretation, inefficiency, and ambiguities. Indeed, one of the outcomes of nearly 35 years' of research into Crew Resource Management is that some aspects of natural language need to be formalized and made more rigorous for efficient crew communications and interactions.)

To some degree, the trade-off is straightforward. Using intent inferencing to save the pilot time and workload in conveying intent to another agent—rather than requiring explicit communication of intent—means that some of the workload required for that communication will be saved. But unless the intent inferencing is perfect, there will also be occasions where the system's pursuit of an incorrect inference will provide disruptions. In general, the time lost to (not to mention the consequences of) an instance of "intent mismatch" can far outweigh the small gains from an instance of correct intent inference. This is greatly compounded if the operator has no explicit insight into what intent the other agent thinks it is assisting, and no explicit ability to override or correct it (both features which exist in explicit intent communications between humans). This is the "Why is it doing that?" and "What will it do next?" phenomena (Sarter and Woods 1995) to an extreme degree.

To the best of my knowledge, no explicit test of these parameters (intent inferencing accuracy, intent mismatch, or the time saved or lost due to either) was made within the PA program or since. While figures of 80% to 90+% accuracy for intent inferencing have been provided in recent years (Pei et al. 2011; Salvucci 2004), these figures alone say little about the relative costs of correct versus incorrect inference and make no comparison to the relative performance implications of accomplishing a task with an agent that is assisting via intent inference versus one assisting via explicit tasking. Anecdotally, it is my impression that, at least in high-criticality domains such as piloting, refinery operation, and so on, performance gains (not to mention operator trust and acceptance) that have been established over a series of hundreds of correct inferences can be wiped out in an instant by one case of intent mismatch in an important or critical event context, especially if the operator has no insight into why the system is doing what it is doing and no ability to alter it.

Finally, even if a FireFox could be created such that it *always* accurately read the pilot's full mental model and, therefore, always accurately inferred what was "in the head" of the pilot, there is reason to believe that the process of articulating that intent is, itself, part of a naturalistic decision-making process that is important to good practice and which brings a number of auxiliary benefits to the human–machine system. This theme will be developed more extensively below, after a discussion of how the realization of this aspect of human–human intent coordination affected the design of the RPA and the benefits we observed from those modifications, followed by an account of how that has shaped our research and design efforts in similar projects since that time.

15.3 THE RPA'S CREW COORDINATION AND TASK AWARENESS DISPLAY

15.3.1 RPA AND ITS COCKPIT INFORMATION MANAGER

The primary goal of the RPA was to reduce operator workload and, thereby, enable better human-system performance, which, in the attack/scout helicopter domain, translated into greater safety and faster operational tempo. The RPA CIM was designed to actively and adaptively manage cockpit displays and controls, as well as some automation functions, to reduce workload for the two interacting pilots. As such, and as with the PA before it, it was intended as a mixed-initiative automation system (Allen et al. 1999) that could take actions it deemed to be in the pilot's best interest without explicit and immediate instructions from the pilot. It was also an example of adaptive automation (Inagaki 2003; Kaber et al. 2001; Opperman 1994; Scerbo 2006) in that it adapted its behavior to perceived needs on the part of the pilot.

The RPA CIM was designed to perform five functions within the cockpit:

1. *Task Allocation*—For authorized tasks, choose who to allocate a task to (between the two pilots and automation) on the basis of who was better able to perform the task given workload, knowledge, preferred assignment, and concurrent tasks.
2. *Page (or Format) Selection*—Given current information needs, what is the best page or format to convey via each display device.
3. *Symbol Selection/Declutter*—Within the selected pages/formats, what level of symbology was most necessary given the current tasks and their information needs, and which symbols could be decluttered.
4. *Window Placement*—The RPA glass-cockpit displays used a rich set of pop-up windows to convey auxiliary and/or more ephemeral information, but placing a pop-up window necessarily obscured some screen symbology. This function involved making decisions about which symbology could most readily be obscured at that point in time.
5. *Pan and Zoom*—The RPA Tactical Situation Display or map display could be automatically panned and zoomed by the CIM which therefore had the job of deciding the most appropriate pan and zoom setting to convey necessary information at an appropriate level of scope and resolution.

CIM had an intent inferencing capability, the Crew Intent Estimator, which worked similarly to that employed in the Lockheed PA, inferring crew activities on the basis of the mission plan and observed actions and world states, and characterizing them as existing within a Task Network—a hierarchical, directed, acyclic network that represented the known human–machine tasks in the domain along with the hierarchically interleaved goals that they served. Figure 15.1 illustrates a portion of the Task Network from RPA. More details on the Crew Intent Estimator can be found in Andes (1997).

The Crew Intent Estimator provided a constantly updated list of tasks that it believed either pilot was doing. Other tasks were triggered directly within the Task Network as needing to be performed (given world conditions). This combined list of tasks was the set for which the CIM managed information presentation. Each task was associated with an abstract set of "Pilot Information Requirements" (PIRs) and the set of active tasks therefore directly provided the set of necessary PIRs, while the priority of the various tasks was used to guide satisfaction of their information requirements.

CIM maintained extensive knowledge about the ability of various display and control formats to satisfy PIRs, the necessary and permissible allocation of display formats to display devices, pilot preferences about where elements should be located,

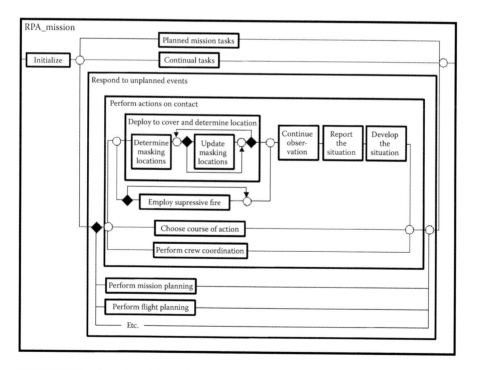

FIGURE 15.1 A portion of the Task Network used in RPA illustrating the actions expected to be taken (with branching alternatives) upon encountering a hostile enemy.

and so on. These were combined in an algorithm, described in detail in Miller (1999) and Miller and Hannen (1999) that strove to optimize and make trade-offs within the arrangement of controls and displays to meet some 11 different goals:

1. Pilot in charge of tasks
2. All needed tasks accomplished
3. Pilot in charge of info presented
4. All needed info provided
5. Stable task allocation
6. Only needed info provided
7. Tasks allocated as expected
8. Info presented as expected
9. Stable info configuration
10. Tasks allocated comprehensibly
11. Only needed tasks active

While generally building on, and more elaborate than, the information management algorithm that had been used in PA, CIM's approach was fundamentally similar. Intent inferencing was used to identify active tasks whose information needs were then satisfied via dynamic configuration of cockpit displays and controls. In one respect, though, the RPA CIM design departed significantly from that used in the PA: it incorporated a means for explicit communication about intent between pilots and the associate system automation during task performance. That is, it did not *solely* rely on intent inferencing, but instead provided the pilots a means of explicitly "talking" about intent with the associate system. This capability was centered in an auxiliary input system known as the Crew Coordination and Task Awareness (CCTA) display and it was motivated by our initial observation that crews naturally and extensively communicated about their intent quite explicitly.

15.3.2 RPA's Crew Coordination and Task Awareness Display Innovation

The CCTA display was intended to provide the pilots some explicit insight into and control over CIM's understanding of their intent. It took advantage of a set of four alphanumeric light-emitting diode (LED) buttons that had been allocated to the glare shield of each pilot's RPA cockpit (see Figure 15.2). These were used to report the following, respectively: (1) the Intent Estimator's inference about the high-level mission context (e.g., what high-level task they were collectively attempting to perform currently); (2) its inference about the highest priority pilot task; (3) its inference about the highest priority co-pilot task; and (4) the highest priority task that the RPA system as a whole was working on at that time. Note that the task names presented on the four buttons were drawn from (and usually abbreviated) the task names in the Task Network. Also, note that the tasks were drawn from different positions in the hierarchy of the task network, with the Mission task in the first button generally representing a parent task at least one level above the child tasks presented for the pilot, co-pilot, and associate. In the example shown in Figure 15.2, the overall

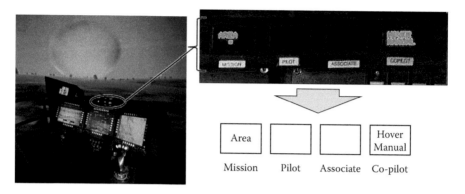

FIGURE 15.2 The RPA cockpit showing the CCTA display (circled on the left and enlarged on the top right, with a schematic view—to compensate for poor image quality—on the lower right). Note that the task inferred for the Mission is "Area" (a type of reconnaissance) and that for the Co-pilot is "Hover Manual."

mission task is Area Reconnaissance, while the associate has inferred that the co-pilot is engaged in a Manual Hover task.

In addition to displaying aspects of the intent the Crew Intent Estimator had inferred, the CCTA also permitted the pilots a form of "talking back" to the associate about their intent. Pushing one of these buttons allowed the pilot to override the current inferred intent for that actor—asserting, in essence, "That is *not* what I am trying to do. Please quit trying to help me do that." Pressing and holding a button scrolled through other tasks at the same level of the Task Network and allowed the pilot to assert a different task to be supported. This corresponded, roughly, to saying "Please quit trying to help me do that, and help me with this instead."

This inclusion of an explicit method of viewing and manipulating inferred intent was motivated by the observation that intent-centered communication and coordination was very much a part of the crew–crew interactions. The thought was that if our "associate" was to truly behave as a member of the crew, it ought to be able to communicate explicitly about intent as well. We note, in retrospect, that it also adheres to and realizes the cognitive systems engineering directive to design for explicit cognitive experiences—to make explicit external representations that design the cognitive environment the user wants or needs to experience. In this case, we were taking a step toward making the cognitive intent states that were internal to the humans and the associate (and which might or might not be the same explicit parts of the designed environment) so that they could be referenced, viewed, and manipulated. Since this inclusion of a direct method of viewing and interacting with intent was a novel approach to interaction with an intent inferencing associate (relative to the prior PA system), we were particularly interested in how it would perform.

15.3.3 THE EFFECTS OF DIRECT INTENT EXPRESSION IN RPA

The CIM, with its Crew Intent Estimator and CCTA, was partially implemented by Boeing Helicopters and tested in both full mission (partial dome) simulation and in an extensive series of flight tests. While there was never a direct comparison between the PA and RPA with regard to the CCTA, and while we have only had access to subjective questionnaire data from early simulation trials, the CCTA seems to have performed well.

Four crews participated in the simulation trials in a full cockpit mockup using the same avionics that were later flown in the Apache flight test vehicle (Miller and Hannen 1999). Crews trained and flew together, as they do in field operations. They received nearly a week of simulator and classroom training in the RPA systems. Crews were given realistic mission briefings and objectives. Crews made their own tactical decisions about how to achieve those objectives. Each crew flew a total of 14 part mission test scenarios of 20 to 50 minutes duration, as well as four full missions with a duration of 1–1.5 hours. Half of these (as a between-subjects design) were with the RPA aiding system and half were with an "Advanced Mission Equipment Package" designed to provide a baseline of advanced sensors and glass displays, but without the associate system's aiding.

The scenarios were constructed to provide numerous opportunities for the CIM behaviors of page selection, window location, symbol selection, and pan and zoom. Because of resource limitations, however, CIM's Task Allocation function was not implemented and the Crew Intent Estimator was only implemented for one, albeit important, parent task with many alternate sub-branches: Actions on Contact.

Working with the RPA and its CIM reduced perceived workload and improved perceived performance on the subscales of NASA's TLX subjective workload measure. Similarly, separate questions about perceived performance on four high-level mission types (Zone and Area Reconnaissance and Deliberate and Hasty Attack) all showed improvements with the RPA (an average of half a point on a 5-point scale, or approximately 12.5%), as indicated in Figure 15.3.

Ratings of CIM behaviors are shown in Figure 15.4. Most CIM behaviors were rated between "of use" and "of considerable use" and all except Window Location

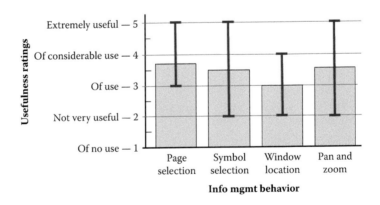

FIGURE 15.3 Perceived usefulness ratings of the four CIM behaviors averaged over pilots. Each bar illustrates the mean rating with the associated range of scores assigned across individuals.

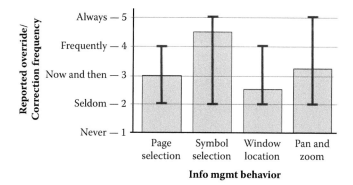

FIGURE 15.4 Perceived ratings for frequency of override/correction for the CIM behaviors. Each bar illustrates the mean rating with the associated range of scores assigned across individuals.

passed the criterion value of 3.5 the team had set on a 5-point scale. On the other hand, we also asked pilots to report their perceived frequency of overrides and corrections of CIM's behaviors. These data are reported in Figure 15.4. It is apparent that pilots felt that they had to "now and then" override CIM's display configuration choices—with symbol selection performing notably better, and window location a bit worse. The Crew Intent Estimator was perceived to be fairly accurate in recognizing crew intent to "Perform Actions on Contact" (average rating of 4.15 on a 5-point scale, roughly corresponding to "Frequently" triggering when the crew intent or mission context made it appropriate). This came at the cost of false positives, however, with CIM "seldom" to "now and then" triggering Actions on Contact when the crew intended to continue past threats (average rating = 2.40).

On the other hand, a final testament to pilot acceptance of CIM design and behaviors came from their full mission simulation trials. In the four missions in which the RPA and CIM functions were available, pilots were given the option of turning off any or all of the CIM behaviors both before and during the missions. Nevertheless, all eight pilots chose to leave all CIM behaviors on throughout their full trials. This seems a sign of trust in, and perceived benefit from, CIM's display management capabilities.

Conceptually, though, these results pose something of a problem. My anecdotal perception, cited above, that one failure of automated aiding could wipe out the benefits of a long series of correct behaviors does not seem in keeping with "now and then" or even "frequently" (in the case of window placement) having to override a behavior that was seen as generally useful.

How did RPA achieve such comparatively high perceived usefulness ratings when "now and then" or even "frequently" needing to be overridden? Although I am far from able to prove this, I believe that the CCTA played a significant role. The overall attempt to make the CIM participate in the cockpit interaction behaviors familiar to teams of human actors, as well as the specific modification of making intent an

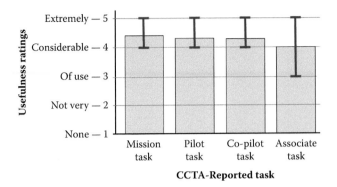

FIGURE 15.5 Perceived usefulness ratings of the CCTA display buttons from full mission simulation trials. Each bar illustrates the mean rating with the associated range of scores assigned across individuals.

observable and manipulable part of the pilot's environment, served to make intent mismatches and their resulting display management "errors" both understandable and easily overridable. When CIM chose to show evasion symbology and pop-up windows instead of attack symbology, this was still "wrong," but by showing that CIM had assumed pilot intent was to evade rather than attack via the CCTA, it showed that what CIM was doing was reasonable given its assumptions—a fact that may have increased pilot trust in the system as a reasonable actor. More importantly, the CCTA gave the pilots a quick and easy way to correct the error at its source—and to get reasonable aiding behavior as a result.

In both cases, these functions mirrored the behavior of a human actor in the domain. By putting intent "out there" in an explicit form in the observable world, it became a part of the pilots' shared observations of the world. Just as human pilots can talk about intent and, thereby, detect mismatches and accept correction, the CCTA enabled the RPA to behave similarly.

In fact, the perceived utility of the CCTA was rated very high by these RPA test pilots. Despite some pilot complaints about inadequate training in their use, most pilots found the inclusion of the LED buttons conveying intent to be "of considerable use," as shown in Figure 15.5.

15.4 DELEGATION AND ADAPTABLE AUTOMATION

Our work on RPA began the realization that explicit intent declarations and negotiation were a significant part of human–human teamwork and, therefore, should generally be a significant part of human–machine teamwork, especially in contexts where the machine has an element of autonomy and substantive complexity and competency. In retrospect, we have realized that this attribute was implicit in Sheridan's initial definition of "supervisory control"—where the human supervisor is presumed to be able to "teach" or program automation in what and how to perform in different contexts (Sheridan 1987). Of course, humans (even in military cockpits) have much more flexible forms of interaction than were supported in the RPA CCTA design.

This realization has led us to a multi-year, multi-project research effort in which we are attempting to develop techniques to enable human–automation interaction that achieves the benefits of a competent assistant. We are also seeking ways of mitigating the costs of disconnect between multiple team members working on the same task and thus, potentially, conflicting with each other. As with the RPA CCTA, we have used human–human teaming interactions as a guide and have generally striven to design environments that make intent an explicit, observable and manipulable part of that environment.

The model we have adopted is one of *delegation*—a relationship in which a human supervisor passes authority, goals, resources, and instructions to one or more automated subordinates. The subordinate is then expected to attempt to follow and perform those instructions, and achieve those goals, with some reporting throughout execution and discussion when compliance with the directives is or will be difficult, impossible, or suboptimal.

Delegation is an approach to "*adaptable*" automation behaviors rather than the *adaptive* automation that the PA (and RPA) and many other approaches since have striven to achieve (Inagaki 2003; Kaber et al. 2001; Scerbo 2006). The terms, and the distinction between them, were originally used by Opperman (1994) to characterize two different kinds of automation—both of which were capable of changing their behaviors to better support their human users in different contexts. Adaptive automation is automation that detects the need to adapt and autonomously takes the initiative to change its behaviors. By contrast, adaptable automation is also capable of adapting to world and user states, but does so only within specific authority and instructions given to it by the user. The user retains the authority over how automation should adapt, though some of that authority may itself be delegated, and although the automation may not autonomously change its behavior, it may propose alternate options.

Adaptable autonomy may also be "flexible" in that it admits a variety of levels and forms of interaction. Human–human delegation is extraordinarily flexible, relying on the full range of natural language expression as well as domain-specific jargon, conventions, and even alternate modalities such as hand gestures. Furthermore, it makes use of the human propensity for "common sense" and fully utilizes shared mental models of the domain and the tasks and operations that humans can perform (perhaps further subdivided by roles) within it. All this provides rich fodder, and multiple challenges, for optimizing human–machine delegation.

In multiple efforts, we have explored enabling humans to express intent and delegate to subordinate automation with the same flexibility possible in human supervisory control (Miller and Parasuraman 2007; Miller et al. 2013). We use the metaphor of a sports team's playbook. Our Playbook® systems allow humans and automation to share a conceptual, hierarchically-structured framework for the goals and methods of a pattern of activity (a "play") and to delegate instructions and discuss performance within that framework.

15.4.1   WHAT IS FLEXIBLE DELEGATION?

In human organizations, supervisors can communicate with trained subordinates with a presumption of shared understanding of the common plans, goals, and methods in the work domain, of the common, culturally determined priorities and limitations on appropriate behavior, and so on. They can talk about activities in detailed natural language, using all available for nuanced intent, urgency, uncertainty, and so on, that humans share. Moreover, they will likely share specialized vocabulary—a jargon or prescribed protocol to reduce errors and improve speed. These factors enable delegation with great flexibility. Supervisors can activate complex, multi-agent behaviors with a word, but they can also shape and refine those behaviors, can create new ones, can task at various time points before and during behavior execution, and can even jump in and physically perform or illustrate behaviors for subordinates if desired— all with reference to this shared vocabulary of tasks in the domain. This is the flexibility we should strive to achieve in human–automation interaction with complex, sophisticated automated agents.

We have argued (Miller and Parasuraman 2003, 2007) that delegation can be viewed as taking place within a "space" of possible domain tasks and this space can be characterized using a hierarchical decomposition (as in Kirwan and Ainsworth 1999) where alternate performance methods are represented as "or" branches, and sequential dependencies, looping, conditional branching, and so on, are captured as a directed graph. Not incidentally, this is essentially the same representation used to structure both intent inferencing and human–machine collaboration in the PA and RPA systems (cf. Figure 15.1).

The selection of a branch for an actor (e.g., a subordinate) to take is, in essence, a function allocation decision—which is to say, a delegation decision when it is made by a superior. It can be made either at design time or left to execution time. If left for execution, this leaves open the question of who gets to make the allocation decision. If automation makes it, then the system can be called adaptive; if the decision is left to a human supervisor, then the system is adaptable (Opperman 1994).

A supervisor interacting with human subordinates could choose to do an entire task alone or could delegate it to a subordinate to plan and execute. He or she could also choose to delegate parts of the task but retain control of the rest and could constrain or stipulate how tasks are to be performed. He or she could offer instructions either holistically or piecemeal, and could jump in to revise or override as needed. He or she could also assemble lower-level tasks to achieve a new and different goal— something that would not have previously existed in the task network, but which could be retained as a new entity if useful. Delegation need not occur at any fixed level within the hierarchy—an entire mission might be assigned to one subordinate, while only a maneuver task to another. Decisions are revisable: after initially delegating full mission execution, the supervisor might later decide to constrain a subtask—or even to perform portions personally. These are all attributes of flexible delegation we have explored in recent programs (e.g., Miller et al. 2013).

Delegation means conveying the supervisor's intent that the subordinate perform an action or achieve a goal using resources that he or she controls or can access. The subordinate then receives some (but not necessarily all) of the supervisor's control

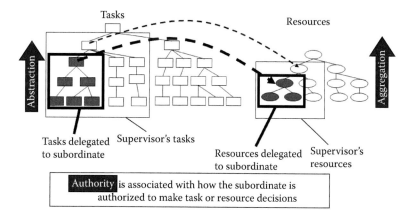

FIGURE 15.6 Key dimensions and entities to be represented in a grammar for delegation or intent expression.

authority to determine how and when to act to achieve the expressed goal. This definition of delegation, as illustrated in Figure 15.6, has brought us to several realizations about the mechanisms of delegation.

For example, we have tended to regard delegation as occurring through a mixture of five expressive methods:

- If the supervisor delegates via a Goal: The Subordinate has the responsibility to achieve the goal if possible; report if incapable
- If via a Plan: Follow plan if possible; report if incapable
- If via a Constraint: Avoid prohibited actions/states if possible; report if incapable
- If via a Stipulation: Achieve stipulated actions/states if possible; report if incapable
- If via a Value Statement: Work to optimize value

The supervisor may express a goal (a world state) to be achieved or a plan (a series of actions) to be performed. Constraints and stipulations on actions, methods, or resources to be used may also be expressed. Finally, less specifically, the supervisor may also express values or priorities. These refer to the relative goodness or badness of states, actions, resource usages, and so on, if they are achieved or used. These methods are rarely mutually exclusive and may be combined to achieve various methods of delegation as appropriate to the domain, and the capabilities of both the supervisor and subordinates. But this set of intent expressions must be applied to specific intent "objects"—some of which will be physical objects in the world, while others will be conceptual objects shared by the team (e.g., task names, spheres of influence, etc.).

While there are specialized vocabularies in different domains, they may frequently and usefully be reduced to those represented in Figure 15.6:

- Tasks to be performed in the domain, arranged in a hierarchical decomposition along what might be termed an *abstraction* dimension (parent tasks representing more abstract, coarse-grained operations)
- Resources used in the performance of tasks. These, too, can be hierarchically arranged in an *aggregation* dimension—with higher-level entities representing more composite, aggregate resources (e.g., an Uninhabited Aerial System—UAS—squadron, as opposed to a single UAS, as opposed to a UAS subsystem)

Actors in the work domain have access and control (which may not be absolute) over some resources in the domain and have responsibility (which, also, may not be complete) for some tasks. An act of delegation involves the handing of some task and resources over which the supervisor has authority to a subordinate, while a request for service involves a request for a task and/or resource to someone who has some authority for those elements. Thus, we need a third dimension for our intent objects grammar:

- Both tasks and resources are owned or shared with a degree of authority that need not be absolute along an *authority* dimension. Sheridan and Verplank's original spectrum of authority levels (Sheridan and Verplank 1978) is one reasonable representation for alternate authority levels with the virtue of simplicity, though they have been criticized and refined repeatedly over the years.

Using intent expressions and intent object representations, a very wide range of intent discussions or delegation actions can be formulated. For example, a supervisor can delegate a task, assert stipulations and constraints on the named subtasks and resources to be used to perform that task with specific authority levels set for using each (e.g., "use at will" vs. "use only with my permission," etc.). Finally, it should be noted that delegation can occur at different time scales with differing implications for authority, responsibility, and the "appropriate" level of delegation.

Delegation always, necessarily, precedes execution of the delegated action, goal, plan, or resources, but the amount of time between the delegation action and its execution can be seconds or years. In a real sense, design itself is a form of delegation in which a designer decides who should do what kind of task with what resources. Similarly, organizational policy creation is also a form of long-term delegation. At the other extreme, a pilot calling out procedure steps to his co-pilot is also "delegating" but doing so in very nearly real time. We have speculated that user willingness to tolerate (and, perhaps, to trust) automation intervention is a function of this timescale and the ability with time lags, or inability with very short ones, to intervene in and override or tune the automation's behavior, as illustrated in Figure 15.7.

Most professional domains in which a repeated set of useful operations are identified codify those operations in jargon for ease of reference. Similar jargon exists for resources, usually emphasizing salient aspects of the aggregation and decomposition dimension. A sports team's "playbook" is an especially well-codified and abbreviated example of this pattern—and is one reason we gravitated toward that

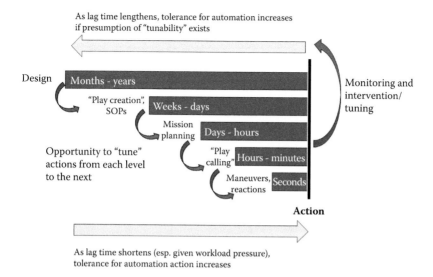

FIGURE 15.7   Delegation at different time scales and the relationship to automation trust and acceptance.

metaphor (see below) as the starting point for a series of projects to develop a "playbook" for human–automation interactions in the control of multiple UASs.

15.4.2   PLAYBOOK® AND DELEGATION

SIFT personnel have been working with the concept of a "playbook" for human interaction with automated systems for nearly 20 years now (Miller and Goldman 1997). Playbook, SIFT's approach to adaptable automation for UASs, has been implemented multiple times, with varying features, in simulation, and versions of it have been flown in demonstration (Flaherty and Shively 2010). All Playbook implementations strive to allow humans to "call a play" that an automated planning and execution control system understands. Once a play is called, perhaps with added stipulation, constraints or values on tasks, resources, or authority by the human supervisor, a planning system expands that play to an executable level and manages its performance, replanning as necessary within the constraints. Plays can be called at a high level, delegating authority over all decisions about alternate subtasks/methods and resource usage to the automation (within the "space" bounded by the play definition itself), or the operator can manipulate the hierarchical structure of the play to offer increasingly specific "instructions" (constraints and stipulations) about exactly how this instance should be performed.

A play is a labeled operation or method that is only partially instantiated; options remain to be "fleshed out" at the time of execution. In that sense, it is a task within the abstraction hierarchy described above. But as used within an organization, plays imply a goal (the expected outcome of the play) and are frequently coupled with embedded or explicit resource usage instructions and may also embed default or editable assumptions about performance. Thus, while a play exists in a task hierarchy, it may also "package" organizational or cultural assumptions about how and why that action is to be performed. A play bounds a "space" of behaviors that are agreed to fall under the play's name as a label. The behavioral space can be thought of as a hierarchical decomposition of alternate tasks, as in task analysis (Kirwan and Ainsworth 1999) and hierarchical task network planning (Nau et al. 1999). Note, that the space does not include all possible behaviors the system can perform, or even all possible behaviors that would achieve the goal; only certain behaviors in certain combinations are agreed to be exemplars of the play.

Sample plays are discussed and illustrated in Miller et al. (2013) and Miller and Parasuraman (2007). One such play we have implemented repeatedly is a "Monitor Target" play that is decomposed into sequentially ordered subtasks (Ingress, Approach, Inspect, etc.). Most of these subtask steps can be skipped, but Inspect is required for a valid instance of Monitor Target. In essence, if you're not "inspecting" something, then you are not doing a "Monitor Target" play—though you might well be doing something else using many of the same subtask components. Each subtask is further decomposed into one or more "Fly Maneuver" sub-subtasks, and "Inspect" also requires a concurrent "Maintain Sensor Focus" sub-subtask. A range of possible maneuvers is the next-level decomposition of the "Fly Maneuver" subtasks. Each maneuver is finally decomposed into specific control actions as operational primitives.

A supervisor interacting with human subordinates could choose to do this entire task alone or could delegate it in a wide variety of ways to a subordinate to plan and execute simply by invoking the name and providing a target. Our Playbook implementations are striving to achieve similar flexibility. Supervisors can choose to delegate parts (e.g., the Ingress portion) but retain control of the rest and can constrain or stipulate how tasks are to be performed (e.g., specific routes to take or avoid, etc.). They can offer instructions either holistically (as a full mission plan) or piecemeal, and can jump in to revise or override as needed. They can provide value statements about how the task as a whole, or specific elements of it, are to be achieved. They could (in principle, though this capability has not been implemented) also assemble lower-level tasks to achieve a new and different goal—a new play that could be retained in the library if useful. Delegation need not occur at any fixed level within the hierarchy—an entire mission might be assigned to one subordinate, while only a maneuver flying task to another. Decisions are revisable: after initially delegating full mission execution, the supervisor might later decide to constrain a subtask—or even to fly portions personally.

Also, as in human–human delegation, specialized vocabularies ("jargon") and even modalities (e.g., hand signals) have their analogs in Playbook implementations. We have developed a wide variety of play-calling methods including graphical user interfaces, spoken voice, touch, and sketch (Miller et al. 2013). We have also developed various mixtures of the five delegation expressions described above for different domains—for example, sometimes including goal-based delegation, sometimes allowing priority/value statements, and so on, according to the different needs of the domains.

It is the underlying, hierarchically structured "map" of tasks or behaviors that provides for adaptation and tuning at execution time and can embed organizational knowledge and conventions about goals, resources, priorities/values constraints and stipulations, and even authorizations, providing a powerful "lingua franca" for expressing, discussing, and even tracking intent—whether from a supervisor to subordinates or among peers, and whether between humans or humans and automation. The lesson learned from PA and RPA is that making this cognitive environment of intent explicit and observable serves to facilitate team action and coordination.

15.4.3 INTENT FRAMES AND THE BENEFITS OF EXPLICIT DELEGATION

While adaptable automation approaches have rarely been compared directly to adaptive ones, we are beginning to amass some evidence for the benefits of explicit delegation via intent declaration. In addition to the benefits I alluded to for the RPA system at the beginning of this chapter, adaptable, delegation approaches have been shown to result in improved overall system performance when examined across unpredictable and/or unexpected contexts (Parasuraman et al. 2005) to reduced human workload relative to adaptive systems in many circumstances (Fern and Shively 2009; Kidwell et al. 2012; Shaw et al. 2010).

In Shaw et al. (2010) using a paradigm pioneered by Fern and Shively (2009), we had users control multiple simulated UASs in an urban environment to monitor ground traffic for occasional military vehicles (HUM-Vs). When HUM-Vs were detected, they had to be tracked. If the HUM-V entered a building and emerged with a mounted weapon, it had to be targeted and destroyed. This simulated environment was used in multiple experiments with varying levels of traditional automation (waypoint control, track mode, etc.) versus play-calling automation to cause the UASs to maneuver to different observation points, track a ground vehicle and reconfigure their monitoring behaviors, call in a weaponized UAS and do target handoff to enable prosecution, and so on. In this environment, we demonstrated a 10%–12% improvement in secondary task performance (designating civilian vehicles) and a nearly 4 second improvement in reaction time (25% of the average) when using plays versus traditional automation in a multi-UAS control and target monitoring task.

Play-based delegation has also proven robust even in situations where the plays do not "fit" the context of use. Using the same testbed described above, we engineered situations in which the available plays occasionally failed to be useable in some situations (e.g., a target to be monitored was not on a list of callable monitoring locations and, therefore, the user had to resort to manual controls to search the area). We had anticipated that such circumstances, especially after a prolonged period in which plays "worked" consistently, might produce workload and performance deficits in a kind of "automation complacency effect" (Parasuraman and Manzey 2010). Instead, we found that performance remained consistently high throughout the "nonoptimal play events" and we found no evidence of complacency—though our longest trials were 30 minutes in duration (Miller et al. 2011).

A more direct test of adaptable versus adaptive automation was provided by Kidwell et al. (2012), albeit not in a play-calling environment. They had users engage in a simulated multi-UAS control task that required monitoring and controlling multiple aspects of the behavior of the simulated squadron. A secondary task involved viewing "sensor images" from the UASs from time to time with a variety of geometric shapes displayed against a terrain-display background. Subjects had to rapidly count or estimate the number of diamond shapes in the presented figure. Three levels of automation were available to assist: low (user had to check a box corresponding to the correct number), medium (automation checked a box indicating its recommended answer but the user had to review and authorize or change that number), and high (automation completed and submitted the count within 20 seconds if the user did not). In the latter two conditions, the automation was 90% accurate in its choices. The primary independent variable, however, was whether the control of automation level was left to the user (adaptable automation) or was itself automatically determined based on the performance of the participant on other UAS control tasks (adaptive automation—increasing after repeated failures, or decreasing after repeated successes). Using this experimental design, the authors showed that users (in a within-subjects design) preferred the adaptable automation to the adaptive, and rated their confidence as significantly higher in that condition. Furthermore, while there was a slight reaction time increase associated with the adaptable automation approach, participants performed significantly better on the image analysis task when using it.

While we do not yet have direct evidence for other benefits of explicit delegation interactions, there are many theoretical reasons to believe they should exist. We argued, in Miller and Parasuraman (2007), that any workload savings from efficient automation design could be devoted to maintaining or achieving better overall situation awareness (SA) about the context of use. SA, as traditionally defined and measured (Endsley 1995), refers to all situational knowledge required to perform one's job. Knowledge specifically about what a subordinate is doing and why is surely part of that set, but it is more specific and subject to different influences. SA reduction specifically about delegated tasks seems to be tolerated, even embraced, in exchange for competency improvements and/or workload reductions resulting from team performance. That said, the reduction in specifically automation-related SA can be subject to different influences in different human–automation interactions. In adaptive or traditional automation, behaviors are disconnected from human intent, and therefore, an additional element of unpredictability enters into the human's experience. This is summed up in Sarter and Woods' (1995) work on "automation surprises"—instances in which automation does unexpected, difficult-to-explain things. One way to alleviate this problem is to require the human to continuously monitor the automation to maintain awareness of what it is doing and, by inference, why. This, though, is akin to having a subordinate whom one has to watch all the time to make sure what he or she is doing is appropriate—perhaps viable, but workload intensive and subject to frequent breakdowns.

Delegation provides another way. By tasking the subordinate, the act of delegation provides an *Intent Frame* that expresses and defines expectations about the subordinate's behavior for both parties. In communication, this frame explicitly details what the supervisor expects the subordinate to do—making it an overt (if historical, in verbal communication) artifact and a reference point for both parties to "view" and compare current actions against. It is not inappropriate to view this as a contract between supervisor and subordinate, or human and automation. Because of this explicit action taken to express intent, aspects of SA relevant to the expressed intent should improve for both parties—if for no other reason than that they have been made explicit in a shared vocabulary.

But delegation should also help awareness and interpretation of observed subordinate behaviors as well because the Intent Frame creates a cognitive expectation about what the subordinate is supposed to do. If a task is delegated, then certain behaviors are expected and others are not. This framing narrows the set of behaviors that need to be attended to by the superior: instead of checking to see what behaviors, from all possible ones, the subordinate is doing, he or she may simply check to see whether the subordinate is doing what was expected or not. Even unanticipated behaviors can be interpreted more directly for whether they are reasonable within the intent instead of for what they could possibly accomplish. In short, explicitly delegated intent shifts the operator's task in monitoring and interpreting automation from one of "what is it doing now?" to a cognitively simpler one of "is it doing what I told it to do?"

The same Intent Frame effect likely has an impact on trust formation. Lee and See (2004) defined trust as "...the attitude that an agent will help achieve an individual's goals in a situation characterized by uncertainty..." (p. 51). The act of delegation as an expression of intent from supervisor to subordinate forms a contract of the form "perform within the space I have delegated to you and I won't blame you."

This contract, then, minimizes one significant component of uncertainty in such relationships—it clarifies what the superior's goals are for both parties. Delegation provides a framework within which trust can be assessed and judged. If I declare a set of intentions as instructions to you, then I can evaluate whether you have performed appropriately within them and, perhaps, even diagnose why not (Did you hear me? Did we share an understanding of the terms? Are you incapable?). Furthermore, an understanding of the requirements of the task (i.e., its subtasks and the capabilities they entail) can lead to improved assessment of the subordinate's reliability for that task both over time, as the supervisor learns about individual subordinates' capabilities, and at delegation time, when the supervisor uses that knowledge to select delegatees.

The supervisor–subordinate relationship establishes the responsibility of the subordinate in general (as in the list of delegation actions with associated subordinate responsibilities above) to follow the instructions provided by the supervisor. In less well-defined relationships, humans must make educated guesses based on affective or analogic cues (Lee and See 2004) as to whether another actor intends to support or further their goals. In supervisory relationships, however, the attitudes that all adherents should adopt are well defined. Moreover, in human–machine delegation (with the human in a supervisory role), one of the main sources of human–human mistrust is defused: machines (to date at least) never try to intentionally mislead a human supervisor into belief that they are complying with delegated instructions while trying to pursue other ends.

While there are certainly edge cases in which a supervisor may not accurately express what he or she intends, or may include instructions that produce adverse consequences, detecting and averting these situations is not the responsibility of the subordinate (though it may well be quite helpful, if possible). Failure to detect these situations is not a matter of disrupting trust of the supervisor for the subordinate (though it might well affect trust of the subordinate for the superior).

Thus, delegation affects trust by making it clear(er) to both parties what the goals and responsibilities are. In Lee and See's (2004) framework (see also Miller 2005) for human–machine trust formation, this provides some of the knowledge required for the deepest and most informed type of trust, analytic trust (i.e., deep structure understanding of how and why the agent behaves the way it does, which enables accurate prediction of future behavior), and thus serves to speed and tune trust formation. While this may or may not lead to increased trust depending on the subordinate's behaviors, it does "sharpen" the test for trust and should, therefore, speed accurate trust tuning (cf. Miller 2005).

These hypothesized effects on trust and automation-specific SA have largely not been tested directly, but there is some indirect support for them. As described above, several studies (Fern and Shively 2009; Kidwell et al. 2012; Miller et al 2011; Shaw et al. 2010) report improved performance and/or faster response times on secondary tasks when using adaptable delegation—which might imply either improved SA or reduced workload or both, though the former was not explicitly measured. The fact that higher subjective confidence ratings (arguably, a loose analog for trust) under adaptable versus adaptive control were observed in Kidwell et al. (2012) can also be seen as support for this hypothesis.

Finally, in their seminal study on the value of "intermediate" levels of automation, Layton et al. (1994) and Smith et al. (1997) had pilots and airline dispatchers interact with three kinds of automated route planning support in a commercial aviation context: a "low" automation level where operators sketched routes and automation computed route details such as fuel consumption and arrival times, a "high" level providing expert system-like support proposing a single complete route to the pilot, and an "intermediate" level where the pilot had to request a route with specific constraints (e.g., "go to Kansas City and avoid Level 3 turbulence") before the decision support tool developed the route. While, in some cases, there could have been overreliance when the software immediately proposed a flight plan before the participants had looked at any of the data, the results found that a label like "over-reliance" was much too simplistic an explanation for most of the participants who ultimately accepted an initial recommendation from the software. In many cases, these participants looked at the pertinent information and explored alternative routes, but their cognitive processes were strongly influenced by the software's initial recommendation. In trials where the software performed suboptimally and automatically recommended a very poor solution (e.g., failed to consider uncertainty in weather predictions), many of the participants selected this poor flight plan *despite the fact* that they looked at all of the relevant data and explored (but rejected) much better alternative plans. By contrast, in conditions with intermediate levels of automation, where pilots were required to identify a priori aspects of a solution that they wanted to see included and "delegate" them to automation to realize if possible, these types of errors were more likely to be avoided. Although situation assessment was not directly assessed, these results caution designers about the impact that solutions (and associated displays) automatically generated by software can have on situation assessment and on the user's perceptual and cognitive processes. (See Chapter 10 for greater detail.)

15.5 CONCLUSIONS

Playbook and other flexible delegation approaches to achieving adaptable automation remain under development and, increasingly, are nearing field testing for unmanned vehicle control in the U.S. Air Force and Army. In almost all cases, at least in initial presentations and encounters (e.g., Calhoun et al. 2012), users find the delegation and supervisor-led adaptation to be more natural and therefore preferable than more adaptive automation approaches that either adapt to world need independently or that adapt, under their own recognizance, to their interpretation of user need.

This is to be expected, of course. Highly trained operators in complex domains rarely relish surrendering control authority and may well have inaccurately positive views of their own competencies. Humans are, after all, not always the best judges of their own workload and capabilities and are not always optimal managers of their own time and resources. Managerial skill, even comparatively simple time and task management skill, does not reside equally in all individuals. For all these reasons, adaptable delegation is not a panacea and will not solve all problems in human–automation interaction.

But it is worth noting that explicit delegation and intent expression is a learned human solution (through human and even primate evolution) to the problem of coordinating the behaviors of multiple agents to achieve more complex goals. As such, we should perhaps have expected difficulties in engineering the sort of high-precision coordination required for an adaptive, intent inference system—something rare enough in human–human interactions to engender comedy.

Instead, humans have much more frequently adopted a strategy of making explicit even the invisible and cognitive aspects of their teamwork and coordination by making overt signals to express, discuss, and negotiate intent. This feature, combined with interaction protocols that define default behaviors and coordinate roles and responsibilities, is what makes collaborative task performance feasible in complex domains. Human–automation interaction should strive to adopt it wherever possible.

ACKNOWLEDGMENTS

The work reported in this chapter has spanned multiple projects and multiple funding sources, and has been conducted with a great many collaborators. My work parallel to the original PA project was funded by the Air Force Research Laboratory under contract F33615-88-C-1739. My work on the RPA program was funded by the U.S. Army's Advanced Aviation Technology Directorate under contract DAAJ02-93-C-0008. Work developing and defining the nature of delegation interactions was partially funded by an Air Force Research Laboratory Small Business Innovation Research contract under SBIR AF04-071 (FA8650-04-M-6477). Data on the workload and resiliency of delegation interactions were funded by another Small Business Innovation Research grant from the U.S. Army's AeroFlightDynamics

Directorate under contract W911W6-08-C-0066. Finally, recent work on the development of user interactions modalities and user acceptance of flexible delegation interactions has been funded by the Air Force Research Laboratory through a subcontract to Ball Aerospace (contract FA8650-08-D-6801). In all cases, however, the opinions expressed are those of the author alone.

REFERENCES

Allen, J.E., Guinn, C.I., and Horvitz, E. (1999). Mixed-initiative interaction. *IEEE Intelligent Systems and Their Applications*, 14, 14–23.

Andes, R. (1997). Assuring human-centeredness in intelligent rotorcraft cockpits: Using crew intent estimation to coordinate RPA functions. *Proceedings of Forum 53, Annual Meeting of the American Helicopter Society* (pp. 73–80). Virginia Beach, VA: American Helicopter Society.

Banks, S.B., and Lizza, C.S. (1991). Pilot's Associate: A cooperative, knowledge-based system application. *IEEE Expert*, 6, 18–29.

Calhoun, G.L., Draper, M., Ruff, H., Miller, C., and Hamell, J. (2012). Future unmanned aerial systems control: Feedback on highly flexible operator-automation delegation interface concept. In *Proceedings of Infotech@Aerospace 2012* (pp. 1–16). Reston, VA: American Institute of Aeronautics and Astronautics.

Carberry, S. (2001). Techniques for plan recognition. *User Modeling and User-Adapted Interaction*, 11, 31–48.

Endsley, M.R. (1995). Measurement of situation awareness in dynamic systems. *Human Factors*, 37, 65–84.

Fern, L., and Shively, R.J. (2009). A comparison of varying levels of automation on the supervisory control of multiple UASs. In *Proceedings of AUVSI Unmanned Systems* (pp. 10–13). Washington, DC: AUVSI Press.

Flaherty, S.R., and Shively, R.J. (2010). Delegation control of multiple unmanned systems. In *SPIE Defense, Security, and Sensing* (pp. 76920B–76920B). Bellingham, WA: International Society for Optics and Photonics.

Hoshstrasser, B., and Geddes, N. (1989). OPAL: Operator intent inferencing for intelligent operator support systems. In *Proceedings of the IJCAI-89 Workshop on Integrated Human–Machine Intelligence in Aerospace Systems* (pp. 53–70). Palo Alto, CA: AAAI Press/International Joint Conferences on Artificial Intelligence.

Inagaki, T. (2003). Adaptive automation: Sharing and trading of control. *Handbook of Cognitive Task Design*, 8, 147–169.

Kaber, D.B., Riley, J.M., Tan, K.W., and Endsley, M.R. (2001). On the design of adaptive automation for complex systems. *International Journal of Cognitive Ergonomics*, 5, 37–57.

Kidwell, B., Calhoun, G., Ruff, H., and Parasuraman, R. (2012). Adaptable and adaptive automation for supervisory control of multiple autonomous vehicles. In *Proceedings of the 56th Annual Human Factors and Ergonomics Society Meeting* (pp. 428–432). Santa Monica, CA: HFES Press.

Kirwan, B., and Ainsworth, L. (1999). *A Guide to Task Analysis*. London: Taylor & Francis.

Layton, C., Smith, P.J., and McCoy, C.E. (1994). Design of a cooperative problem-solving system for en-route flight planning: An empirical evaluation. *Human Factors*, 36, 94–119.

Lee, J.D., and See, K.A. (2004). Trust in computer technology: Designing for appropriate reliance. *Human Factors*, 46, 50–80.

List of M*A*S*H characters. (n.d.). Wikipedia. Retrieved from http://en.wikipedia.org/wiki/List_of_M*A*S*H_characters#Radar_O.27Reilly.

Miller, C. (1999). Bridging the information transfer gap: Measuring goodness of information "fit". *Journal of Visual Languages and Computing*, 10, 523–558.

Miller, C. (2005). Trust in adaptive automation: The role of etiquette in tuning trust via analogic and affective methods. In Schmorrow, D. (Ed.), *Foundations of Augmented Cognition* (pp. 551–559). Mahwah, NJ: Erlbaum.

Miller, C., Draper, D, Hamell, J., Calhoun, G., Barry, T., and Ruff, H., (2013). Enabling dynamic delegation interactions with multiple unmanned vehicles; flexibility from top to bottom. In Harris, D. (Ed.), *HCI International Conference on Engineering Psychology and Cognitive Ergonomics, LNCS vol. 8020*, (pp. 282–291). Heidelberg: Springer.

Miller, C., and Goldman, R. (1997). Tasking interfaces: Associate systems that know who's the boss. In *Proceedings of the 4th Joint GAF/RAF/USAF Workshop on Human–Computer Teamwork*. Farnborough, Hampshire: Defence Research Agency.

Miller, C., and Hannen, D. (1999). The Rotorcraft Pilot's Associate: Design and evaluation of an intelligent user interface for cockpit information management. *Knowledge-Based Systems*, 12, 443–456.

Miller, C., and Parasuraman, R. (2003). Beyond levels of automation: An architecture for more flexible human–automation collaboration. In *Proceedings of the 47th Annual Meeting of the Human Factors and Ergonomics Society* (pp. 182–186). Santa Monica, CA: HFES.

Miller, C., and Parasuraman, R. (2007). Designing for flexible interaction between humans and automation: Delegation interfaces for supervisory control. *Human Factors*, 40, 57–75.

Miller, C.A., Shaw, T., Emfield, A., Hamell, J., Parasuraman, R., and Musliner, D. (2011). Delegating to automation performance: Complacency and bias effects under non-optimal conditions. In *Proceedings of the 55th Human Factors and Ergonomics Society Annual Meeting* (pp. 95–99). Thousand Oaks, CA: Sage Publications.

Morgan Jr, B.B., Salas, E., and Glickman, A.S. (1993). An analysis of team evolution and maturation. *The Journal of General Psychology*, 120, 277–291.

Nau, D., Cao, Y., Lotem, A., and Munoz-Avila, H. (1999). SHOP: Simple hierarchical ordered planner. In *Proceedings of the 16th International Joint Conference on Artificial Intelligence*, 2 (pp. 968–973). Burlington, MA: Morgan Kaufmann.

Opperman, R. (1994). *Adaptive User Support*. Hillsdale, NJ: Erlbaum.

Parasuraman, R., Galster, S., Squire, P., Furukawa, H., and Miller, C. (2005). A flexible delegation-type interface enhances system performance in human supervision of multiple robots: Empirical studies with RoboFlag. *IEEE Systems, Man and Cybernetics— Part A*, 35, 481–493.

Parasuraman, R., and Manzey, D.H. (2010). Complacency and bias in human use of automation: An attentional integration. *Human Factors*, 52, 381–410.

Pei, M., Jia, Y., and Zhu, S.C. (2011). Parsing video events with goal inference and intent prediction. In *Proceedings of the 2011 International Conference on Computer Vision* (pp. 487–494). Piscataway, NJ: IEEE.

Salvucci, D.D. (2004). Inferring driver intent: A case study in lane-change detection. In *Proceedings of the Human Factors and Ergonomics Society Annual Meeting* (pp. 2228–2231). Santa Monica, CA: HFES.

Sarter, N.B., and Woods, D.D. (1995). How in the world did we ever get into that mode? Mode error and awareness in supervisory control. *Human Factors*, 37, 5–19.

Scerbo, M. (2006). Adaptive automation. In Parasuraman, R. and Rizzo, M. (Eds.), *Neuroergonomics: The Brain at Work* (pp. 238–251). Oxford, UK: Oxford University Press.

Shattuck, L.G., and Woods, D.D. (1997). Communication of intent in distributed supervisory control systems. In *Proceedings of the Human Factors and Ergonomics Society Annual Meeting* (pp. 259–263). Santa Monica, CA: HFES.

Shaw, T., Emfield, A., Garcia, A., deVisser, E., Miller, C., Parasuraman, R., and Fern, L. (2010). Evaluating the benefits and potential costs of automation delegation for supervisory control of multiple UAVs. In *Proceedings of the 54th Annual Human Factors and Ergonomics Society Meeting* (pp. 1498–1502). Santa Monica, CA: HFES.

Sheridan, T. (1987). Supervisory control. In G. Salvendy (Ed.), *Handbook of Human Factors* (pp. 1244–1268). New York: John Wiley & Sons.

Sheridan, T., and W. Verplank, (1978). *Human and Computer Control of Undersea Teleoperators*. Technical Report. Cambridge, MA: MIT Man–Machine Systems Laboratory.

Smith, P.J., McCoy, E., and Layton, C. (1997). Brittleness in the design of cooperative problem-solving systems: The effects on user performance. *IEEE Transactions on Systems, Man and Cybernetics*, 27, 360–371.

Woods, D., and Roth, E. (1988). Cognitive systems engineering. In M. Hellander (Ed.), *Handbook of Human–Computer Interaction* (pp. 3–43). Amsterdam: Elsevier.

16 A One-Day Workshop for Teaching Cognitive Systems Engineering Skills

Gary Klein, Laura Militello, Cindy Dominguez, and Gavan Lintern

CONTENTS

16.1 INTRODUCTION

The goal of this chapter is to describe a method for teaching cognitive systems engineering (CSE) to non-specialists. We designed a workshop to teach the activities required to accomplish CSE when designing a system. This introductory section explains how this method, using a workshop structure, came into existence and what it was intended to achieve. Section 16.2 describes the framework and content of the workshop, which we eventually presented at three different conferences. Section 16.3 details the workshop activities. Section 16.4 provides reflections on our overall project to support the teaching of CSE.

Our effort began in 2003 as a simple exploration of similarities and differences between different CSE approaches. We were concerned about the fragmentation of the field. Hoffman et al. (2002) had used a concept map to show the different facets of CSE. It revealed the diversity of the field. Researchers and practitioners had developed contrasting CSE approaches and methods, and competing brands such as Cognitive Work Analysis, Applied Cognitive Work Analysis, Situation-Awareness Oriented Design, Decision-Centered Design, and Work-Centered Design. System developers seemed confused about what CSE was supposed to achieve and which particular approach to adopt.

In response to this confusion, we initiated a series of lunch meetings at a favorite Vietnamese restaurant, taking advantage of our being co-located at the time in

Dayton, Ohio. Gavan Lintern, who was then with General Dynamics in Dayton, was a staunch advocate for Cognitive Work Analysis. Laura Militello and Gary Klein, both working at Klein Associates at the time, were equally staunch advocates for Decision-Centered Design. Cindy Dominguez, then a Lieutenant Colonel at the Air Force Research Laboratory at Wright–Patterson Air Force Base, was more interested in seeing CSE applied broadly in practice than in advocating for any specific approach. So we had two bases covered, Cognitive Work Analysis and Decision-Centered Design, plus a "referee."

These meetings resulted in a generic description of the activities required for a CSE effort to succeed. Different approaches might employ different methods—Abstraction Hierarchies for Cognitive Work Analysis, Cognitive Task Analyses for Decision-Centered Design—but the game plan was largely the same. We published two articles describing this generic account of CSE (Militello et al., 2009a,b) but then wondered whether we might develop an introductory workshop for CSE. Even if we never ran the workshop, the process of designing it might prove instructive, and so we embarked on a discussion around what approach and content would best communicate this material.

We agreed that the workshop should be experiential, with minimal lecturing. Other CSE practitioners had offered workshops that explained specific methods and compared different approaches. We did not think workshop participants at an introductory level would care about the nuances that differentiated one approach from another. We imagined they would want to gain a sense of why CSE could be useful and how to plan and accomplish a CSE effort. We agreed that we wanted to get the participants to practice CSE skills by the end of the workshop and therefore we would need to illustrate a full CSE cycle within the time available.

We settled on a single-day workshop largely because it would fit the full-day conference formats that were commonly in use (e.g., at meetings of the Human Factors and Ergonomics Society). But in addition, we believed that if we spilled into a second day, it would be too easy to defer tough decisions about what to include. We believed that the discipline of a single-day format would help us maintain focus and priorities. Thus, we decided to run the participants through an entire CSE effort in a single day—a whirlwind tour of the topic. In describing the results here, we hope to provide a view of CSE as a field and provide insights into how to familiarize people with its practice.

16.2   DESIGNING THE WORKSHOP

The core of the workshop would have to be an example application that called for CSE. We considered several different possibilities. One option was to design a kitchen. Another option was to redesign a GPS device for helping rental car customers navigate in unfamiliar cities. We initially selected the kitchen design exercise as one that would allow participants to immerse themselves in the design problem without any prior need for familiarization with the problem domain. We anticipated that workshop participants would be able to link kitchen design issues to the workshop exercises that were to be introduced through the remainder of the day.

We chose to focus on Decision-Centered Design because we were familiar with it and were confident that, in contrast to other approaches we considered (e.g., Cognitive Work Analysis), we could fit the essentials into a one-day workshop. We believed it important that participants should leave the workshop with a level of usable skill that would allow them to initiate a CSE effort when they returned to their workplace. Decision-Centered Design offers a comprehensive story that fits well into a one-day format without straying into complexities and subtleties that would distract participants from the unifying theme of CSE.

From a theoretical perspective, Decision-Centered Design is closely aligned with the concepts of macrocognition, emphasizing the study of cognition in real-world settings (e.g., decision making, sensemaking, planning, and coordinating) involving multiple players, shared resources, risk, and often competing goals (Klein et al., 2003). Decision-Centered Design uses Cognitive Task Analysis interviews to identify the most important and challenging cognitive functions involved in performing a task. One aim is to capture the factors that make it difficult to execute these cognitive functions and the types of common errors people make. Another aim is to identify the critical cues needed to execute these cognitive functions. As a result, design activities can then be oriented toward supporting important and challenging cognitive functions such as decision making, sensemaking, planning, coordinating, reducing execution barriers, and highlighting critical cues (Militello and Klein 2013).

With time at a premium, we avoided discussion of the relative merits of different CSE approaches. We judged that such a comparison would have added unnecessary complexity to our message and detracted from it. We considered ways to fit Cognitive Work Analysis into the day, but Gavan Lintern, our Cognitive Work Analysis champion, suggested that we should avoid it. He thought that it alone would be too challenging to bootstrap people to a working level in one day, and that putting it adjacent to the ideas of Decision-Centered Design in an introductory workshop would confuse many of our participants.

The initial workshop, conducted by the four authors plus Corey Fallon, was entitled "The Road to Cognitive Systems Engineering," and was presented in 2008 at the annual meeting of the Human Factors and Ergonomics Society. It went fairly well. There were 16 attendees. In rating the overall workshop quality on a 5-point scale (1 = Poor: of no redeeming value and 5 = Excellent: among the best workshops offered anywhere), 3 participants rated it a 5, 11 rated it a 4 (Good: Very important information packaged to meet participants' needs and presented clearly) and 2 rated it a 3 (Acceptable: conveyed key issues clearly). On average, the participants rated the workshop 4.0 (out of 5) for overall quality.

We believed that the workshop could be improved. In particular, we thought that the kitchen design exercise was not as effective as we had anticipated, mainly because we had not given participants an opportunity to engage actively in kitchen work during the workshop, and in particular, participants had not teamed with other people in a kitchen setting prior to the workshop.

We revised the workshop by substituting an information management exercise that had participants either working in small teams or observing the small teams. This exercise had the particular advantage of allowing workshop participants to gather their own activity data that they could then use in the subsequent workshop exercises throughout the day.

The next opportunity to present the workshop was at the International Symposium on Aviation Psychology in April 2009. There were 14 participants. This conference used a 7-point evaluation format (with 7 being the highest). Our workshop received one rating of 7, eight ratings of 6, four ratings of 5, and one rating of 4, for an average of 5.7 out of 7. These evaluations suggested that the workshop needed additional development.

We presented the CSE workshop for a third time at the Human Factors and Ergonomics Society meeting in October 2009. Our review of the feedback from the second workshop suggested that we needed to make this more interactive than it already was. We trimmed some of the didactic material but, in particular, sought to set the scene for more interaction early by encouraging participant involvement in the Critical Decision Method interview. In both earlier workshops, we had conducted the entire interview with participants observing. For this third workshop, we used a hybrid approach, in which the facilitator began the interview and then opened it up to participants, encouraging them to ask additional questions. This new approach gave participants an opportunity to observe the facilitator's questioning style and strategy for structuring the interview, but also gave them an opportunity to practice firsthand from the interviewer perspective. We also closed the cycle by having the participants finish the workshop with a review of the initial mission statement and the data collection activities. The participants had a chance to reflect on how they could have designed a better set of CSE tactics, which helped them solidify their learning.

This time, the workshop received extremely high evaluation scores—the highest any of us had ever experienced. There were 12 attendees. In rating the overall workshop quality on a 5-point scale (1 = Poor: of no redeeming value and 5 = Excellent: among the best workshops offered anywhere), eight participants rated it a 5, three rated it a 4, and one rated it a 3. On average, the participants rated the workshop 4.6 (out of 5) for overall quality, 4.3 for usefulness, 4.3 for value, and 4.7 for quality of the instructors.

Section 16.3.1 describes some details of approach we took as we boiled the CSE process down to its essentials. For further reference, we have made the workshop slides available on the Workshops page of www.cognitivesystemsdesign.net.

16.3 WORKSHOP ACTIVITIES

This description of the workshop focuses on the third offering, the one that seems to have been the most effective.

Introduction (30 minutes). In the workshop introduction, our intent was to engage participants immediately to signal to them that the workshop would be an interactive rather than a passive learning experience. We started by asking participants to introduce themselves and to describe a project or context in which they might apply CSE methods. From there, we provided a brief definition of CSE and a quick overview of the five most commonly used frameworks: Cognitive Work Analysis (Rasmussen et al., 1994; Vicente, 1999), Applied Cognitive Work Analysis (Elm et al., 2003), Situation-Awareness Oriented Design (Endsley et al., 2003), Decision-Centered Design (Hutton et al., 2003), and Work-Centered Design (Eggleston 2003).

Although we were tempted to begin with a lecture that would thoroughly lay the groundwork (i.e., theoretical background, methods, commonalities, and differences across approaches, labels, definitions, etc.), we chose to present just enough information to orient the participants to the essentials. The slides contained key representations from each CSE approach and references for further read. We explained that because the presenters were most familiar with Decision-Centered Design, we would use methods closely associated with Decision-Centered Design throughout the workshop.

After these orienting slides, we advised the participants that they would be asked, over the course of the workshop, to conduct a CSE project to support team decision-making and other cognitive work in a command and control environment. Figure 16.1 describes this task. We knew from previous experience that grounding methods in even a hypothetical project makes it easier for participants to think critically about how the methods will work. Having a goal in mind encourages participants to think more concretely about how each phase (data collection, analysis, and design) informs the next. We limited ourselves to 30 minutes for these introductory materials, so that we could move quickly into the first interactive exercise.

Information Management Exercise (1 hour). We employed an exercise designed to pull participants deeply into a moderately stressful team situation, so that they could use themselves as subject matter experts when they applied the CSE methods we intended to teach in the workshop. This exercise also broke down barriers among participants who did not know each other and facilitated a personal understanding of what it means to experience stressful time-critical team decision making under uncertainty, as well as what it means to study and understand how people act in such situations.

To conduct this exercise, we first asked for four to six volunteers to act as military decision makers. This group was organized in a hierarchical set of roles and was told that it was their job to (a) understand from the information provided where the friendly and enemy forces were, (b) predict future enemy actions, and (c) command

---

Mission statement:
Supporting small team decision-making in a command
and control task

You have been commissioned to undertake a CSE project involving the research and design of technology and training concepts that support team decision making and other cognitive work in a command and control environment.

An aerospace engineering company (that wishes to remain anonymous) has failed miserably in its first attempt to do so, and has engaged HFES to conduct a workshop in October 2009 to gain fresh insights and recommendations to support this problem space.

FIGURE 16.1   CSE workshop project.

friendly forces based on this knowledge. One volunteer was asked to assume the role of team leader.

The exercise was facilitated by a workshop instructor who had practiced the management of the tone and pacing of instructions, to bring about a somewhat realistic (i.e., time-pressured) experience. The team leader was given a fast-paced set of message injects, called spot reports, along with a map of the space where friendly and enemy forces were active, as shown in Figure 16.2. The team was required to act on their assessment of the situation based on these messages. The spot reports might say that a UAV detected two red vehicles in sector D3 or that a friendly force requires resupply. Example messages are shown in Table 16.1.

Fourteen messages were handed to the team leader by the facilitator, then 3 minutes elapsed, and then another set of messages was delivered, followed by multiple sets of messages delivered according to a predetermined schedule.

As these reports rapidly accumulated, the team leader realized the need to set up a process for triaging message priority and updating the situation on the map, as well as the need to allocate roles to team members and to better organize the team. All the

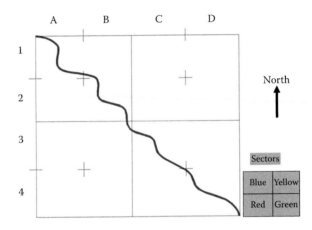

FIGURE 16.2 Information management exercise map.

TABLE 16.1
Example Message Injects for Information Management Exercise

To/From, Time	Subject	Message
00:00 To: Sunray, From 1st Platoon	Spot report	1st Platoon relocated to A2
00:00 To: Sunray, From 2nd Platoon	Spot report	2nd Platoon relocated to A3
00:00 To: Sunray, From Hawk	Spot report	Hawk reports 2 unknowns in D4
03:00 To: Sunray, From DIVMAIN	Movement	Automated message: Sensors detect movement

while, the facilitator continued to pressure the leader and the group by delivering spot reports at short intervals regardless of whether the team had finished reviewing and sharing previous injects. The pacing of the spot reports ensured that the team would be falling behind and would be deliberating about older reports while newer ones piled up unread. Some spot reports provided solid situation awareness while others were designed as distractors or were ambiguous in the information provided, thereby increasing the uncertainty and adding to the chaos.

While this information management activity was underway, the attendees who were not playing a role on the team were assigned to two observer groups. One group was instructed to just watch the information management team and note events of interest, while the other group was instructed to watch for specific macrocognitive functions and processes: the decisions the team made, shifts in the situation awareness of the team, and problems and workarounds in coordinating the team.

We paused the exercise midway through to allow the information management team to reflect on the way it was pursuing the task. The observers got to listen as the team identified the problems it was facing and crafted a new organizational structure to better handle the volume and pacing of information. Then the exercise resumed.

After all of the messages were delivered, the team leader was asked for a solution (recommended action to higher headquarters about where to direct missile attacks).

We then debriefed the exercise. Each observer group and the decision makers in the small team described decision challenges, assessment challenges, sensemaking challenges, and organizational issues. Observers in each observer group then compared notes on whether having a structure helped them to see and record in different ways than having no structure.

Next, the entire group shared experiences across teams, with the facilitator noting the aspects of team cognition that were already covered, and adding items from the set described in the previous paragraph. (The set of topics can and should change in future workshops depending on the particular interests of the sponsors.)

After this exercise, workshop participants were observed to be more relaxed, more communicative, and more engaged than when the workshop began. More importantly, each participant had experienced or observed a situation with characteristics similar to those faced by real-world high-stakes decision makers and observers, a situation similar to those that CSE practitioners need to understand and support in their everyday work.

Critical Decision Method overview and demonstration (75 minutes). Next, we introduced the Critical Decision method as a technique for uncovering decision requirements. The Critical Decision method (Crandall et al., 2006) is an incident-based method for capturing subtle aspects of expertise. It provides an understanding of how people think in the context of a lived event. We highlighted the value of incident-based methods for increasing recall, facilitating discussion, encouraging a first-person perspective, and evoking detailed memories. We also provided a task aid. Figure 16.3 summarizes the four sweeps of the Critical Decision method. Then, we conducted a demonstration interview with a volunteer from the small team that had participated in the information management exercise.

The workshop facilitator elicited an incident (sweep 1) and drew a timeline (sweep 2). Then, the facilitator asked workshop participants to follow up with

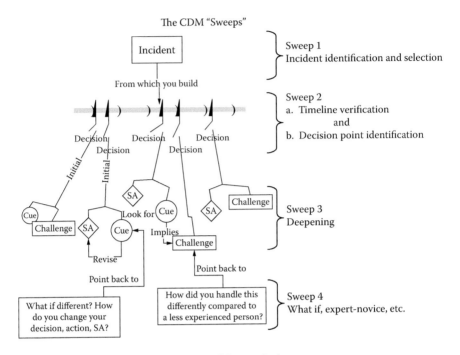

FIGURE 16.3   Job aid for the Critical Decision method.

deepening questions to probe the challenges, cues and situation awareness (sweep 3), and hypotheticals such as ways that a novice might blunder and differences between experts and novices (sweep 4). All participants were asked to take notes during the interview to inform the design of a small team decision-making tool as requested in the work statement (Figure 16.1).

Macrocognition lecture (15 minutes). After encouraging participants to collect both observation and interview data using the information management exercise, we recapitulated the goals of CSE and introduced macrocognition (see Figure 16.4) as a framework to guide CSE. We included a brief discussion of how the macrocognitive functions and processes link to the cognitive work observed in the information management exercise. Our intent was, again, to limit the amount of theoretical lecture, while providing orienting information that would help participants identify useful decision requirements to support small team decision making in the context of command and control.

Decision Requirements Table (1 hour). A Decision Requirements Table is used to organize cues and information as well as strategies and practices that support performance of important cognitive functions. As shown in Figure 16.5, a Decision Requirements Table documents specific difficulties associated with performance challenges and identifies potential pitfalls and errors as well as possibilities for intervention in the form of design ideas.

To transition from data collection to data analysis, we asked participants to refer to their observation notes from the information management exercise and the follow-up

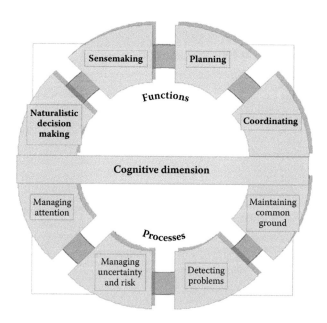

FIGURE 16.4   Macrocognition as a framework for CSE.

Decision/ Assessment	Why difficult?	Critical cues/ Anchors	Potential errors	Design ideas

How do you fill in the table?

FIGURE 16.5   Decision Requirements Table.

interview to develop a Decision Requirements Table. After a few minutes of independent work developing their own version of a table, they moved into small groups to share findings and develop an integrated table. Figure 16.6 shows how a Decision Requirements Table captured the ideas that emerged from the information management exercise.

Design Exercise (1 hour). For the design exercise, we reminded participants that their project goal was to design technology and training supports for team decision making and other cognitive work in a command and control environment

Decision/Assessment	Why difficult?	Critical cues/Anchors	Potential errors	Design ideas
Use message injects to locate friendly and enemy forces in order to report situation assessment to HQ	Identify and resolve anomalies Build teamwork Assemble coherent situation picture	Enemy location and estimates of speed and direction of movement Frequency and clarity of team exchanges Positions of critical entities within the battle space	May not link same entities from different spot reports Inexperienced members may not coordinate well HQ may misinterpret verbal descriptions	Electronic tracking system Preliminary team briefing Shared electronic map

FIGURE 16.6 Decision Requirements Table: Information management.

(see Figure 16.1) and asked them to spend an hour in their small groups discussing concepts for a tool to facilitate team decision making in the context of command and control. The tool should directly support the decision requirements that had been identified in the exercises earlier in the day.

We provided large sheets of paper and pens to help the groups articulate and illustrate their design concepts. Groups were informed that they would have the opportunity to brief their design concepts at the end of the hour.

Cognitive Performance Indicators (25 minutes). No design process is complete without some type of assessment. We introduced participants to a tool called Cognitive Performance Indicators, a set of heuristics for assessing how well a technology supports cognitive performance. The Cognitive Performance Indicators were identified by interviewing experienced CSE practitioners about their own experiences in developing and evaluating cognitive support technologies (Long and Cox 2007; Wiggins and Cox 2010). In the CSE workshop, we gave each participant a job aid handout that listed the cognitive indicators shown in Figure 16.7. We then explained each of the nine indicators and provided real-world examples of how they act as heuristics for identifying strengths and weaknesses in a system's support for cognitive work.

As an example, the "transparency" indicator proposes that a support tool should provide access to the data it uses and show how it arrives at processed data. All of the Cognitive Performance Indicators emerged from interviews with CSE leaders; we have chosen "transparency" to illustrate the Cognitive Performance Indicator construct because it was one of the indicators that emerged from an interview with David Woods, reflecting the importance he places on observability and understandability. Transparency is important in operational work in which robotic (and other) systems use multiple algorithms to shape the displayed information. Operators may or may not trust the numbers or assessments that are displayed, depending on whether the algorithms are transparent. For example, if a map display shows a geographic area as red, indicating the area is not safe to traverse, but an operator has been ordered into the area, the reason *why* the system has coded the area red would provide transparency to help that navigator assess risk. (See Wiggins and Cox 2010, for a complete description of this approach.)

Heuristics for cognition:
Cognitive indicators

1. Option workability
2. Cue prominence
3. Fine distinctions
4. Direct comprehension
5. Transparency
6. Historic information
7. Situation assessment
   - Enabling anticipation
8. Directability
9. Flexibility in procedures
   - Adjustable settings

FIGURE 16.7   Cognitive Performance Indicators.

We provided the workshop participants with details and examples of how to apply the indicators in practice to assess support for user cognition and gave participants the opportunity to offer their own examples of how these indicators manifest in systems they have evaluated. Participants' examples were insightful and deepened the understanding of the group in a meaningful way.

To make a clear linkage to the workshop goals and the conceptual stance underlying the Decision-Centered Design approach to CSE, we tied the Cognitive Performance Indicators back to macrocognition functions and processes by describing how the indicator tool was developed from a review of the literature behind these functions and processes and how the tool was validated in hospital and National Weather Service settings. We emphasized that the approach should be tailored to a particular domain—only a subset of the indicators might be chosen to apply to any given system, depending on that system's functions and goals. For example, a technology that does not include algorithms or the need for transparency should not be evaluated based on that indicator. Also, we outlined options for using Cognitive Performance Indicators at the start of a CTA research effort in order for the researchers to clearly understand what issues the users experience with an existing system. Finally, we outlined how Cognitive Performance Indicators could be used to assess the effectiveness of the cognitive support provided by the new interface developed following the CSE analysis (see Nielsen and Molich 1990).

Redesign (25 minutes). To encourage reflection on this CSE process, we asked participants to consider how they might redesign the data collection process. We asked our group to use hindsight to find ways to strengthen the observation and interview approach used in the workshop. Another group, with hindsight, worked on redesigning the initial mission statement shown in Figure 16.1. This discussion topic elicited thoughts about both the number of observations/interviews and the content of each. Building on their experience developing Decision Requirements Tables and design concepts, the workshop participants identified new questions and information gaps.

16.4   REFLECTIONS

In our retrospective analysis of the success of this enterprise, we reflected on three issues: (1) whether the desire for this sort of experience within the human factors and cognitive engineering community was at the level we imagined, (2) whether the participants in the workshop remained engaged and interested throughout the day, and (3) whether our workshop format allowed workshop participants to develop a self-sustaining level of skill.

Was there a perceived need? The workshops were well-attended and the participants expressed immediate needs for learning more about CSE; therefore, we would answer yes.

Did the participants remain engaged and interested throughout the day? We recalled our own experiences as participants in workshops where we have been promised a good deal of class activity but where that promise remained unfulfilled because the workshop presenters spent most of the available time doing the talking. In reflecting on those experiences, we resolved to make our workshop heavily interactive. Reviewing the time allocations, we spent 205 minutes on participant activities versus 145 minutes on lecture, and some of the lecture included give-and-take discussions. Possibly, we might further reduce the lecturing but we think we were approaching a limit. The participants needed some guidance for the activities they performed.

The initial exercise was challenging. Every participant had a job to do and, as far as we could tell, became deeply involved. After this initial exercise, further analysis and design exercises throughout the day drew on the participants' experiences of this first exercise, thereby maintaining a narrative through the day that was only briefly interrupted by short presentations of didactic material. Indeed, the demeanor of the workshop participants throughout the day and their assessments showed that they enjoyed the day and remained engaged with the material. We believe that beginning with the initial exercise and, subsequently, the frequent use of data from that exercise in analysis and design exercises throughout the day was a key to this continuing engagement.

Did the participants learn? As a team of five, we distributed ourselves around the room during exercises and monitored progress. By our own judgment, we saw participants improving in their skill throughout the day. All of the exercises were successful and all participants seemed to be contributing to the quality of the work. In our judgment, workshop participants generally acquired a level of skill with the methods that make it possible to apply them to rather straightforward problems and, by that method, gain sufficient experience that they would be able to apply them effectively in their ongoing projects.

The experiential exercises also seem crucial to the success of the workshop. We had discussed several types of exercises. The one we chose for the second and third workshops had a game-like quality in which the exercise team had to process disorganized (and sometimes distracting) information to develop a plan. It had a military theme that seemed to work very well, although we suspect that specialist groups such as in health care or process control might respond better to an exercise that captured some elements of their specific domain.

Our goal for the amount of material that we would introduce was somewhat modest, but at the end of the day, we could see that our workshop participants were satisfied. We believe that any attempt to pack more ideas into the day would diminish the experience.

Our workshop was stimulated by the thought that CSE is a relatively unfamiliar discipline (outside of the CSE community) and that very few university programs teach its theories and methods. Practitioners of CSE have become proficient by reshaping their existing capabilities, often drawn from human factors science, cognitive science, or engineering, to the practice of CSE. Most practitioners have essentially learned on the job.

In the development of this workshop, we imagined that some of those who had learned on the job would appreciate an introductory experience that laid out a foundation and some basic methods in an organized and engaging fashion. We pondered the alternatives. Returning to university for a couple of semesters would be entirely impractical for all but a very few. A program of several days duration, such as the one conducted by the University of Michigan for human factors, could fill the need, but as far as we know, no such program exists for CSE and may not be economically viable. There is, of course, a large amount of literature, but little of it provides a gentle introduction and much of it offers competing claims. We saw the possibility that a one-day workshop could fill a need by framing the issues and instructing in straightforward techniques that flowed from knowledge elicitation through knowledge representation into design and then into evaluation.

Our main goal was to provide a form of experience in which participants would develop a level of comfort with some basic methods and acquire an overview perspective of the CSE enterprise. Given the constraint that this had to be done in a single day, we believed we should focus on a small set of tools with an intuitively plausible rationale and a framework that would establish a context.

Decision-Centered Design as supported by the Critical Decision Method and the Decision Requirements Table satisfied that constraint and fulfilled the requirement of supporting a progression through the complete sequence of knowledge elicitation, knowledge representation, and design. The addition of the material on Cognitive Performance Indicators completed the sequence through the evaluation stage. The material on macrocognition provided the conceptual framework that would allow workshop participants to view these tools as an integrated suite.

In presenting this material, we noted other frameworks but avoided any evaluative judgment regarding relative effectiveness within the CSE enterprise. Additionally, we kept away from arguments about effectiveness of CSE, assuming that those in attendance would already be favorably inclined, and we minimized theoretical discussion in the belief that this workshop would be successful if it were focused on the development of self-sustaining level of skill with basic tools.

REFERENCES

Crandall, B., Klein, G., and Hoffman, R.R. (2006). *Working Minds: A Practitioner's Guide to Cognitive Task Analysis*. MIT Press: Cambridge, MA.

Elm, W.C., Potter, S.S., Gualtieri, J.W., Roth, E.M., and Easter, J.R. (2003). Applied cognitive work analysis: A pragmatic methodology for designing revolutionary cognitive affordances. In E. Hollnagel (Ed.), *Cognitive Task Design* (pp. 357–387). Mahwah, NJ: Erlbaum.

Eggleston, R.G. (2003). Work-centered design: A cognitive engineering approach to system design. Proceedings of the Human Factors and Ergonomics Society, 47th Annual Meeting (pp. 263–267). Santa Monica, CA: Human Factors and Ergonomics Society.

Endsley, M.R., Bolté, B., and Jones, D.G. (2003). *Designing for Situation Awareness: An Approach to User-Centered Design*. New York: Taylor & Francis.

Hoffman, R.R., Feltovich, P.J., Ford, K.M., Woods, D.D., Klein, G., and Feltovich, A. (2002, July/August). A rose by any other name...would probably be given an acronym. *IEEE Intelligent Systems*, 72–29.

Hutton, R.J.B., Miller, T.E., and Thordsen, M.L. (2003). Decision-centered design: Leveraging cognitive task analysis in design. In E. Hollnagel (Ed.), *Handbook of Cognitive Task Design* (pp. 383–416). Mahwah, NJ: Erlbaum.

Klein, G., Ross, K.G., Moon, B.M., Klein, D.E., Hoffman, R.R., and Hollnagel, E. (2003, May/June). Macrocognition, *IEEE Intelligent Systems*, 81–85.

Long, W., and Cox, D.A. (2007). Indicators for identifying systems that hinder cognitive performance. Paper presented at the Eighth International Conference on Naturalistic Decision Making, K. Mosier and U. Fischer (Eds.), Pacific Grove, CA (June 2007), CD Rom.

Militello, L.G., Dominguez, C.O., Lintern, G., and Klein, G. (2009a). The role of cognitive systems engineering in the systems engineering design process. *Systems Engineering*, 13(3), 261–273.

Militello, L.G., and Klein, G. (2013). Decision-centered design. In J. Lee and A. Kirlik (Eds.), *Oxford Handbook of Cognitive Engineering* (pp. 261–271). New York: Oxford University Press.

Militello, L.G., Lintern, G., Dominguez, C.O., and Klein, G. (2009b). Cognitive systems engineering for system design. *INCOSE Insight Magazine Special Issue on Cognition*. April, 12(1), 11–14.

Nielsen, J., and R. Molich. (1990). *Heuristic Evaluation of User Interfaces*. New York: ACM Press.

Rasmussen J., Petjersen, A.M., and Goodstein, L.P. (1994). *Cognitive Systems Engineering*. New York: John Wiley & Sons.

Vicente, K.J. (1999). *Cognitive Work Analysis: Toward Safe, Productive, and Healthy Computer-Based Work*. Mahwah, NJ: Erlbaum.

Wiggins, S.L., and Cox, D.A. (2010). System evaluation using the cognitive performance indicators. In E.S. Patterson and J. Miller (Eds.), *Macrocognition Metrics and Scenarios: Design and Evaluation for Real-World Teams* (pp. 285–302). London: Ashgate.

17 From Cognitive Systems Engineering to Human Systems Integration
A Short But Necessary Journey

Lawrence G. Shattuck

CONTENTS

17.1 INTRODUCTION

I first stepped onto The Ohio State University campus in the summer of 1992. I was there to begin a PhD program in cognitive systems engineering (CSE). Serving in the United States Army, I had just completed a 3-year assignment in Germany, which included a 6-month deployment to Desert Shield/Desert Storm. The U.S. Army would pay for my degree and then assign me to the U.S. Military Academy at West Point where I would serve as an academy professor for the next decade.

The process of applying to Ohio State was completed while I was in Germany. At some point in the application and admission processes, I was asked for my preference for a particular advisor. I had learned in my master's degree program the importance of having the right advisor, so I reviewed the credentials and research activities of the Industrial and Systems Engineering (ISE) faculty closely and decided that Dr. Philip Smith's interests aligned most closely with mine. But, when I processed into the ISE Department, I was informed that Dr. David Woods would be my advisor. My military training taught me to salute smartly and carry on. Thus began my relationship with

David. For nearly 25 years, he has been a mentor, colleague, and friend. I continue to be amazed and inspired by his passion and intellect, and his creativity and intuition.

This chapter hinges on just a few of the countless ideas he has inspired and encouraged that have resulted in improved cognitive systems. And, it is about pushing CSE beyond the "cognitive" and broadening "systems" to address the entire life cycle of a system.

My introduction to the field of cognitive engineering began with Woods and Roth (1988). In that article, the authors described this nascent, interdisciplinary endeavor and the reason it came into being: powerful technological systems were transforming the nature of cognitive work. The authors proposed cognitive engineering as a way to understand how to support problem solvers in this new era. Nearly 30 years later, cognitive engineering has become a respected discipline with hundreds of practitioners who self-identify as cognitive engineers or cognitive systems engineers. The field of cognitive engineering now has its own journal. Several universities offer graduate degrees and support cognitive engineering laboratories. Numerous research program announcements refer to challenges for cognitive engineers to engage in both field and laboratory work.

By these measures and others, cognitive engineering appears to be a success. This chapter, however, suggests that there is more work to be done and that work will require CSE practitioners to expand their vision. This chapter will advocate that CSE has some inherent limitations that can be mitigated by adopting a perspective similar to another young discipline—human systems integration (HSI). This expanded vision is necessary if the putative benefits of CSE are to be fully realized in the design, manufacturing, operation, maintenance, and support of complex systems.

The chapter begins with a brief overview of CSE followed by a description of HSI. A model of human–technology integration (HTI) is introduced that describes how these two system components are joined within the design, engineering, and manufacturing processes. This model illustrates how gaps between human capabilities and limitations and technology affordances and constraints can be closed to achieve optimal total system performance. The chapter concludes with a discussion of how CSE and HSI practitioners can work together for the benefit of those who will ultimately use the systems they help design.

17.2   COGNITIVE SYSTEMS ENGINEERING

Hollnagel and Woods (1983) introduced the world to CSE. They proposed it as a novel and necessary approach to the design and study of "man–machine systems." Traditional approaches were not sufficient to capture the complexities and nuances of cognitive systems that include both human and machine agents. These approaches lacked the tools, concepts, and models needed to observe, analyze, and make improvements to man–machine systems. Hollnagel and Woods stated that human–machine systems work best when man's understanding of the machine and the machine's model of the man align with one another. They concluded that for CSE to be successful, practitioners would need to develop methods for cognitive task analysis that would reveal underlying cognitive models that could be used to design better man–machine systems.

In Woods and Roth (1988), we learned that cognitive engineering is focused on human work in complex worlds and the importance of designing technologies to support that behavior. Cognitive engineering is also ecological in that it studies how humans solve problems with tools in operational settings. We came to understand that tool-driven approaches to problem solving ignore the all-important semantics of a field of practice. And we were taught that cognitive engineering seeks to improve performance in systems composed of multiple cognitive agents.

By the mid-2000s, Hollnagel and Woods had evolved their thinking about man–machine systems and had begun referring to "joint cognitive systems" (Hollnagel and Woods 2005; Woods and Hollnagel 2006). They began to focus on the performance of a joint cognitive system rather than simply considering the cognitive functions embedded in a technology. Rather than focusing on the cognitive aspects of multiple system components and their interaction, they redrew the boundaries to include all agents—human and machine—that comprised the work system. Studying the performance of this joint system became their goal.

The following are the three themes that have been present in CSE: understanding how cognitive work systems cope with complexity, how to design complex cognitive work systems in which human and machine agents interact with one another, and how artifacts shape the nature of the work that is performed. Fundamental to CSE is the belief that "all work is cognitive" (Hollnagel and Woods 2005, p. 24). Some activities may be so well-practiced that they require minimal cognitive resources, but even those are initiated and controlled by our central nervous system.

Another fundamental principle of CSE is its focus on the intersection of the people engaged in work, the technologies they employ, and the nature of the activities they undertake. CSE practitioners employ three general methodological principles: (1) they focus on aspects of complex systems where problems exist or where they may arise; (2) they study the circumstances surrounding actual or potential problems to better understand their etiology; and (3) they propose practical solutions to these problems rather than to satisfy themselves with theoretical models and speculative explanations (Hollnagel and Woods 2005).

Hollnagel and Woods (2005) acknowledged that theories, models, and methods of disciplines such as human factors engineering and ergonomics had evolved a great deal over the past five decades. This progress was applied during the emergence of newer fields of study, including human–computer interaction, cognitive ergonomics, and computer-supported cooperative work, and, of course, CSE. But Hollnagel and Woods also admitted that there is more work to be done to establish the scientific foundations of CSE. Many CSE researchers simply investigate problems and propose solutions. These researchers do not actually build machines, write code, or manufacture components because they are more often researchers than engineers, designers, or acquisition professionals. (See Chapters 10, 14, and 15, as well as Guerlain et al. 1999 and Smith et al. 2005, for examples of exceptions to this.)

Hollnagel and Woods recognized that if CSE is to reach its potential, it needs to consider the entire life cycle of a human–machine work system. "It is becoming clear that systems must be considered over a much longer time span—beginning with system design and ending with maintenance, or perhaps even decommissioning....

Many of the problems at the sharp end have their origin in the choices made during the design phase" (2005, p. 179).

One manifestation of the problems that arise when CSE practitioners do not participate throughout the entire system life cycle is the likelihood that the mental models of designers and users will differ. The model the designer has of the work system and its users will form the basis for the design of technology system, and hence the work method that is imposed by the technology. The technology's image— that is, the look and feel of the technology—is what is presented to the user and is the predominant determinant of the user's model of the system (Hollnagel and Woods 2005; Norman 2013).

Even if the designer and user models are perfectly aligned in the planning phase, there is no assurance that the operational work system will replicate the models that were intended. The design and production of a large-scale work system can involve hundreds of designers, engineers, programmers, program managers, and a host of other personnel responsible for manufacturing budgeting, contracting, maintenance, and support. These people will be employed by different organizations, each with its own processes, policies, and procedures, many of which may be proprietary and inaccessible by other organizations. Any one decision made by anyone involved in the system life cycle—especially early in the process—could permeate it in ways that are implicit or explicit, expected or unexpected (see Hoffman and Elm 2006; Neville et al. 2008).

How can CSE practitioners use what they have studied and observed over the last few decades to affect such an extraordinarily convoluted process? This is where CSE can benefit from HSI. As will be described below, HSI embeds itself in the system life-cycle process, seeks to work closely with program managers and system engineers, uses a variety of methods (found within the HSI domains) to solve problems, and works to make explicit the impact of poor decisions or trade-offs before they are implemented to mitigate unintended consequences.

17.3   HUMAN SYSTEMS INTEGRATION

More so than CSE, HSI has struggled with its identity. Booher (2003) suggested that HSI is a concept. Tvaryanas (2010) advocated for HSI as a philosophy. Winters and Rice (2008) pondered HSI as a form of speech. Is it a noun, verb, adverb, or preposition? Deal (2007) collected 40 definitions from various publications and reports in an attempt to find a suitable definition for the International Council on Systems Engineering. Three definitions are presented here.

Handbook of Human Systems Integration (Booher 2003, p. 4):

Human systems integration is primarily a technical and managerial concept, with specific emphasis on methods and technologies that can be utilized to apply the HSI concept to systems integration. As a concept, the top-level societal objectives of HSI are to significantly and positively influence the complex relationships among: (1) People as designers, customers, users, and repairers of technology; (2) Government and industrial organizations that regulate, acquire, design, manufacture, and/or operate technology; (3) Methods and processes for design, production, and operation of systems and equipment.

International Council on Systems Engineering:

[HSI is an] interdisciplinary technical and management processes for integrating human considerations within and across all system elements; an essential enabler to systems engineering practice. (Human Systems Integration Working Group, n.d.)

Naval Postgraduate School: The Naval Postgraduate School offers both resident and distance learning master's degrees in HSI and a graduate certificate. The definition used in those curricula is:

Human Systems Integration acknowledges that the human is a critical component in any complex system. It is an interdisciplinary approach that makes explicit the underlying tradeoffs across the HSI domains, facilitating optimization of total system performance in both materiel and non-materiel solutions to address the capability needs of organizations. (Human Systems Integration Program, n.d.)

Most definitions of HSI include or infer a list of domains that are essential to achieving total system integration. The Department of Defense (DOD) describes the following seven domains (DOD 2015). Manpower is the number of people who will operate, maintain, support and provide training for a system. This domain also concerns the mix of uniformed military personnel, DOD civilians, and defense contractors required by the system.

The personnel domain identifies the requisite characteristics of the users and establishes selection criteria for each occupational specialty. The training domain determines requirements for individual, collective, and joint training for those who will operate, maintain, and support the system. These requirements include the appropriate training methods, the necessary equipment and technologies, the length of training, and the methods for assessing the effectiveness of the training.

The DOD description of the human factors engineering domain incorporates both ergonomics and cognitive engineering and sets it within the systems engineering process across the entire life cycle. The stated purpose of HFE is to develop "effective human–machine interfaces and to meet human systems integration requirements" (DOD 2015, p. 118). Habitability practitioners focus their efforts on the physical work and rest environments, personnel services, and living conditions. This domain is particularly important when personnel will live on the system (e.g., ships, submarines, spacecraft, etc.). Failure to design for habitability will have a negative effect on total system performance, quality of life, and morale; recruitment and retention will also be adversely affected. Practitioners in the compound domain of safety and occupational health seek to "minimize the risks of acute or chronic illness, disability, or death or injury to operators and maintainers" and to "enhance job performance and productivity of the personnel who operate, maintain, or support the system" (DOD 2015, p. 119). Finally, force protection and survivability aims to mitigate the risks of harm to personnel from environmental (including enemy) threats through system design.

The conceptual model in Figure 17.1 illustrates how these domains interact with one another and with other aspects of the system acquisition process.

HSI happens within the system acquisition life cycle. This life cycle will differ from one industry or agency to another, but in general, the process includes the identification of a need for a new work system or revision to an existing work system, planning and design, manufacturing, testing, distribution, operation, and disposal. The model shows manpower, personnel, training, and human factors engineering as primary or input domains that are the result of decisions made early in the acquisition process. The decisions may be conscious or unconscious, good or bad, right or wrong. These input domain decisions are either enabled or constrained by acquisition factors such as the cost associated with the input domain decisions (and the budget available to the program manager), the schedule implications for the domain decisions (sufficient or insufficient time available), and the risk incurred by the acquisition program as a result of the domain decisions (tolerable or not).

The combination of the input domain decisions; the enablers or constraints of cost, schedule, and risk; and the other aspects of the design and engineering processes will result in a human–technological system that has the potential for some level of total system performance. Measuring total system performance of large systems with any degree of precision is extremely challenging, at best. However, there are intermediary or first-order outcomes that are both measurable and quantifiable. Several of these first-order outcomes are listed in Figure 17.1. Among the outcomes listed are HSI domains of habitability, safety, and survivability. These outcomes can be used to infer and estimate total system performance. If the estimated total work system performance is less than needed, the feedback loop indicates that input domains can be tweaked,

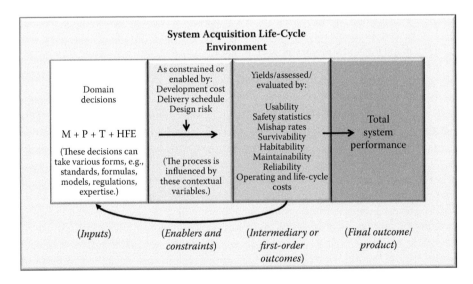

FIGURE 17.1 Conceptual model of HSI within the system acquisition life-cycle environment.

perhaps in combination with changes to the enablers and constraints, to yield more acceptable intermediary outcomes and ultimately better total system performance.

This model tells only part of the story. For human and non-human elements of a work system to be integrated in a manner that has the potential to achieve optimal total system performance, the focus must be on the capabilities and limitations of the humans who will operate, maintain, and support the system and the affordances and constraints that are inherent in the system. The HTI model discussed below illustrates how human and non-human elements should be integrated throughout the entire system life cycle.

17.4 OVERVIEW OF THE HTI MODEL

The HTI model is based on the work of O'Neil (2014) and illustrates the manner in which humans and technology can be integrated during the design, engineering, manufacturing, and testing processes to create a work system that optimizes performance while minimizing risk and cost. The bottom portion of Figure 17.2 shows an acquisition process. It could be any acquisition process, but in this case, it is the DOD acquisition process. This establishes the HTI model's context. For more generic acquisition and design models, see Blanchard and Fabrycky (2006), Pew and Mavor (2007), Kirwan and Ainsworth (1992), or Parnell et al. (2008). HTI occurs within the context of the acquisition process. Broadly speaking, the acquisition process spans the entire life cycle of the system—from the inception of the idea through disposal when it becomes obsolete or is replaced by another system. In the HTI model, just above the acquisition process is something that resembles a zipper or merging railroad tracks. On the left side are icons that represent humans and technology. Represented in this idealized HTI model, as an acquisition life-cycle advances to product maturity, the

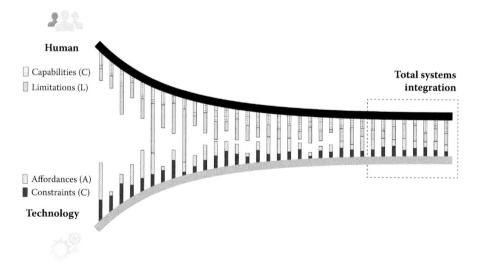

FIGURE 17.2 The HTI model. Image produced by the Naval Postgraduate School's Center for Educational Design, Development, and Distribution (CED3).

alignment between humans and technology becomes closer and more seamless. The intended end result is total systems integration.

17.5   THE DOD ACQUISITION PROCESS

An acquisition process begins by identifying an unfulfilled need. In the DOD, needs are identified by conducting a capability-based assessment. The capability-based assessment is a formal way to consider what must be accomplished, what is needed to accomplish it, and what is currently available to accomplish it. The difference between what is needed and what is available is referred to as a capability gap. A capability-based assessment often uncovers a multitude of gaps, and it is the responsibility of military leaders to prioritize the gaps and decide how they should be closed. A gap could be closed by implementing a non-materiel solution (e.g., assigning medical personnel to all forward-deployed ground units) or by acquiring a new technological system (e.g., a robotic system to evacuate wounded Soldiers from the battlefield). This decision point is the Materiel Development Decision and initiates the materiel acquisition process.

During the Materiel Solution Analysis phase, which resides between the Materiel Development Decision and Milestone A, several approaches are considered. For example, battlefield medical evacuations could be conducted using manned aircraft systems or manned ground systems, or unmanned aircraft systems or unmanned ground systems. Each alternative will have different HSI implications, especially for the domains of manpower, personnel, training, and human factors engineering. It is imperative that HSI practitioners use their expertise to inform decision makers of how their choices will affect those who will operate, maintain, and support the system. Choosing an unmanned ground evacuation system may reduce the number of medical personnel that accompany units into combat, but the number of personnel needed to supervise, operate, maintain, and support the system will not result in any significant reduction to overall manpower requirements.

At Milestone A, a senior military official determines whether all the tasks specified for the Materiel Solutions Analysis phase have been accomplished successfully and is presented with a recommendation on the type of technological system that should be developed. This decision has tremendous implications for all HSI domains and the decision maker would be wise to consider the implicit and explicit trade-offs being made that will affect the entire life cycle of a set of machines and software. In the case of DOD, a technological system's life cycle can easily span more than 60 years!

A decision made at Milestone A to pursue an unmanned ground system for casualty evacuation leads to the Technology Maturation and Risk Reduction phase. In Figure 17.2, this phase fits between Milestones A and B. As its name implies, this phase provides time and opportunity for researchers, designers, and programmers from multiple defense contractors to mature the technologies they plan to include in their system. The more mature the technologies, the lower the risk associated with including them in the system. During this phase, HSI practitioners should be attuned to the emerging affordances and constraints, ensuring they will align with the human capabilities and limitations of those who will operate, maintain, and support the system.

At Milestone B, the senior military official reviews the various unmanned casualty evacuation ground system designs and selects the alternative that closes the capability gaps with the least amount of risk and cost and has the best chance of meeting the acquisition timeline. At this point in the acquisition process, a majority of the design decisions have been made and it would be costly to modify them. Figure 17.3 conceptually shows the difference in cost and performance when investing in HSI early or late in the acquisition process. Investing in HSI earlier costs less and results in much greater improvement to total system performance throughout the system life cycle.

Beyond Milestone B is the Engineering and Manufacturing Development phase. The objective of this phase is to move from a prototype design into a design that is producible on a large scale and verify that the manufactured system is able to meet all essential requirements, also referred to as key performance parameters. Many modifications to the design of the hardware and software are still being made during this phase to accommodate the manufacturing processes. Any one of these changes could affect one or more of the HSI domains. It is vital that HSI practitioners stay informed about these changes and communicate to program managers how these changes will affect total system performance.

At Milestone C, if the senior military official is convinced that the system can be manufactured as required, the technology progresses into the Production and Deployment phase. In this phase, test and evaluation activities determine whether the production version of the system can achieve the total system performance necessary to close the capability gaps. HSI personnel should be fully engaged in the test and evaluation process. Ideally, this testing takes place in an operational setting with personnel who are representative of the Warfighters who will actually use the system in combat and who have received the same training as that which is planned for the Warfighters. In addition, the test and evaluation activities should assess maintenance and support functions related to the system rather than focusing exclusively on the operational aspects of the system. If the system achieves the minimally acceptable performance criteria, a full rate production decision is made and the number of systems budgeted for is built and distributed to the designated organizations. Included in the acquisition process is a study of the impact of introducing the new technology into an organization and the changes needed to ensure a smooth transition and a successful adoption. Areas considered include changes in doctrine, organizational structure, training, leadership, personnel, facilities, and policies.

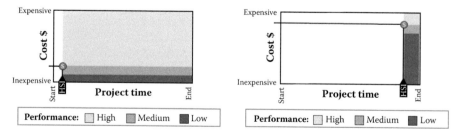

FIGURE 17.3 The impact of investing in HSI early (left) or later (right) in the acquisition process.

The final phase of the acquisition process is the Operation and Support phase. This is, by far, the longest and most costly phase. While technologies may have a planned service life of 25 years, defense budget constraints frequently require service life extensions for decades beyond that. These timelines often necessitate numerous upgrades and modifications to hardware and software that affect every domain of HSI, but especially manpower, personnel, training, and human factors. This portion of the acquisition process is yet another period in which HSI practitioners need to be decisively engaged to ensure any changes made do not negatively affect operators, maintainers, or supporters. These upgrades and modifications also present an opportunity to introduce HSI changes that will improve both human and total system performance.

The preceding description of the acquisition process establishes the context for the discussion that follows. Section 17.6 describes the two major elements of a complex system: the technological elements composed of hardware and software and the human elements consisting of those who will operate, maintain, and support the system. The manner in which these two major components are integrated will determine the success or failure of the total system.

17.6   HUMANS AND TECHNOLOGY

The goal of the acquisition process depicted in Figure 17.2 is to achieve total work system integration by combining or integrating the human and technological components. These components differ in many ways. One of the biggest differences is that over the course of a work system acquisition, technology will mature from just an idea or a drawing to something that can actually be used by Warfighters. During that same time, though, the humans who will operate, maintain, and support the technology will not likely change in any significant manner. Their vision and hearing will remain relatively consistent. Their cognitive processes won't improve or decline dramatically. Their limbs and their joints will still function about the same. All of these human attributes are measurable. Many of these factors are measured with more or less precision when someone joins the military. In fact, there are requirements that must be met before people can join the military. Candidates have to be a certain height and weight. They have to meet certain standards for vision, hearing, and intelligence. Taken as a whole, the range of attributes for a military population can be thought of as capabilities and limitations.

In the HTI model, a capability is something that a human is able to do (without the aid of any tools or technological systems), whereas a limitation is something that the human cannot do—or at least the limitation establishes the boundary between what the human can and cannot do. Technology presents affordances and constraints. Gibson (1979), who coined the term "affordance," described it as follows. "An affordance is not bestowed upon an object by a need of an observer and his act of perceiving it. The object offers what it does because it is what it is" (p. 130). Decades later, Norman provided a more relational view of an object's affordances and the agents who use them. "An affordance is a relationship between the properties of an object and the capabilities of the agent that determine just how the object could possibly be used" (Norman 2013, p. 13). A constraint is a design feature that prevents certain actions from being taken. Many cars are designed so that the key cannot be locked inside. The car's sensors detect the presence of the key and prevent the driver from locking the car until the key has been removed from the vehicle.

A technological design feature can, at the same time, present both affordances and constraints (Bennett and Flach 2011). For example, many smartphones have a concave button near the bottom of the device. In the hands of a human, the design affords pushing in the button with a finger but constrains just about any other action. The user cannot rotate it like a dial, flip it like a switch, or pull it out. The design at once affords the one action that is intended by its designers and constrains every other action. It is quite elegant.

Humans and technology are integrated best when the capabilities and limitations of the humans map to—or align with—the affordances and constraints of the technology. The capabilities and limitations of humans for a given subset of the population such as U.S. military personnel are measurable (qualitatively if not quantitatively), relatively stable (within an individual over time and across a specific subpopulation), and generally predictable (e.g., fatigue slows reaction time and increases errors). The best time for designers, engineers, and programmers to bound the technological possibilities only to those that will encompass the full range of human capabilities and limitations of the user population is early in the acquisition process, before there is any commitment to the type of materiel solution that will be built or the characteristics that the technology will have. The rationale for this approach is that there are many more degrees of freedom early in the acquisition process for the affordances and constraints of the technological system than for the capabilities and limitations of human operators, maintainers, and supporters of the system.

A close look at many major DOD acquisition programs reveals that system designs were not sufficiently bound by the human capabilities and limitations of the envisioned operators, maintainers, and supporters. As a result, well into the acquisition process, substantial HSI-related shortcomings have been discovered. A ship has too few crew members for the work that must be done. Only a small portion of the envisioned user population can eject from a cockpit due to the weight of the helmet and ejection forces experienced by the crew member. A proposed hardware design significantly increases the time needed for maintainers to access components that reduce system availability. See Hoffman et al. (2016) for a detailed description of HSI-related problems in the Patriot missile system and the Predator unmanned aerial system.

The preceding discussion is captured in the following six principles:

- Humans have capabilities and limitations; technology offers affordances and constraints (Hoffman and Woods 2011). Capabilities and limitations bound what the human can do. Affordances and constraints affect what can be done with the technological system.
- Total system performance is enhanced when the technology capabilities and limitations map well to (or align with or are constrained by) human capabilities and limitations. This mapping happens within the design and acquisition processes.
- Human capabilities and limitations are variable within a specific range and are measureable for a given subset of the population. Technology affordances

and constraints have more degrees of freedom, especially early in the design and acquisition processes.

- The mapping of technology affordances and constraints to human affordances and constraints will ebb and flow throughout the design and acquisition processes. The mapping can be close at one point but a bad decision could result in a wide gap later in the process.
- Those who design and build the technological systems must ensure that the technology affordances and constraints map well to human capabilities and limitations.
- The success of a system is a function of the extent to which human capabilities and limitations are mapped appropriately to technology affordances and constraints to enhance total system performance.

When a gap is identified at any point in the acquisition process, the nature of that gap should be assessed. Is it a minor or a major gap? What are the consequences if the gap is not closed? And, what alternatives are available for closing the gap? Virtually every alternative will involve modifications to the technology's hardware, software, or to one or more of the HSI domains. Section 17.7 examines how to close these gaps.

17.7   CLOSING THE GAP

Figure 17.2 shows that the gap between humans and technology begins wide and then decreases as the acquisition progresses from inception to the final stage. This may or may not be the case. In reality, the gap could be narrow or wide throughout the process or the gap could widen and narrow throughout the process (Hoffman and Woods 2011).

The gap between humans and technology will be wide at the beginning of the acquisition process when early discussions and conceptualizations of technological solutions do not appear to be bounded by the capabilities and limitations of those who will operate, maintain, and support the system. Alternatively, the initial gap could be narrow if, in those early formulations of the system, designers, engineers, and programmers fully appreciate the human capabilities and limitations of the users and use them to bound the solution space for their system design.

The size of the gap will change throughout the acquisition process because the process is dynamic. Thousands of design, engineering, manufacturing, programming, contracting, scheduling, and budgetary decisions are made over the life cycle of a work system, each of which has the potential to narrow or widen the gap and, as a result, positively or negatively affect work system performance. Such decisions are often made to resolve local problems, without complete knowledge or full consideration of the consequences across the entire life cycle. A capable HSI practitioner will review the alternatives available and make explicit the immediate and long-term impacts of each as they relate to the operators, maintainers, and supporters of the work system. To do so, HSI practitioners have to be given full access to acquisition information and be involved throughout the entire acquisition process.

HSI practitioners are essential to the task of closing the gap between the humans and technology. When problems arise and decisions must be made, HSI practitioners partner with the engineers and acquisition professionals by making explicit the implications of the alternatives and suggesting possible solutions and trade-offs. But importantly, HSI practitioners must rely on the engineers and acquisition professionals to implement the decisions.

Engineers and acquisition processionals have a vast array of alternatives available to them in the acquisition process, but the number of viable alternatives decreases quickly—and the cost skyrockets—as the technology progresses through the acquisition process. HSI practitioners also have a larger set of solutions they can employ earlier in the process rather than later. The best way to achieve human–technology integration is to have HSI practitioners work together with those responsible for the technological alternatives to find solutions that will close the gaps as early as possible in the acquisition process (see Chapter 14).

Figure 17.2 is intended to suggest that some gaps are easier to close than others. Some can be closed earlier in the process while others may not be fully closed until much later. Also important to note is that sometimes the gaps are closed when technology affordances and constraints meet human capabilities and limitations in the middle. Other gaps are closed when the technology side must make greater accommodations than the human side and extend beyond the midpoint. Still other gaps may exist in which there is little flexibility in the hardware and software and, in those cases, the human side may have to go more than halfway.

On the human side, the HSI practitioners' best tools for closing the gaps are the domains of manpower, personnel, training, and human factors engineering (see Figure 17.1). A shortcoming on the technology side can be overcome by modifying the characteristics of any one of these domains, some combination of multiple domains, or a combination of HSI domains and accommodations on the technological side. This calls for HSI practitioners to employ their trade-off and negotiation skills but also requires they have data to support their solutions.

A thorough work analysis early in the acquisition process could reveal that the total workload for the number of personnel assigned to operate, maintain, and support the system will exceed their collective cognitive or physical limits. Human modeling can identify what missions, functions, and crew members are at risk. For these analyses to be valid, they must take into account the conditions under which the activities must be carried out (e.g., time constraints, physical stress and psychological stress, adverse environmental conditions, etc.). One way to close the gap between the human and the technology side would be to increase the number of personnel assigned to the work system and to redistribute the workload until models show that no one is overloaded.

While adding manpower would seem like a purely human-side solution, it is not. On a Navy ship, for example, adding sailors could require not only additional workstations but also more berthing spaces, showers, toilets, and laundry facilities. Military services look to reduce manpower whenever possible because manpower costs are much higher than hardware or software across the entire life cycle of the work system. (See Tvaryanas 2010, for a detailed history of the Army's HSI program [formerly known as MANPRINT] and its efforts to reduce manpower through the development of more highly automated systems.)

Increasing manpower is just one way to close the gap between the human and technology sides. Personnel and training domain solutions can also be employed. Selecting people who are more capable to operate, maintain, and support a system may mitigate the need for additional manpower because they are able to accomplish the required work with fewer cognitive and physical resources and in less time. In a similar manner, providing additional training and opportunities to practice the necessary skills will reduce the cognitive and physical demand of tasks, resulting in personnel being able to accomplish more tasks for a sustained period. (See Hoffman et al. 2016, for examples of training and retraining failures that created excessive cognitive demands on users.) But relying on higher personnel selection standards and additional training to close gaps is not without cost. The pool of candidates decreases in size as the selection criteria increase. And more qualified personnel expect greater compensation. Additional training can be costly as well in terms of both training resources and manpower. Many program managers are enticed by the notion of using training to close the human–technology gap because training costs are not borne by the program office responsible for design and manufacturing the system. Instead, the costs are borne by the organizations that operate, maintain, and support the system.

CSE must work to **close or reduce** the gaps that cannot be closed by the other HSI domains. A high-performance fighter aircraft may be required to perform aerial combat maneuvers that would subject both the aircraft and the pilot to extreme gravitational forces. Increasing manpower, altering selection criteria, or extending and enhancing the training will do little to improve a pilot's G-force tolerance. However, anti-G suits that prevent the blood from pooling in the lower portion of the pilot's body and positive pressure breathing systems are human factors engineering solutions that will permit pilots to tolerate higher forces. Implementing a human factors engineering solution, however, may necessitate changes to other HSI domains. G-force tolerance, even with anti-G suits, will vary among the subpopulation and calls for more discriminating selection criteria. Pilots will need anti-G suit training to use them properly. And the aircraft's hardware and software will need to be modified to accommodate the suit.

The other HSI domains (safety, survivability, occupational health, and habitability) are the result of the extent to which the primary domains (manpower, personnel, training, and human factors engineering) were properly addressed in the technology acquisition process. A work system acquisition program in which there may be insufficient manpower, inappropriate personnel selection criteria, inadequate training, or poor human factors engineering will lead to safety, occupational health, and habitability issues. It is for this reason that we refer to the "work system" as being what is referenced in the acquisition, and not just machines and software. And these issues can lead to a multitude of challenges once the system is put into operation. Morale, retention, and the ability to attract and recruit new personnel are just a few of the activities that can suffer in a system where the human and technology gap is not closed completely or correctly.

Important activities for HSI practitioners throughout the acquisition process include identifying the domain (or domains) that can most effectively **close or reduce** the gap between the human and technology sides, making trade-offs explicit in terms of cost (both immediate and total life cycle), risk, schedule, and performance, and advocating with program managers and other acquisition professionals for decisions that are within the human capabilities and limitations of solution space. The success of HSI practitioners will depend on several things:

- *Data*—Are data available and persuasive enough to convince program managers that the proposed trade-offs will enhance total system performance?
- *Perspective*—Do the program managers have a total life-cycle perspective or are they more disposed to solving immediate problems without regard for long consequences?
- *Status of HSI in the acquisition process*—Is HSI valued by program managers or are HSI practitioners buried so deep in the acquisition organizational structure that HSI issues never reach program managers?
- *Funding of HSI activities*—How are acquisition programs' HSI activities funded? In most cases, HSI activities are funded by program managers. Historically, program managers allocate very little funding to HSI, which affects the quantity and quality of HSI activities.
- *Timing*—Where in the acquisition process are the trade-offs considered? As noted previously, HSI trade-offs implemented earlier in the acquisition have a much greater chance of success at a lower cost than if they are implemented later in the process.
- *Mandates*—What are the forcing functions that drive decisions and activities to close gaps? In the commercial sector, consumers have options. Poorly designed products are forsaken for those with smaller human–technology gaps. Warfighters, however, do not get a choice in the equipment they use. Therefore, acquisition policies are what dictate how much (or little) HSI activities are included in a work system's life cycle.

Closing the gaps between humans and technology that inevitably arise in any work system is challenging. As the six items above suggest, closing gaps requires planning, resources, and organizational commitment. It also requires knowledgeable practitioners who understand how to negotiate the research, development, and acquisition communities (Endsley 2016). The following section addresses how CSE and HSI practitioners can join together to ensure work systems are appropriately designed, manufactured, fielded, and sustained.

17.8   FROM CSE TO HSI

More than three decades ago, David Woods and his colleagues rejected the *status quo* of an information processing paradigm and the traditional view of human factors and ergonomics when they formulated the field of CSE. They focused their efforts on the intersection of people, the technology they use, and the work they perform. Their unit of analysis was not the interaction of the human and the machine but the joint cognitive system. They sought to understand how cognitive systems cope with complexity, how to design better complex systems, and how artifacts shape the nature of the work activities.

Over time, it became clear to these innovators that for CSE to succeed, it had to expand along two dimensions. The first dimension is the time span; the second is the influences on performance. They acknowledged the need for CSE to consider more of the work system life cycle and to investigate distal as well as proximal influences of cognitive work system performance. But there is little evidence that CSE has had much success in expanding as needed. (See Chapter 14 and Cooke and Durso 2008, for examples of CSE successes. Unfortunately, far too many systems fail compared to those that succeed.) However, HSI practitioners do address the entire work system life cycle and consider both distal and proximal influences. CSE practitioners can benefit from the progress that HSI practitioners have already made in these important areas.

HSI works best when practitioners are engaged early and often throughout the system life cycle. Within DOD, HSI practitioners are able to participate throughout the life cycle for several reasons. First, HSI is mandated by government policy (DOD 2015), but the current language is both vague and weak. Program managers are required only to "consider" HSI in the acquisition process. More forceful and successful language would be that the delivery of a work system shall be accompanied by convincing empirical evidence from an operational context that the technology is usable, learnable, useful, and understandable. Such wording would put HSI issues on par with other critical acquisition priorities. Second, HSI practitioners are trained to speak the same language as program managers and system engineers. Third, HSI practitioners understand the acquisition process well enough to know when and where to observe, influence, and intervene in the process to ensure decisions that are made do not widen the gap between the human capabilities and limitations and the technology's affordances and constraints. And fourth, HSI practitioners understand how to employ trade-off analyses among the domains to make explicit the impact of acquisition decisions and to advocate for decisions that favor those who will operate, maintain, and support the complex system.

17.9 CONCLUSION

Both HSI and CSE have much to offer in the design of advanced technologies and human–machine work systems. It is not a competition but an opportunity for cooperation and collaboration. It is time for each to embrace what the other has to offer. HSI can and does extend the work of CSE practitioners. CSE acknowledges the system life cycle but does not engage with it in a holistic and continuous manner from beginning to end. Total engagement with the life-cycle process is absolutely essential in order to ensure that what CSE practitioners have designed will be implemented.

Hollnagel and Woods (1983) wrote, "If the designer is to build an interface compatible with human cognitive characteristics rather than force the human to adapt to the machine, he must be provided with a clear description of these characteristics and with tools and principles that allow him to adapt machine properties to the human" (p. 597). This is where HSI practitioners can partner with CSE practitioners. HSI practitioners, through their intimate knowledge of, and their daily contact with, the system life cycle, can leverage domain trade-offs to close the gap between humans and technology that will ensure total system integration and optimal system performance. This, after all, is the ultimate goal of both CSE and HSI.

REFERENCES

Bennett, K.B., and Flach, J.M. (2011). *Display and Interface Design: Subtle Science, Exact Art.* Boca Raton, FL: CRC Press.

Blanchard, B.S., and Fabrycky, W.J. (2006). *Systems Engineering and Analysis.* Upper Saddle River, NJ: Pearson Prentice Hall.

Booher, H.R. (2003). Introduction: Human systems integration. In H.R. Booher (Ed.), *Handbook of Human Systems Integration* (pp. 1–30). Hoboken, NJ: John Wiley & Sons.

Cooke, N.J., and Durso, F.T. (2008). *Stories of Modern Technology Failures and Cognitive Engineering Successes.* Boca Raton, FL: CRC Press.

Deal, S. (2007). Definition of human systems integration. An unpublished white paper prepared to set the context for discussions at the International Council on Systems Engineering International Workshop, Albuquerque, New Mexico.

Department of Defense. (2015, January 7). *Operation of the Defense Acquisition System* (DOD Instruction 5000.02). Washington, DC.

Endsley, M. (2016, January–February). Building resilient systems via strong human systems integration. *Defense AT&L Magazine,* 6–12.

Gibson, J. (1979). *The Ecological Approach to Visual Perception.* Boston: Houghton Mifflin.

Guerlain, S., Smith, P.J., Obradovich, J.H., Rudmann, S., Strohm, P., Smith, J.W., Svirbely, J., and Sachs, L. (1999). Interactive critiquing as a form of decision support: An empirical evaluation. *Human Factors,* 41, 72–89.

Hoffman, R.R., Cullen, T.M., and Hawley, J.K. (2016). The myths and costs of autonomous weapon systems. *Bulletin of the Atomic Scientists,* doi: 10.1080/00963402.2016.1194619

Hoffman, R.R., and Elm, W.C. (2006). HCC implications for the procurement process. *IEEE Intelligent Systems,* 21(1), 74–81.

Hoffman, R.R. and Woods, D.D. (2011, November/December). Beyond Simon's slice: Five fundamental tradeoffs that bound the performance of macrocognitive work systems. *IEEE: Intelligent Systems,* 67–71.

Hollnagel, E., and Woods, D.D. (1983). Cognitive systems engineering: New wine in new bottles. *International Journal of Man–Machine Studies,* 18, 583–600.

Hollnagel, E., and Woods, D.D. (2005). *Joint Cognitive Systems: Foundations of Cognitive Systems Engineering.* Boca Raton, FL: Taylor & Francis Group.

Human Systems Integration Program. (n.d.). Retrieved January 05, 2017, from https://my.nps .edu/web/dl/degProgs_MHSI.

Human Systems Integration Working Group. (n.d.). Retrieved January 05, 2017, from http:// www.incose.org/ChaptersGroups/WorkingGroups/analytic/human-systems-integration.

Kirwan, B., and Ainsworth, L.K. (1992). *A Guide to Task Analysis.* London: Taylor & Francis.

Neville, K., Hoffman, R.R., Linde, C., Elm, W.C., and Fowlkes, J. (2008). The procurement woes revisited. *IEEE Intelligent Systems,* 23(1), 72–75.

Norman, D. (2013). *The Design of Everyday Things.* New York: Basic Books.

O'Neil, M.P. (2014). *Development of a Human Systems Integration Framework for Coast Guard Acquisition.* Monterey, CA: Naval Postgraduate School.

Parnell, G.S., Driscoll, P.J., and Henderson, D.L. (2008). *Decision Making in Systems Engineering and Management.* Hoboken, NJ: Wiley-Interscience.

Pew, R.W., and Mavor, A.S. (2007). *Human–System Integration in the System Development Process: A New Look.* Washington, DC: National Academics Press.

Smith, P.J., Klopfenstein, M., Jezerinac, J., and Spencer, A. (2005). Distributed work in the National Airspace System: Providing feedback loops using the Post-Operations Evaluation Tool (POET). In B. Kirwan, M. Rodgers, and D. Schaefer (Eds.), *Human Factors Impacts in Air Traffic Management.* Aldershot, UK: Ashgate.

Tvaryanas, A.P. (2010). *A Discourse in Human Systems Integration*. Monterey, CA: Naval Postgraduate School.

Winters, J., and Rice, J. (2008). Can you use that in a sentence: What part of speech is 'HSI' anyway? (PowerPoint presentation). *Proceedings of the 60th Department of Defense Human Factors Engineering Technical Advisory Group*, Washington, DC.

Woods, D.D., and Hollnagel, E. (2006). *Joint Cognitive Systems: Patterns in Cognitive Systems Engineering*. Boca Raton, FL: Taylor & Francis.

Woods, D.D., and Roth, E.M. (1988). Cognitive engineering: Human problem solving with tools. *Human Factors*, 30(4), 415–430.

Part IV

Integration

18 Future Directions for Cognitive Engineering

Karen M. Feigh and Zarrin K. Chua

CONTENTS

Over 30 years have passed since Dave Woods and Eric Hollnagel introduced cognitive systems engineering (CSE) as "New Wine in New Bottles"; it is right to summarize what we have learned and to contemplate the future. CSE has made enormous strides and established itself as a separate discipline of engineering complete with unique vocabulary, theory, and methods.

The goal of CSE is to improve sociotechnical work systems by enabling analysis, design, implementation, validation, and certification. This chapter explores the future of CSE. The chapter has two parts: the first summarizes the well-known weaknesses within CSE and how to address them as the field matures, and the second part is more speculative. It is written from the perspective of what the world might be like if the tenants of Dave Woods and colleagues were to be truly embraced by CSE specialists, system designers, engineers, and computer scientists alike. In doing so, this chapter will ruminate broadly and boldly in the hopes that these visions will come to pass.

18.1 FINISHING WHAT DAVE WOODS STARTED

For all of its accomplishments, CSE has many elements that are currently a work in progress. As practitioners continue the work that Dave Woods and colleagues started, a number of specific issues stand out as needed for CSE to take its place as an accepted science whose goals, methods, and contributions are understood and valued.

Specifically, at least five outstanding challenges can be identified. First, explanatory and predictive models of cognition are still under development and those that exist are in need of better integration across work domains and levels of abstraction and explanation. Second, technology that adapts to context is missing along with the associated methods to design it. Third, the further migration of technology design toward more holistic design is needed. Fourth, the further integration of CSE into the systems engineering design process must continue. Fifth, our understanding of individual cognition must expand to encompass teams, groups, and organizations. These issues are not new to the CSE community, but they remain open challenges.

18.2 DEVELOPMENT AND INTEGRATION OF PREDICTIVE MODELS OF VARIOUS ASPECTS OF COGNITION

The aim of any science is the creation of explanatory models sufficiently detailed as to enable explanation and prediction. Furthermore, the aim of any engineering discipline is the application of scientific models to the design and creation of things. We seek, as Dave has argued, not to simply inform design but to use our tools and methods to create products. Traditional physics- or chemistry-based engineering disciplines focus almost exclusively on the creation of engineered systems or technology through the application of models of the appropriate physical or chemical mechanisms. Cognitive engineering applies psychology and cognitive and computational sciences to the design of work, procedures, and technology. There is a fundamental need, therefore, to develop predictive models of various aspects of cognition. From the microcognitive elements (such as perception, attention, and memory) to macrocognitive components (such as judgment, decision making, and ideation) to meta cognition (such as situation awareness and problem solving), and through macrocognition, which stretches across individuals, explanatory models having predictive power are needed in order to anticipate and design for cognitive work.

18.3 DEVELOPMENT OF TECHNOLOGY THAT ADAPTS TO CONTEXT

One of the main themes through the history of CSE has been to document and prove the importance that task context plays in human cognition, learning, control, and subsequent behavior (Hollnagel and Woods 2005). CSE has always treated context in its broadest terms, as any mediating construct that influences human cognition and behavior. Feigh et al. (2012) suggested five context categories for human–machine systems: system state, world state, task/mission state, spatiotemporal dimensions, and human state. In judgment and decision-making literature, particularly in psychology, the impact of a wide range of contextual features context has been investigated and determined to be significant (Hollnagel and Woods 2005; Woods and Hollnagel 2006). Appropriately, research on the design of automated systems that can adapt to context has been extensive. The literature can be categorized along four main themes: (1) determining which elements of cognition are best replaced/supplemented by technology; (2) distinguishing which is better for overall work system performance, resilience, and combating complacency and overreliance on technology: human-initiated or technology-initiated changes to function allocation; (3) understanding which triggers (performance and physiological) are the most appropriate for technology-initiated changes in machine functionality; and (4) establishing guidelines for changes in technology by deconstructing exemplar machines that have known defects. Improving the design of such technology merits several requirements. First, we require better methods to document and model the influence of context on all aspects of cognition. Second, we require methods to design technology such that their activities are observable to and predictable by their human co-workers. Third, we require comprehensive methods to test and evaluate such adaptive or adaptable work systems. And fourth, if goals 1–3 are achieved, we will need to change the way we train people to interact with intelligent technologies that now operate much more like humans.

18.4   THE FURTHER MIGRATION OF TECHNOLOGY DESIGN TOWARD A MORE HOLISTIC WORK DESIGN

Presently, much of technology design is done in isolation from the work design. Work design, in its loosest sense, encompasses the function allocation (the distribution of work between the human and the machine), procedures, and the responsibilities of the different personnel interacting with the system (users, operators, maintenance, etc.). Many products are designed with the explicit assumption that the new technology will not alter the wider work in any way; the input/output, the individuals and their competencies, and the responsibility and authority should remain the same even though the introduction of technology has been shown to implicitly alter work (Woods 1985). As work system technologies grow more complex and they evolve from user–tool combinations to sociotechnical systems consisting of multiple users, tools, and multiple interactions, the pretense that the impact on work will be negligible ceases to hold and work design must be carefully considered in tandem with the technology design as to maintain performance, safety, and transparency through all facets of the system design. The future of technology design, especially for safety critical applications, will also require work redesign that may not be technology centric, including the creation of new procedures, or new requirements for personnel or training for existing staff. The complete work design must support all cognitive work—not just one small aspect of it—and the work of all stakeholders. As new breakthroughs continue to occur in each of these aspects, we must continue to progress in a few key areas before technology design and work design become synonymous, specifically the identification of key context features and their impact, the creation of parsimonious predictive models for work design, and the creation of metrics for system-level constructs such as robustness and resilience.

18.5   THE CONTINUED INTEGRATION OF COGNITIVE ENGINEERING INTO SYSTEMS ENGINEERING

Historically, the role of the cognitive engineering community, as part of a broader human factors or human systems integration group, was primarily limited to the role of testing and evaluation or ensuring that designs met constraints outlined by professional or military standards. For decades, the cognitive engineering community has petitioned for inclusion early and often in the design process and has outlined the most effective roles and applications of cognitive engineers (Bailey 1989; Byrne 2003; Chapanis 1965; Wickens 2003). Standards and government regulations now require the inclusion of cognitive engineering to maintain safety (Ahlstrom and Longo 2003, MIL-STD-1472F, 1999, NASA Procedural Requirements; O'Hara et al. 2011). Yet, these regulations remain weak and underspecified. There persists a lack of clarity of how to integrate CSE and frustration with few methods to approximate and predict workload, errors, training requirements, and other human–automation interaction issues. The outstanding issues indicate that this is an area for active research. To facilitate in the exploration of design space, the trade-offs between cognitive engineering metrics and other engineering metrics must be understood, yet key measures (the cognitive equivalents of physical units of weight, power, stability, and performance) and methods to approximate those metrics early in the design process are missing. Finally, we must educate engineering students about the general methods and metrics used by CSE.

18.6 UNDERSTANDING COGNITION OF TEAMS, GROUPS, AND ORGANIZATIONS

While understanding the cognitive process of an individual is critical to the development of sociotechnical systems and working environments, most macrocognitive work is conducted in teams, groups, and organizations. Future CSE will need to better understand groups and teams. Specifically, three areas are promising: (1) improving our understanding of how cognition changes from individuals to teams and groups both at the individual construct level such as team decision making and communication and at the overall cognition level as in team or shared cognition; (2) incorporating knowledge from other fields such as management and organizational psychology; and (3) improving methods for observing and measuring shared cognition.

CSE has made enormous strides, bringing much needed research and new insight into the challenge of designing effective joint human–machine systems. CSE continues to gain followers interested in solving hard design problems and creating systems that are not only helpful but also a pleasure to use. The five themes postulated here are areas of open research needed to continue what Dave Woods began.

18.7 A WOODS-INSPIRED FUTURE

The second half of this chapter speculates about how the world could be if Dave Woods' lessons and suggestions were taken to heart and implemented widely. To illustrate this imagined future, we will use a more narrative form and peek into the design process at an engineering design firm. Here we meet Tim, a new employee fresh out of his graduate studies in computer science. Tim is on his way to a design kickoff meeting where he will discuss the design of a new joint human–machine system with two others: Ian, a more senior colleague who has grand ideas for how to change current work practices through the introduction of new technology and procedures, and Paula, a practitioner from the field. Paula is also a senior figure who has been in the work domain of interest for over 20 years. Her role in the meeting will be to bring a user's perspective and to make sure that whatever is designed actually improves the joint human–machine system.

In this scenario, we can think of Tim as our technologist, Ian as our innovator, and Paula as our practitioner—the CSE team triumvirate that Roesler et al. (2005) suggested as necessary for all design processes. Tim has been preparing for this meeting for a little over 2 weeks now. He's been studying all of the advance work that preceded the decision to proceed with the design project. Ian has overseen a preliminary ethnographic study of the work domain of interest—which is where the company found Paula. The study included on-site observations, review of documents and standard operating procedures, a focus group, and several minimally obtrusive interviews with a variety of practitioners in the work domain. The results were captured in a series of models describing the work domain. The CSE team's goal was to gain apprentice-level proficiency as rapidly as possible.

As this is Tim's first major design assignment, he's glad that he stayed for his graduate work. Although he had a class on human–machine systems in his undergraduate program, it was his graduate work that really exposed him to the theory behind CSE. His classes have familiarized him with how each of the models in his preparation packet was produced and what aspect of the work domain each one describes (the people doing the work, the technology designed to help, and the context in which the work will be accomplished). He especially likes the Abstraction Hierarchy from Work Domain Analysis (Rasmussen 1985; Vicente 1999) where the entire work domain is decomposed into multiple levels of abstraction. This helps Tim, as the one responsible for suggesting the technological specifics and mechanisms, to better understand the big picture and explain "why" certain functionality in the final design is needed. Before finding this model, Tim would often find himself not understanding why his suggested solutions to requirements documents were not well received by the eventual users. The Abstraction Hierarchy gave him a fuller understanding of what was missing in conventional requirements documents. Another model that Tim is a particular fan of is the information flow model from contextual design, which helped him see the exchange of information between the human and the system.

Also in his packet is a series of design scenarios put together by Paula and Ian. These scenarios describe the work and context that are of particular interest to the new design. Tim can already envision how some tweaks to work practice and the inclusion of some technology might help support the users and ease some of the identified trouble points. What was really helpful about the scenarios is that they also included some alternative scenarios that outlined a "structured set of alterative scenarios (to sample) the difficulties, challenges, and demands of a field of practice" (Roesler et al. 2005). The design team will use these as they begin their work to identify quickly via mental simulation what types of technology and process interventions might be helpful across a range of scenarios. Tim knows how easy it is to design for a typical day, but that those designs can often be too brittle to handle off-nominal days (Smith et al. 1997).

Tim enters the meeting that is being held in a design room. The room has been prepopulated by the work models and design scenarios that Tim has been reviewing in his packet. The room is equipped with digital recording equipment so that the team members are free to focus on their creative task safe with the knowledge that any insight uttered will be captured for future reference if needed. In this meeting, Ian, Tim, and Paula are beginning the process of designing a new work flow that will likely, but not necessarily, involve the creation and integration of new technology.

They are designing for an organization that is heavily dependent on expert-level knowledge and highly proficient performance at complex tasks. The reasons for the effort include unanticipated and radical changes in the situations to which the organization must respond and a recognition that their legacy technology and work methods not only create barriers to effective performance but also make the work system nonadaptive and nonresilient. However, not all of the technology and work methods were ineffective, and those should be preserved. This CSE project would be a mix of legacy system redesign and envisioned world design.

Here, Ian's job is to evaluate the usability of the system that results from the conceptual design phase. Paula's job is to evaluate the usefulness of the system proposed, calling on her extensive knowledge of both nominal and off-nominal operations. And Tim's job is to evaluate the anticipated changes and suggest ways to bring them into the world.

Ian begins the meeting by briefly reviewing their goals and inviting Paula to briefly go over the models and findings of the ethnography. The team quickly settles into a good working rhythm where Ian and Paula do most of the talking. Tim is happily nodding along when Ian stops and asks for his opinion. Tim interjects that he's been thinking that a key to the new design would be an improvement in situation awareness and that he'd like to add that as a design goal/requirement (Woods and Sarter 2010). Ian agrees that situation awareness is indeed an important aspect of this design, but that situation awareness is a desired performance *improvement* (Woods and Sarter 2010), rather than a requirement, that comes about because the system is directing the user's attention between different scales: from the individual, physical world and the interconnected digital world, against the fluctuations of system functionality, and across multiple people in different roles. Paula adds that they cannot ignore the scope and scale of the project (Woods and Cook 1999; Woods and Sarter 2010).

Clearly having hit upon a topic both Ian and Paula feel strongly about, she continues to remind Tim that while improved situation awareness is easy to say, it is vague notion that doesn't translate well to the design process. Remembering years of dealing with strong and silent automation, Paula points out that the system cannot just redirect the "user's attention" without some sort of reasoning or logic that the user can understand (Potter et al. 2006; Woods et al. 2004). And not necessarily just at the end, but a continued logic throughout the work environment (Woods and Patterson 2001) as to have shared understanding of unfolding events and clear cooperation. Otherwise, there is risk for automation surprise, or a miscommunication between the human and the technology (Woods and Sarter 2000). Additionally, she firmly insists, the human should always have the mechanism to direct the system's attention to other activities. Tim nods, appreciating this subtlety, and the meeting continues.

Ian muses that the interface will need to provide new feedback, in modes other than the traditional auditory and visual, perhaps haptic and even natural language? His eyes gleam at the last idea. Here, Tim finally feels he can contribute something.

Tim points out that while not impossible, natural language is more difficult to implement and, given the large range of accents present in the user population, may simply serve to frustrate a large segment of users. Paula concludes that whatever the modality (including many nonverbal cues) (Murphy and Woods 2009), the system needs to think in terms of high-level goal-oriented patterns, to have an eye out for the future and any possible event transitions, and to not just be limited to reporting on the actual situation—trying to avoid building clumsy automation at all costs. Tim draws the link that a more adaptive system is likely to increase safety (Woods 2015).

As Tim finishes making his point, the company's resilience officer, Rebecca, steps into the room. Rebecca makes her apologies, having been caught up in a meeting with the County regulator's resilience team. Rebecca was originally hired to reevaluate the company's safety policies. A well-known fan of the Q4-Balance Framework (Woods et al. 2015), she and her team developed a portfolio of the company's safety proactive indicators, mapped out how the company used this kind of information, and started redirecting the company's "safety energy," or the resources used to proactively maintain safety. Since her arrival, the company was able to substantially reduce its liability insurance and cash reserves.

Rebecca makes it a priority to sit in on the tri-specialty design kickoff meetings. She wants to keep abreast of how the work flow is likely to change as a result of the new design. Additionally, she feels that if she's present with the CSE design team from the beginning, she can intervene quickly and early should the proposed new design weaken the work domain's resilience. She also keeps tabs of future difficult decisions involving the trade-off between safety and economics. A large part of her role is helping to make those critical decisions. Ian quickly recaps the meeting for Rebecca. Rebecca tells Paula that she will coordinate with her to help develop the competence envelope for this work domain and sources of possible perturbations that would push beyond this envelope. She also double checks with Ian that he's got an assessment of the work system performance with at least two measurables from each of the five bounds of performance: ecology, cognizance, perspectives, responsibility, and effectiveness (Hoffman and Woods 2011).

This sounds familiar to Tim as he remembers learning about the four laws of cognitive work (Woods 2002): laws of adaptation, laws of models, laws of collaboration, and laws of responsibility. He makes a comment on his meeting notes to remind his team of them as they begin their work. As the meeting draws to a close, they are one step closer to finalizing the design and are in agreement about the goals of the new system. Everyone is familiar with the context in which the system will be used, the culture of the users, the information available, and the work to be done. They have about three new ideas for how the work could be modified to better achieve its goals. They have already begun to identify not just existing difficulties with the work flow but also those that are likely to persist/be created in the new work flow if care is not taken with the new design.

Before the next meeting, several things will happen. Tim and Ian (the members of the CSE team having significant human factors experience) will evaluate the current technologies of the work system, focusing on interface design. Their experience has proven that individuals possessing human factors experience can examine existing tools, interfaces, and work methods, and expeditiously identify interface elements that are deficient in terms of such features as understandability, usability, and usefulness. This is especially true for graphical user interfaces and web pages, which are typically user-hostile. Ian and Paula will conduct retrospective structured interviews to develop (1) models of a "day in the life" of a number of key roles/ positions in the work system and (2) rich descriptions of how the workers and their work process adapted when they were confronted by challenging cases that fell outside their usual experience. The CTA questions address both the legacy work issues and current design challenges. The information from the interviews will augment the preliminary ethnographic study conducted by Ian before the first CSE team meeting.

Following the collection of this information will come the integration phase. A more in-depth Abstraction Decomposition analysis will take place. They are looking for responsibility gaps that will likely be observable as process bottlenecks. When they find these gaps, they will need to determine at what echelon and place of cognitive work authority and responsibility should come together. If this role does not currently exist, then it will need to be created and included in the envisioned work. The three designers will be developing a series of hypotheses of how the redesigned system should work, which will eventually progress to prototypes before finally progressing into scenarios each of which will be evaluated for robustness to nominal and off-nominal scenarios (Roesler et al. 2005). For now, Tim, Ian, and Paula will return to their respective teams. For the next meeting, Ian has a deliverable of going deeper into the concepts that were discussed and what capabilities will be needed to support the new design concept. Tim will be working closely with Ian, and his team will be continuing to put together different pieces of technology that will be capable of bringing the envisioned system to a functional reality. It is Tim's job to make sure that the technology proposed is capable of delivering on the requirements; it is a difficult task integrating it all. Tim also knows that while a bit unusual for someone in his position, he also has the responsibility to justify the use of all technology. Technology often changes the nature of work and can make a system more brittle, so its inclusion must "buy its way on" and not simply be used for its own sake. Also, he's well aware of the substitution myth, that human work can't just be replaced by automation (Sarter et al. 1997). Paula has one of the hardest jobs. She must oversee the creation of multiple scenarios in which the system will be evaluated and verified. And her scenarios must test the system in nominal and off-nominal situations not only once the full prototypes are ready but also earlier when the new design is still in its conceptual phase. Ian will be working with her as well, to map out how specific cues in the environment, sometimes simply lumped under the heading "context," in these scenarios influence user decisions and to explore possible intervention points where technology can be injected and adapt to the unfolding situation. Ian will also help bridge Tim and Paula's work, by suggesting solutions to overcoming system brittleness to changes of the work environment.

As professionals, they are fully aware of the delicate balance between designing a robust, proactive, adaptive joint human–machine system and meeting stakeholder demands without surpassing the budget. Conversely, they must guard against simply injecting technology to seemingly cut costs unless they can be certain that the new work flow will maintain or improve the work domain's resilience.

Following the initial activities and the integration of a "cardboard mock-up," Tim, Ian, and Rebecca were able to create a set of new workstations and descriptions of roles and responsibilities, both legacy and new. The team placed the mock-ups in a room adjacent to the room where the actual work takes place and the workers were invited to work through the redesign for both their legacy work and for the envisioned work. Paula organized several scenarios for the evaluations and the workers were encouraged to express their desired functionalities in their own words. The first evaluation took about a week. The results were used by everyone in the CSE team to generate requirements for Tim's team. Tim's presence during the early design phase, evaluation, and requirements writing was critical as he would serve as the liaison between the CSE team and the software and hardware engineers creating envisioned technologies.

The process of evaluation was repeated. This time, the mock-ups had become prototypes with limited capabilities. Using another set of scenarios created by Paula, the workers were again asked to use the technology and clarify any desires they still had to support the legacy work and/or the envisioned work. The process again took about a week, which felt like three to Tim and his team, who worked late into the night every night to incorporate small tweaks requested during the day's trials. The results enabled the CSE team to refine their requirements documents and Tim to begin a second wave of implementation. The new prototypes were in the experimentation room within a week—the fast turnaround impressed and excited the workers. The workers were now able to perform many of their job tasks using the new work system. Its functionalities were more apparent, and the work method was more usable, useful, and understandable. The interfaces were much better. The workers began to prefer working with new work system and started spending considerable time in the experimentation room. The system was working very well for routine tasks.

The final evaluation requires seeing how the joint human–machine system performs in some off-nominal and very demanding situations. This will hopefully be the last round of testing before finalizing the design. Emotions are running high, especially for Tim. The previous round of testing—using Paula's off-nominal scenarios—illustrated some weaknesses in the design. Tim's team spent all of last month implementing Rebecca's remediation plan so that the human–machine system with the latest design iteration would be able to deal with the unanticipated perturbations that Paula's latest off-nominal scenario will create. Rebecca (and everyone on the CSE team really) wants to see that the joint human–machine system can adapt to the situation as it changes. Additionally, there was a recent accident with a similar human–automation system. The situation had begun to escalate and both the technology and the user were unable to compensate for the dynamically changing scenario. Luckily, no one was injured, but there was significant damage to the physical plant. Investors were blaming the problem on human error, arguing that the user should have made the logical connections and that the user training program should include methods on combating this same scenario. Rebecca tells Tim that this is a typical knee jerk response and that it is not the proper way to conduct error assessment—there is always more to the story (Woods 1990, 2002).

Rebecca had been named to the committee investigating the accident. She reexamined the accident from the user's point of view and determined how the system had led the user to think a certain way. There were a lot of late nights, but she had been using those lessons learned to improve Tim, Paula, and Ian's new system, especially to combat escalation. She was happy to see that their system was already fairly robust to the type of situation that caused the accident. They had previously mapped how user expertise and knowledge come into play during these sorts of situations during their initial design process. This allowed them to understand the augmentation of resources to account for the increased need in monitoring and attention and included better methods for efficiently educating and informing the user and others outside the joint human–machine system when such human intervention is needed for problem solving (Woods and Patterson 2001). And between Ian and Paula, they had also assessed the new joint human–machine system against the pitfalls of adaptive system failure: decompensation, working at cross-purposes, and getting stuck in outdated behaviors (Woods and Branlat 2011).

After a week of testing and numerous small tweaks, the verdict is in, and the system is cleared for production. From the perspective of the client company and its workers, this experience was very different from past designs in which the workers would only get to see the "final deliverable," to which the workers were only asked how well they "liked" it. Additionally, this design process, which included more prolonged and persistent access to the prototype systems, resulted in not just a work system redesign but also a cohort of workers who are already comfortable using the new work methods, enabling immediate productivity gains as they trained the remainder of the workforce.

As the team celebrates their success, Tim has mixed emotions. He's glad to see his first design project finalized, but he knows that his job is never really over. The design cycle never stops. He'll be keeping an eye on the product over the course of its life. If any issues come up, he is very likely to be called on to help set it right. But there's no time to dwell on their success. Each member of the CSE team is already working on a few other projects at various phases of design. Tim is, however, looking forward to a few weeks of well-deserved vacation followed by a quick stop at his old research lab where he's been invited to give a talk to the new crop of graduate students about what it's like to be on a CSE team in the "real" world.

BIBLIOGRAPHY

Ahlstrom, V., and Longo, K. (2003). *Human factors design standard (HF-STD-001)*. Atlantic City International Airport, NJ: Federal Aviation Administration.

Bailey, R.W. (1989). *Human Performance Engineering: Using Human Factors/Ergonomics to Achieve Computer System Usability*, 2nd edition. Hillsdale, NJ: Prentice Hall.

Branlat, M., and Woods, D.D. (2010). How do systems manage their adaptive capacity to successfully handle disruptions? A resilience engineering perspective. In *2010 AAAI Fall Symposium Series*.

Byrne, M.D. (2013). 27 Computational cognitive modeling of interactive performance. In John D. Lee and Alex Kirlik (Eds.), *The Oxford Handbook of Cognitive Engineering*, Oxford, UK: Oxford University Press, 415–423.

Chapanis, A. (1965). On the allocation of functions between men and machines. *Occupational Psychology*, 39(1), 1–11.

Christoffersen, K., and Woods, D.D. (2002). How to make automated systems team players. *Advances in Human Performance and Cognitive Engineering Research*, 2, 1–12.

Feigh, K.M., Dorneich, M.C., and Hayes, C.C. (2012). Toward a characterization of adaptive systems a framework for researchers and system designers. *Human Factors*, 54(6), 1008–1024.

Hoffman, R.R., and Woods, D.D. (2005). Toward a theory of complex and cognitive systems. *Intelligent Systems, IEEE*, 20(1), 76–79.

Hoffman, R.R., and Woods, D.D. (2011). Beyond Simon's slice: Five fundamental trade-offs that bound the performance of macrocognitive work systems. *Intelligent Systems, IEEE*, 26(6), 67–71.

Hollnagel, E., and Woods, D.D. (2005). *Joint Cognitive Systems: Foundations of Cognitive Systems Engineering*. Boca Raton, FL: CRC Press.

Johnson, M., Bradshaw, J.M., Hoffman, R.R., Feltovich, P.J., and Woods, D.D. (2014). Seven cardinal virtues of human–machine teamwork: Examples from the DARPA Robotic Challenge. *IEEE Intelligent Systems* (6), 74–80.

Klein, G., Pliske, R., Crandall, B., and Woods, D.D. (2005). Problem detection. *Cognition, Technology and Work*, 7(1), 14–28.

Klein, G., Woods, D.D., Bradshaw, J.M., Hoffman, R.R., and Feltovich, P.J. (2004). Ten challenges for making automation a "team player" in joint human–agent activity. *IEEE Intelligent Systems* (6), 91–95.

Lee, D.S., Woods, D.D., and Kidwell, D. (2005). Escaping the design traps of creeping featurism: Introducing a fitness management strategy. In *Usability Professionals' Association Conference (UPA 2006)*. Broomfield, CO, USA (June 13–16, 2006).

Murphy, R.R., and Woods, D.D. (2009). Beyond Asimov: The three laws of responsible robotics. *Intelligent Systems, IEEE*, 24(4), 14–20.

O'Connor, B., and Chief, S. (2011). *Human-Rating Requirements for Space Systems*. Office of Safety and Mission Assurance, NASA Report NPR, 8705.

O'Hara, J., Higgins, J., and Fleger, S. (2011, September). Updating human factors engineering guidelines for conducting safety reviews of nuclear power plants. In *Proceedings of the Human Factors and Ergonomics Society Annual Meeting* (Vol. 55, No. 1, pp. 2015–2019). Thousand Oaks, CA: Sage Publications.

Potter, S.S., Woods, D.D., Roth, E.M., Fowlkes, J., and Hoffman, R.R. (2006, October). Evaluating the effectiveness of a joint cognitive system: Metrics, techniques, and frameworks. In *Proceedings of the Human Factors and Ergonomics Society Annual Meeting* (Vol. 50, No. 3, pp. 314–318). Thousand Oaks, CA: Sage Publications.

Rasmussen, J. (1985). The role of hierarchical knowledge representation in decision making and system management. *IEEE Transactions on Systems, Man and Cybernetics (2)*, 234–243.

Roesler, A., Woods, D.D., and Feil, M. (2005, March). Inventing the future of cognitive work: Navigating the "Northwest Passage." In *Proceedings of the 6th International Conference of the European Academy of Design*.

Sarter, N.B., Woods, D.D., and Billings, C.E. (1997). Automation surprises. *Handbook of Human Factors and Ergonomics*, 2, 1926–1943.

Smith, P.J., McCoy, C.E., and Layton, C. (1997). Brittleness in the design of cooperative problem-solving systems: The effects on user performance. *IEEE Transactions on Systems, Man and Cybernetics, Part A: Systems and Humans*, 27(3), 360–371.

Vicente, K.J. (1999). *Cognitive Work Analysis: Toward Safe, Productive, and Healthy Computer-Based Work*. Boca Raton, FL: CRC Press.

Wickens, C.D., Lee, J.D., Liu, Y., and Gordon-Becker, S. (2003). *An Introduction to Human Factors Engineering*. Hillsdale, NJ: Prentice Hall.

William J. Hughes Technical Center (1999). Human Engineering Design Criteria for Military Systems, Equipment and Facilities, Military Standard MIL-STD-1472D, Notice 3, U.S. Department of Defense.

Woods, D.D. (1985). Cognitive technologies: The design of joint human–machine cognitive systems. *AI Magazine*, 6(4), 86.

Woods, D.D. (1990). On taking human performance seriously in risk analysis: Comments on Dougherty. *Reliability Engineering and System Safety*, 29(3), 375–381.

Woods, D.D. (2002). Steering the reverberations of technology change on fields of practice: Laws that govern cognitive work. In *Proceedings of the Annual Meeting of the Cognitive Science Society* (Vol. 10). Fairfax, VA.

Woods, D.D. (2006). Essential characteristics of resilience. In N. Leveson, E. Hollnagel and D.D. Woods (Eds.), *Resilience Engineering: Concepts and Precepts* (pp. 21–34). Aldershot, UK: Ashgate.

Woods, D.D. (2009). Escaping failures of foresight. *Safety Science*, 47(4), 498–501. Elsevier

Woods, D.D. (2015). Four concepts for resilience and the implications for the future of resilience engineering. *Reliability Engineering and System Safety*, 140 (September), 5–9.

Woods, D.D., and Branlat, M. (2011). Basic patterns in how adaptive systems fail. In *Resilience Engineering in Practice* (pp. 127–144). Boca Raton, FL: CRC Press.

Woods, D.D., Branlat, M., Herrera, I., and Woltjer, R. (2015). Where is the organization looking in order to be proactive about safety? A framework for revealing whether it is mostly looking back, also looking forward or simply looking away. *Journal of Contingencies and Crisis Management*, 23(2), 97–105.

Woods, D.D., and Cook, R.I. (1999). Perspectives on human error: Hindsight biases and local rationality. In F.T. Durso (Editor) and R.S. Nickerson, R.W. Schvaneveldt, S.T. Dumais, D.S. Lindsay, M.T.H. Chi (Associate Editors), *Handbook of Applied Cognition*. West Sussex, UK: John Wiley & Sons.

Woods, D.D., and Cook, R.I. (2002). Nine steps to move forward from error. *Cognition, Technology and Work*, 4(2), 137–144.

Woods, D.D., and Hollnagel, E. (2006). *Joint Cognitive Systems: Patterns in Cognitive Systems Engineering*. Boca, Raton, FL: CRC Press.

Woods, D.D., and Patterson, E.S. (2001). How unexpected events produce an escalation of cognitive and coordinative demands. In P. Hancock and P. Desmera (Eds.), *Stress Workload and Fatigue* (pp. 290–304). Hillsdale, NJ: Erlbaum.

Woods, D.D., and Sarter, N.B. (2000). Learning from automation surprises and going sour accidents. In N. Sarter and R. Amalberti (Eds.), *Cognitive Engineering in the Aviation Domain* (pp. 327–353). Boca Raton, FL: CRC Press.

Woods, D. D., and Sarter, N.B. (2010). Capturing the dynamics of attention control from individual to distributed systems: The shape of models to come. *Theoretical Issues in Ergonomics Science*, 11(1–2), 7–28.

Woods, D.D., Tittle, J., Feil, M., and Roesler, A. (2004). Envisioning human–robot coordination in future operations. *IEEE Transactions on Systems, Man, and Cybernetics, Part C: Applications and Reviews*, 34(2), 210–218.

Author Index

Subject Index